Wildland Fire Smoke in the United States

David L. Peterson · Sarah M. McCaffrey ·
Toral Patel-Weynand
Editors

Wildland Fire Smoke in the United States

A Scientific Assessment

Springer

Editors
David L. Peterson
School of Environmental and Forest Sciences
University of Washington
Seattle, WA, USA

Sarah M. McCaffrey
Rocky Mountain Research Station
USDA Forest Service
Fort Collins, CO, USA

Toral Patel-Weynand
Research & Development
USDA Forest Service
Washington, DC, USA

Research & Development, USDA Forest Service

ISBN 978-3-030-87044-7   ISBN 978-3-030-87045-4   (eBook)
https://doi.org/10.1007/978-3-030-87045-4

This is a U.S. government work and not under copyright protection in the U.S.; foreign copyright protection may apply 2022. This book is an open access publication.
Open Access This book is licensed under the terms of the Creative Commons Attribution 4.0 International License (http://creativecommons.org/licenses/by/4.0/), which permits use, sharing, adaptation, distribution and reproduction in any medium or format, as long as you give appropriate credit to the original author(s) and the source, provide a link to the Creative Commons license and indicate if changes were made.
The images or other third party material in this book are included in the book's Creative Commons license, unless indicated otherwise in a credit line to the material. If material is not included in the book's Creative Commons license and your intended use is not permitted by statutory regulation or exceeds the permitted use, you will need to obtain permission directly from the copyright holder.
The use of general descriptive names, registered names, trademarks, service marks, etc. in this publication does not imply, even in the absence of a specific statement, that such names are exempt from the relevant protective laws and regulations and therefore free for general use.
The publisher, the authors and the editors are safe to assume that the advice and information in this book are believed to be true and accurate at the date of publication. Neither the publisher nor the authors or the editors give a warranty, expressed or implied, with respect to the material contained herein or for any errors or omissions that may have been made. The publisher remains neutral with regard to jurisdictional claims in published maps and institutional affiliations.

Cover illustration: The smoke plume in this photo is from the Fishlake National Forest, Utah, USA. This burn was part of the Fire and Smoke Model Evaluation Experiment (FASMEE) on 7 November 2019. Photo by John Zapell (retired), Fishlake National Forest.

This Springer imprint is published by the registered company Springer Nature Switzerland AG
The registered company address is: Gewerbestrasse 11, 6330 Cham, Switzerland

Foreword

The Wildland Fire Smoke Science Assessment documents the state of smoke-related science from past to present and provides insights into future needs. The assessment was motivated, at least in part, by recent wildfire years with substantial smoke impacts in the continental USA. From concept to publication, heavy smoke from wildfires and bushfires has been observed in Alaska, Canada, South America, Australia, Greenland, Europe, and Siberia. Transport of the smoke from these wildfires sent emissions around the world and exposed millions of people to unhealthy levels of fine particulates and other pollutants for extended periods. These recent events have had significant effects on the broad field of smoke science and future directions for research and collaboration.

In the USA, particulate matter with aerodynamic diameter < 2.5 μm ($PM_{2.5}$) from all sources, excluding wildland fire (wildfire and prescribed fire), has been declining over the last several National Emissions Inventories conducted by the US Environmental Protection Agency (USEPA). In many places, wildland fire smoke has displaced all other sources, including mobile and industrial sources, as the most significant source of $PM_{2.5}$, accounting for 43% of total $PM_{2.5}$ in the USA in 2017.

The last five years have seen greater focus on smoke and its effects on a range of social values including impacts on mortality, morbidity, and susceptibility to infection and disease; impacts on the economies of communities adjacent to wildlands; and disruption of life when contending with smoke for weeks at a time. While the USA grapples with these mostly adverse smoke effects, the ecological need for periodic fire also highlights the need for research that can help to understand how to balance the objectives of resilient landscapes with the desire for good air quality.

In recent years, the severity and duration of wildfire smoke impacts across the western coast of the USA has been growing: 2020 saw numerous 24-hour periods of $PM_{2.5}$ above USEPA thresholds considered healthy, which led to media and researchers claiming thousands of individuals had potentially died as a result of wildfire smoke. The year 2020 also brought significant focus on the potential compounding effects of air pollution on human health during the COVID-19 pandemic, especially for respiratory health and the potential adverse effects of long-term air pollution on increased susceptibility and severity of COVID-19 infections.

The pandemic also increased public awareness of the importance of personal protective equipment to protect respiratory health (e.g., N95 filtering face respirators). Concerns about the combined effects of smoke exposure and COVID-19 also elevated the issue of wildland firefighter health within the fire management community.

Growing recognition of wildfire smoke impacts has contributed to increased research attention from federal agencies, universities, and the National Academy of Sciences which have explored wildland fire smoke since the 1970s. This growth in interest and research has been building for the last 20 years. Initially, a by-product of other fire science endeavors, smoke science has emerged as a discrete scientific focus, with stand-alone international symposia that draw worldwide attention and engagement. This growth in smoke research will need coordination for improved data sharing mechanisms, technology transfer, and consensus on research needs and priorities.

US Forest Service Research & Development has been a leader in the development of a substantial portion of the science found in the assessment. Initial investment into smoke science as an agency began in the late 1960s and 1970s, a period of growing awareness of the importance of environmental health. The commitment of the USA to the value of clean air was encapsulated in the Clean Air Act of 1970 and the formation of the USEPA in the same year. And with that awareness and national environmental goal, the challenge of balancing the desire for clean air with the need for use of fire for the health of many ecosystems in the USA became evident.

Historically, many wildland fire smoke research efforts have been driven by air quality regulatory efforts to predict and manage prescribed fire smoke. The 1970s was a busy period of research on prescribed fire smoke, and to some extent wildfire smoke, including prescribed fire and wildfire activity levels, prescribed fire emissions and smoke management, and how to reduce prescribed fire emissions. Emphasis was placed on the prescribed fire side of the wildland fire equation based on a misguided view that little could be done about wildfire smoke, which was perceived as "natural" and uncontrollable. National rules supported the idea that areas affected by wildfire smoke should logically not be counted in the calculation of whether an area was meeting National Ambient Air Quality Standards (NAAQS). This interpretation suppressed research on wildfire smoke and operational response for decades, although the prescribed fire focus did provide tools that have recently had utility for addressing wildfire smoke. The assessment highlights the need for future research specific to wildfire smoke, as it has emerged as a serious health risk for the public and firefighters.

Area burned by wildfire has grown since the 1980s, the result of decades of fire suppression, increased fuel loads, and recent periods of severe drought. At the same time, the reality of wildfire smoke as a growing public health issue has also spawned both greater focus and greater collaboration. In the 1970s, the intersection of prescribed fire smoke impacts and wildfire smoke impacts with the NAAQS was virtually nil, because the standards were exceeded infrequently. However, the cumulative impacts of multiple concurrent prescribed fires, and the need to avoid nuisance impacts of any prescribed fire, began to drive development of regulatory smoke management programs in the Northwest and Southeast, with other regions soon to follow. The research community also responded to this regulatory challenge with Forest Service Research & Development leading the charge and working in concert with the USEPA, a collaborative partnership that endures today.

A critical early driver of smoke research in the USA has been the multiagency-funded Joint Fire Science Program (JFSP). Since its establishment in 1998, the JFSP has funded 67 smoke-related research projects, with over $25 million invested in smoke emissions, models, smoke perceptions, and firefighter smoke exposure. One of the hallmarks of these projects has been the creation of new partnerships among federal agencies and academia, thus helping develop new scientists in the field. In addition, the JFSP regional Fire Science Exchange Network has helped to put emerging smoke science into the hands of users and facilitate the transition of research models into operational decision support tools, an important ongoing challenge as noted in the assessment chapter on management needs.

Over time, smoke research has expanded to investigate smoke production and dispersion, more closely address operational needs, and better understand the global dynamics of smoke. Most of the recent large wildfire seasons around the world have been captured on diverse remote sensing platforms and satellite systems; for example, the global transport of wildfire emissions during the 2019–2020 bushfires in Australia was clearly documented. This technology also provides support for previously modeled worldwide mortality of 180,000 per year due to smoke from biomass burning. Long-lived and aging smoke, which may have distinct health risks compared to less aged smoke, has been recognized as a new scientific challenge. Recognition of the need to reduce wildfire risk while allowing fire to fulfill its critical ecosystem function is driving interest in greater use of prescribed fire. In turn, this has raised interest in understanding how to manage smoke from prescribed fire, minimize emissions when possible, and effectively communicate smoke concerns, from social science and medical science perspectives.

Prioritizing research in the future will rely on fully understanding the state of the science, as represented by this assessment, determining lines of research, and then developing the resources and capability to execute new research efforts. This assessment facilitates understanding of critical gaps in research regarding wildland fire smoke impacts and notes the challenge of managing, quantifying, and mitigating effects of prescribed fire smoke. It also underscores the importance of research on wildfire smoke emissions, prediction of impacts, and public and firefighter health effects as the USA begins to address climate change and its effects.

As land managers—federal, state, tribal, and private—struggle with addressing the growing wildfire risks through fuels management and increased use of prescribed fire, it increases the importance of understanding how to minimize and mitigate smoke effects on human health, especially for those most at risk, and prevent adverse smoke effects to roadway visibility. In addition, while much of the smoke that the US public and firefighters breathed in 2020 was from burning vegetation, consumption of fuels was not limited to biomass; it also included human structures and infrastructure, with more than 4,000 structures lost in Oregon alone. The research community is faced with a challenge of understanding the constituents of this source of smoke and what their effects are on humans through both local and downwind exposure. Motivated by almost continuous wildfire years and impacts on public health and the economies of the rural and urban West, the U.S. Congress has also been engaging with questions about wildland fire smoke, with hearings on wildfire risk, prescribed fire and fuel treatments, and the effects of smoke. Issues and questions that emerge at the national level will likely motivate further work on the health and economic effects of smoke and how to best mitigate them.

In recent years, research has expanded into the sources, chemistry, and physics of smoke, how smoke disperses, and the worldwide burden of smoke impacts, including how wildfire smoke that spans the globe is a factor in climate change as well as a source of short-term air pollution. The wheels are now in motion for increased research funding and collaboration in multiple programs, such as the international multiagency and multi-academic institution development of the FIREX-AQ aerial platform spearheaded by National Oceanic and Atmospheric Administration and National Aeronautics and Space Administration, with support from the Forest Service. The level of investment in this project and number of scientists from around the world were unprecedented, as aircraft flew smoke plumes across the USA in 2019. The Forest Service has also stepped into a "big science" role with the ongoing Fire and Smoke Model Evaluation Experiment, which brings individuals together from academic institutions and federal agencies across the country to analyze data during active research burns. This large-scale experiment is ongoing and will help advance smoke science through sharing of data and transfer of new information to users.

The need for more extensive understanding of human health effects, both physiological and mental, of smoke exposure is also coming into focus. Mental health effects of smoke were not a topic of discussion until recently, but months of dark skies, as experienced in the western USA for multiple summers, have prompted many new questions. Concerns for outdoor workers who must conduct their work in the high smoke levels downwind of recent wildfires have generated new protective regulations and the need for improved smoke forecasting to reduce exposure. This concern for health impacts of smoke is also driving greater investment in addressing wildland fire personnel exposure and response.

A number of scientific efforts are addressing the need for better wildland fire smoke information. Investments in smoke modeling, predictions, and forecasting have increased, both as a research area and in support of wildfire response efforts. Ensuring that the public is apprised of the risks from high levels of smoke in the appropriate language and with clear guidance and availability of information is critical. Science to inform outreach efforts that help the public to take appropriate protective actions could keep many branches of research and agencies engaged for years to come. International efforts such as the World Meteorological Organization's Vegetation Fire and Smoke Pollution Warning and Advisory System, including its regional smoke modeling hubs, expand this concept across the globe, matching the scale of the challenge of wildfire and smoke.

The assessment also documents growth in technology and scientific advancement of the field of smoke science associated with collaborations and investments worldwide. This work has provided a basis for today's smoke science and operational smoke management efforts. The impact of technology has progressed rapidly over the last 10 years, building on the foundation of older science with new and more powerful analytical approaches. This is especially evident in the smoke emissions field with recent research efforts utilizing novel ground-based technology, unmanned aerial vehicles, and aerial platforms that can track smoke for many kilometers downwind.

Much of the science found in the pages of this assessment will be integral to meeting the domestic and global challenge of balancing the fire needs of natural ecological systems with a world striving to balance carbon, greenhouse gas emissions, and clean air objectives. Effective emission reduction techniques and accurate smoke impact forecasting will be critical to meet the challenge of smoke management in a warmer climate. Improving our ability to translate smoke science into operations and to inform policy is urgently needed.

There is no shortage of research and management questions in all chapters of the assessment. Maintaining the many collaborations and partnerships of agencies and academia documented here will provide a foundation for answering these questions and informing public policies, management directions, and fire operations. Wildland fire smoke science and its many facets have developed into a new and important research field, one at the front and center of the world stage.

Peter W. Lahm
Fire & Aviation Management
State & Private Forestry,
U.S. Forest Service
Washington, DC, USA

Acknowledgments

Although the National Wildland Fire Smoke Science Assessment documented in this book was led by the US Forest Service, the participation of many other agencies, universities, and other organizations was critical to the success of the project—we are grateful for their collaboration. We are also grateful to participants in the national smoke science workshop that took place in Seattle, Washington, in June 2019. Discussions and planning at the workshop created the vision for this book. We thank authors of all chapters for their dedication, their responsiveness to the review process, and their persistence in writing high-quality chapters despite COVID-19 and other distractions during 2020. We thank Colin Hardy, Sarah Henderson, and Talat Odman who reviewed the entire book manuscript, providing insightful reviews that improved the quality and coherence of the final product. We also thank the following reviewers of individual chapters: Wei Min Hao, Kevin Hiers, Amara Holder, Leda Kobziar, Elizabeth Noth, Susan O'Neill, Jeff Pierce, Don Schweizer, Morgan Varner, and Adam Watts. Yvonne Shih played a key role in the assessment process by helping with workshop details, compiling Appendices B and C, and working on various aspects of editing and communication. Linda Geiser made a major contribution in planning the workshop and other aspects of the assessment and led the development of Appendix A.

Disclaimer

The findings and conclusions in this publication are those of the authors and should not be construed to represent official USDA or US Government determination or policy.

Contents

1 **Assessing the State of Smoke Science** 1
Daniel A. Jaffe, David L. Peterson, Sarah M. McCaffrey,
John A. Hall, and Timothy J. Brown

2 **Fuels and Consumption** 11
Susan J. Prichard, Eric M. Rowell, Andrew T. Hudak,
Robert E. Keane, E. Louise Loudermilk, Duncan C. Lutes,
Roger D. Ottmar, Linda M. Chappell, John A. Hall,
and Benjamin S. Hornsby

3 **Fire Behavior and Heat Release as Source Conditions
for Smoke Modeling** 51
Scott L. Goodrick, Leland W. Tarnay, Bret A. Anderson,
Janice L. Coen, James H. Furman, Rodman R. Linn,
Philip J. Riggan, and Christopher C. Schmidt

4 **Smoke Plume Dynamics** 83
Yongqiang Liu, Warren E. Heilman, Brian E. Potter,
Craig B. Clements, William A. Jackson, Nancy H. F. French,
Scott L. Goodrick, Adam K. Kochanski, Narasimhan K. Larkin,
Peter W. Lahm, Timothy J. Brown, Joshua P. Schwarz,
Sara M. Strachan, and Fengjun Zhao

5 **Emissions** 121
Shawn P. Urbanski, Susan M. O'Neill, Amara L. Holder,
Sarah A. Green, and Rick L. Graw

6 **Smoke Chemistry** 167
Matthew J. Alvarado, Kelley C. Barsanti, Serena H. Chung,
Daniel A. Jaffe, and Charles T. Moore

7 **Social Considerations: Health, Economics, and Risk Communication** 199
Sarah M. McCaffrey, Ana G. Rappold, Mary Clare Hano, Kathleen M. Navarro, Tanya F. Phillips, Jeffrey P. Prestemon, Ambarish Vaidyanathan, Karen L. Abt, Colleen E. Reid, and Jason D. Sacks

8 **Resource Manager Perspectives on the Need for Smoke Science** 239
Janice L. Peterson, Melanie C. Pitrolo, Donald W. Schweizer, Randy L. Striplin, Linda H. Geiser, Stephanie M. Holm, Julie D. Hunter, Jen M. Croft, Linda M. Chappell, Peter W. Lahm, Guadalupe E. Amezquita, Timothy J. Brown, Ricardo G. Cisneros, Stephanie J. Connolly, Jessica E. Halofsky, E. Louise Loudermilk, Kathleen M. Navarro, Andrea L. Nick, C. Trent Procter, Heather C. Provencio, Taro Pusina, Susan Lyon Stone, Leland W. Tarnay, and Cynthia D. West

Appendix A: Regional Perspectives on Smoke Issues and Management 279

Appendix B: Smoke Monitoring Networks, Models, and Mapping Tools 311

Appendix C: Abbreviations and Acronyms 339

Chapter 1
Assessing the State of Smoke Science

Daniel A. Jaffe, David L. Peterson, Sarah M. McCaffrey, John A. Hall, and Timothy J. Brown

Abstract Recent large wildfires in the USA have exposed millions of people to smoke, with major implications for health and other social and economic values. Prescribed burning for ecosystem health purposes and hazardous fuel reduction also adds smoke to the atmosphere, in some cases affecting adjacent communities. However, we currently lack an appropriate assessment framework that looks past the planned versus unplanned nature of a fire and assesses the environmental conditions under which particular fires burn, their socio-ecological settings, and implications for smoke production and management. A strong scientific foundation is needed to address wildland fire smoke challenges, especially given that degraded air quality and smoke exposure will likely increase in extent and severity as the climate gets warmer. It will be especially important to provide timely and accurate smoke information to help communities mitigate potential smoke impacts from ongoing wildfires, as well as from planned prescribed fires. This assessment focuses on primary physical, chemical, biological, and social considerations by documenting our current understanding of smoke science and how the research community can collaborate with resource managers and regulators to advance smoke science over the next decade.

D. A. Jaffe (✉)
University of Washington, Bothell, WA, USA
e-mail: djaffe@uw.edu

D. L. Peterson
University of Washington, Seattle, WA, USA
e-mail: wild@uw.edu

S. M. McCaffrey
U.S. Forest Service, Rocky Mountain Research Station, Fort Collins, CO, USA
e-mail: sarah.m.mccaffrey@usda.gov

J. A. Hall
Boise State University, Division of Research and Economic Development, Boise, ID, USA
e-mail: johnhall440@boisestate.edu

T. J. Brown
Division of Atmospheric Sciences, Desert Research Institute, Reno, NV, USA
e-mail: tim.brown@dri.edu

This is a U.S. government work and not under copyright protection in the U.S.; foreign copyright protection may apply 2022
D. L. Peterson et al. (eds.), *Wildland Fire Smoke in the United States*,
https://doi.org/10.1007/978-3-030-87045-4_1

Keywords Emissions · Fire behavior · Fuels · Health effects · Smoke chemistry · Smoke plume · Smoke management · Wildland fire

1.1 Recent Trends

Data from the National Interagency Fire Center show that annual area burned by wildfires in the USA has increased in recent decades (NIFC 2021). Smoke generated from these fires is of particular concern because it is harmful to human health and can have significant economic implications for nearby communities. In recent years, air quality impacts due to wildfires in the USA have exposed tens of millions of people to elevated and sometimes hazardous concentrations of particulate matter, specifically particulate matter with aerodynamic diameter ≤ 2.5 μm ($PM_{2.5}$), the smoke pollutant of most concern in relation to human health (Chap. 7).

Smoke can affect broad geographic areas, well beyond the actual wildfires. In 2017, numerous large wildfires in the western USA generated smoke plumes that were transported across North America and resulted in $PM_{2.5}$ concentrations that reached unhealthy to hazardous levels (based on the USEPA Air Quality Index(AQI) in many areas (Fig. 1.1). Although US air quality has been improving for decades, largely due to implementation of the Clean Air Act, the effects of wildfires in the past decade have been acute, and in some regions, wildfire smoke has led to a reversal in the general trend toward cleaner air (McClure and Jaffe 2018). Periodic pulses of high $PM_{2.5}$ from smoke are typically much higher than ambient $PM_{2.5}$ concentrations otherwise seen in both rural and urban areas. In 2017 and 2018, many cities in the western USA experienced their all-time highest $PM_{2.5}$ concentrations due to the number of wildfires burning simultaneously (Laing and Jaffe 2019). Very high $PM_{2.5}$ concentrations can also occur in the southeastern USA, although less frequently than in the western USA.

Although most smoke is associated with wildland fires[1] within the USA, fires in other countries can also affect US air quality. In 2017, high $PM_{2.5}$ concentrations in the Pacific Northwest were associated with large fires in British Columbia, Canada (Laing and Jaffe 2019). These same fires were associated with smoke transport to Europe and, locally, strong thunderstorm–pyrocumulonimbus activity, which injected smoke into the stratosphere (Baars et al. 2019). In addition, large fires in Quebec, Canada, have significantly affected air quality in the northeastern USA (DeBell et al. 2004); smoke from fires in Mexico and Central America can affect Texas (Mendoza et al. 2005; Kaulfus et al. 2017); and fires in Siberia can affect air quality in the western USA (Jaffe et al. 2004; Teakles et al. 2017).

---

[1] Throughout this document "wildland fire" is used to encompass both wildfires and prescribed fires. The individual terms are used only when they specifically refer to that specific source of smoke.

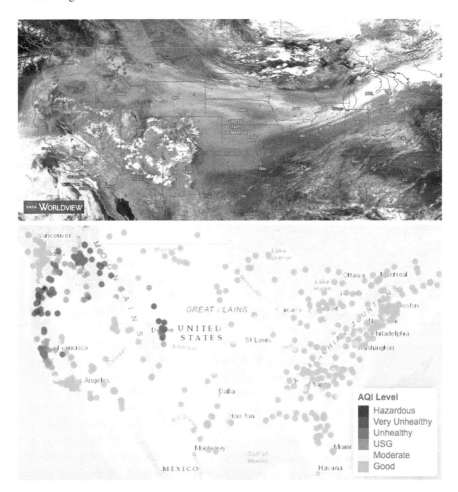

Fig. 1.1 Observed smoke on September 4, 2017. NASA Worldview (https://worldview.earthdata.nasa.gov) image (upper) showing fire hotspot detections from the VIIRS and MODIS satellite instruments, along with visible satellite imagery from the VIIRS instrument between 1200 and 1400 local time. Bright white areas are clouds; grayer areas are smoke. 24-h average $PM_{2.5}$, shown as the corresponding air quality index (AQI) category colors (lower), based on surface PM sensors collected in the USEPA AirNow system (https://www.airow.gov) (From Jaffe et al. (2020))

These increasingly broad and adverse effects of wildland fire smoke have led to growing interest in (1) assessing the state of science in relation to smoke and (2) improving smoke science in order to develop information and tools that can better inform management decisions (e.g., forest treatments and prescribed burning) and mitigate potential smoke impacts of future wildland fires.

1.2 Environmental and Social Context

Wildland fire is an essential ecological process that influences the structure and function of most North American ecosystems. The scale of fire phenomena differs across the nation, with consequences for both emissions and effects of smoke. Wildland fire smoke can affect at least some part of the USA throughout the year (Fig. 1.2). In winter, fires are found mainly in the Southeast, typically as prescribed, low-intensity understory burns to rejuvenate grasses and forbs and prepare seed beds for new tree seedlings, as well as reduce understory growth in pine forests. As spring approaches, fire detections move north and west, with increased prescribed fires on rangelands in the central USA. In Alaska, the wildfire peak is typically in May and June, and summer is the peak wildfire season for the western USA. Late fall can be a time of many wildfires in California and the Southeast. This progression of fire throughout the seasons and ecosystems across the USA has implications for the overall quantity, duration, and human impacts of the emitted smoke (Table 1.1).

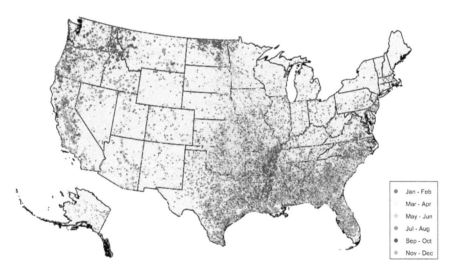

Fig. 1.2 Progression of fires throughout the year using 2017 MODIS hotspot fire detections. (*Source* U.S. Forest Service, from Jaffe et al. (2020))

Table 1.1 Summary of wildland fire for different regions in the USA. Adapted from Jaffe et al. (2020)

Region[a]	Typical fire season	Wildfire characteristics
Alaska	May–Jun	Mostly lightning- caused; high interannual variability in fire depending on the occurrence of dry weather; largest fires >100,000 ha
Eleven western contiguous states, minus California, Arizona, and New Mexico	Jun–Sep	Mostly lightning- caused in mountains; high fuel loadings in many dry forests can facilitate intense fires; largest fires >100,000 ha
California	Oct–Nov Jun–Sep	Many lightning- caused in Sierra Nevada, mostly human-caused elsewhere; high fuel loadings in many dry forests can facilitate intense fires; largest fires >100,000 ha
Arizona and New Mexico	May–Jun	Combination of lightning- and human-caused; fires often driven by interannual variation in fuel production (e.g., grasses); largest fires >100,000 ha
Great Plains	Apr–Jul	Mostly human- caused, some lightning-caused; largest fires are rarely >10,000 ha
Midwest and Northeast	Apr–Jun	Mostly human- caused; dependent on dry spring weather; fires are small
Southeast	Feb–Nov	Mostly human- caused, some lightning-caused; largest fires are usually <10,000 ha, although fires in 2016 burned more than this

[a] Hawai'i and USA-affiliated areas are not included because they comprise a very small portion of fire and smoke occurrence

Humans have a long history of using fire and it is difficult to separate human influence from the natural occurrence of fire on the landscape (Pyne 1997). For centuries, Native Americans used fire as a tool for multiple purposes, including agriculture, managing wildlife habitat and hunting grounds, and cultural practices. As a result of lightning fires and Native American burning, as well as agricultural clearing fires by European settlers, dense and extended periods of smoke were a fairly common occurrence prior to 1900 in many places in the USA. In the 1800s, smoke from wildland and agricultural fires in Oregon hindered navigation on the Columbia River and was credited with contributing to increased illness (Pyne 1997).

The practice of suppressing most wildfires was introduced in the late 1800s. Over time, this policy has contributed to elevated fuel loadings that are one factor contributing to increasing fire size in recent years (Ryan et al. 2013). Fire suppression (and other forms of fire exclusion [e.g., agriculture]) have meant that up until about 1990, less fire has occurred on the landscape than in pre-European settlement times (Leenhouts 1998), resulting in less smoke in the air (Brown and Bradshaw 1994). Recent episodes of smoke across the USA in the last two decades have been driven by large wildfires, and this may be, to some extent, a return to conditions that have not existed since the implementation of widespread fire suppression. A key challenge for forest managers therefore is how to address the fuel accumulation that has occurred as a result of fire suppression (Calkin et al. 2015), while addressing the potential impacts of smoke on a growing human population.

Although 98% of wildfires are currently suppressed before reaching 120 ha (Calkin et al. 2005), annual area burned by wildfires is increasing (Dennison et al. 2014). In the decade between 1991 and 2000, wildfires burned an average area of 1.46 million ha y^{-1}, whereas in the most recent decade (2011–2020) wildfires burned an average area of 3.04 million ha y^{-1} (NIFC 2021). This is mainly due to an increase in large fires that are difficult to control (Dennison et al 2014).

One study has suggested that climate change is contributing to the increased size of wildfires in the western USA (Abatzoglou and Williams 2016), although this study did not consider how fuels and other factors affect wildfire (Dennison et al. 2014). Rising temperatures affect fuel moisture and the length of the fire season (Jolly et al. 2015; Freeborn et al. 2016; McKenzie and Littell 2017). The effects of climate change on area burned will differ by ecosystem and fuel conditions (Littell et al. 2009), with larger areas burned by wildfire in some regions and longer durations of poor air quality due to smoke (Pechony and Shindell 2010; Vose et al. 2018). Changes in fuel composition, loading, and areal extent (Chap. 2) may lead to regional variability that alters the effects of climate change, especially after mid-century. For example, if large wildfire patches comprise an increasing proportion of the landscape, they may limit fire spread.

Prescribed fire—*planned ignition in accordance with applicable laws, policies, and regulations to meet specific objectives* (NWCG 2020)—is an important land management tool that can be used for several management objectives including fuel reduction and ecosystem health. All potential smoke production from such burning must be considered in the context of human health and air quality standards (Chap. 7). Prescribed fires occur under environmental conditions more amenable to fire control (Chaps. 2 and 8) and, depending on the state, may need to be permitted under a smoke management plan to ensure that smoke exposure will not exceed air quality standards or affect sensitive populations.

The ability to plan for when and where a prescribed burn will happen provides some control over the duration, overall amount, and spatial extent of smoke production, although unexpected atmospheric conditions (e.g., a change in wind direction) can result in smoke dispersion into nearby communities (Chap. 4). When a large number of prescribed fires are planned to occur simultaneously, they can create accumulated smoke impacts, making collaboration among burners advisable (Chap. 8).

A final challenge in relation to wildland fire smoke is that wildland fires do not occur in a vacuum. Rather, they occur in landscapes with expanding human populations, increasing the potential for social impacts for both rural and urban areas. Although health impacts are usually the primary concern, smoke can adversely affect a range of social values beyond health (e.g., transportation and tourism) (Chap. 7) and affect areas far beyond the fire perimeter. For example, in 2016, the Chimney Tops fire near Gatlinburg, Tennessee, a major tourism center, caused 15 deaths and burned 2500 homes. It also exposed large populations beyond the immediate area to severely degraded air quality for weeks: monitors in many cities in the southeastern USA had daily $PM_{2.5}$ averages that exceeded 100 $\mu g\ m^{-3}$, a level of exposure that greatly increases risk for people with compromised respiratory function and other medical conditions (Jaffe et al. 2020; Chap. 7). In addition, fires that burn human infrastructure may produce toxins from building materials into the smoke (Chap. 6).

1.3 Overview of This Assessment

This assessment builds on previous integrated analyses of wildland fire and smoke (e.g., Sommers et al. 2014). To better address the growing societal impacts discussed above, an improved understanding of smoke dynamics is needed to more accurately predict the location, extent, and likely effects of smoke, as well as how to effectively mitigate any adverse effects. Because understanding how fire influences air quality is a complex process due to high variability among fires in the quantity and composition of emissions, this will require the compilation of knowledge from diverse scientific disciplines.

Emission characteristics vary as a function of the amount and type of fuel, meteorology and burning conditions (Chap. 2), fire behavior (Chap. 3), and smoke dispersal (Chap. 4); therefore, emissions (Chap. 5) for individual fires are often uncertain and difficult to predict. In addition to $PM_{2.5}$, smoke contains numerous gaseous compounds, some of which are harmful to people, including nitrogen oxides, carbon monoxide, ozone, methane, and hundreds of volatile organic compounds (Chap. 6). This chemical complexity makes wildfire smoke different from typical industrial pollution. In addition, once emitted, wildland fire smoke undergoes chemical transformations in the atmosphere, which alter the mix of compounds and generate secondary pollutants, such as ozone and secondary organic aerosols (Chap. 6); some of these secondary compounds appear to be more toxic than the primary emissions (Wong et al. 2019).

Ultimately, given that the social impacts of smoke are the foundation for these scientific needs, a better understanding of the full range of human health and economic costs of smoke is needed (Chap. 7). Complex interactions among wildland fires, climate change, and other factors mean that the different disciplines of smoke science need sufficient integration to ensure credible and consistent projections of physical phenomena and human impacts through space and time. Clear linkages between what resource managers and regulators need and what is being produced through scientific research is also critical (Chap. 8).

The technical capability of smoke measurement and modeling has increased significantly over the past decade. Our understanding of acute human health effects has also increased, partially in response to big smoke events and partially in response to concerns about effects on wildland firefighters who are exposed to smoke for weeks at a time during the course of their work. This scientific knowledge is encouraging, but greater accuracy is needed in all aspects of smoke science to better mitigate future health and economic impacts.

To that end, we are now at a critical point in the development of smoke science. Several large-scale field projects, focused on comprehensive measurements and modeling (detailed in subsequent chapters) have been recently completed or will be completed within the next few years (e.g., FASMEE; Prichard et al. 2019). These experiments include simultaneous satellite-, aircraft-, drone-, and ground-based sensors which, along with fuel measurements, should significantly improve our knowledge about a number of smoke phenomena.

Accompanying this potential wealth of data will be the need to develop new assessment frameworks through which we can compare and evaluate characteristics of different types of fires, their smoke consequences, and opportunities for planning and managing fires to reduce smoke impacts. However, this information will be meaningful only with a better understanding of the health and economic effects of smoke and identification of which actions most effectively mitigate those effects. Williamson et al. (2016) articulated the principles of a potential framework for guiding scientific and management needs associated with fire and smoke, but more effort is needed to develop this framework.

Poised on the cusp of a new wave of technically advanced smoke research and a surge in new data, it is imperative that we summarize the current state of science for wildland fire smoke as a foundation for integration of new information. The subsequent chapters of this book assess that state of science as follows:

- Fuels and consumption (Chap. 2)
- Fire behavior and heat release (Chap. 3)
- Smoke plume dynamics (Chap. 4)
- Emissions (Chap. 5)
- Smoke chemistry (Chap. 6)
- Social Considerations: Health, Economics, and Risk Communication (Chap. 7)
- Resource manager perspectives on the need for smoke science (Chap. 8).

Chapters 2 through 6 focus on physical, chemical, and biological factors that affect fire and smoke. Chapter 7 examines the existing knowledge on key impacts, particularly human health, all of which rely on a better understanding of the physical and chemical nature of smoke, as well as on improved knowledge of human sensitivities and responses to smoke to understand social and economic consequences. We note here that the social costs of smoke are significant and include documented increases in cardiovascular issues, premature mortality, and direct health costs in the billions of dollars annually (Fann et al. 2018). A summary of management and regulatory issues related to smoke science is presented in Chap. 8, which can be used to inform research and facilitate science-management collaborations in the future.

Although this assessment is, by necessity, divided into the primary components of smoke science, authors of the above chapters have integrated among components as much as possible. This assessment emphasizes recent discoveries, linking to projects and lines of inquiry that are in progress or soon will be. Recommendations for future research are included in each chapter.

This is an exciting time for the science and management of smoke in the USA and other parts of the world, and we anticipate rapid progress in the years ahead. As smoke will likely become a more pervasive issue in a warmer climate with more extensive wildfires, it is also a critical time for the smoke science community to continue to make progress. Our hope is that collaboration at all levels will improve effectiveness of the research process and timeliness of integration into useful applications, ultimately benefiting the health and welfare of all communities affected by smoke.

References

Abatzoglou JT, Williams AP (2016) Impact of anthropogenic climate change on wildfire across western US forests. Proc Natl Acad Sci USA 113:11770–11775

Baars H, Ansmann A, Ohneiser K et al (2019) The unprecedented 2017–2018 stratospheric smoke event: decay phase and aerosol properties observed with the EARLINET. Atmos Chem Phys 19:15183–15198

Brown JK, Bradshaw LS (1994) Comparisons of particulate-emissions and smoke impacts from presettlement, full suppression, and prescribed natural fire period in the Selway-Bitterroot Wilderness. Int J Wildland Fire 4:143–155

Calkin DE, Gebert KM, Jones JG, Neilson RP (2005) Forest Service large fire area burned and suppression expenditure trends, 1970–2002. J For 103:179–183

Calkin DE, Thompson MP, Finney MA (2015) Negative consequences of positive feedbacks in US wildfire management. For Ecosyst 2:9

DeBell LJ, Talbot RW, Dibb JE et al (2004) A major regional air pollution event in the northeastern USA caused by extensive forest fires in Quebec Canada. J Geophys Res Atmos 109:D19305

Dennison PE, Brewer SC, Arnold JD, Moritz MA (2014) Large wildfire trends in the western USA, 1984–2011. Geophys Res Lett 41:2928–2933

Fann N, Alman B, Broome RA et al (2018) The health impacts and economic value of wildland fire episodes in the US: 2008–2012. Sci Total Environ 610:802–809

Freeborn PH, Jolly WM, Cochrane MA (2016) Impacts of changing fire weather conditions on reconstructed trends in U.S. wildland fire activity from 1979 to 2014. J Geophys Res 121:2856–2876

Jaffe DA, Bertschi I, Jaegle L et al (2004) Long-range transport of Siberian biomass burning emissions and impact on surface ozone in western North America. Geophys Res Lett 31:L16106

Jaffe DA, O'Neill SM, Larkin NK et al (2020) Wildfire and prescribed burning impacts on air quality in the USA. J Air Waste Manag Assoc 70:583–615

Jolly W, Cochrane M, Freeborn P et al (2015) Climate-induced variations in global wildfire danger from 1979 to 2013. Nat Commun 6:7537

Kaulfus AS, Nair U, Jaffe D et al (2017) Biomass burning smoke climatology of the USA: implications for particulate matter air quality. Environ Sci Technol 51:11731–11741

Laing JR, Jaffe DA (2019) Wildfires are causing extreme PM concentrations in the western USA. In: EM—the magazine for environmental managers (July). Air and Waste Management Association, Pittsburgh

Leenhouts, B. (1998). Assessment of biomass burning in the conterminous USA. Conserv Ecol 2(1)

Littell JS, McKenzie D, Peterson DL, Westerling AL (2009) Climate and wildfire area burned in western U. S. ecoprovinces, 1916–2003. Ecol Appl 19:1003–1021

McClure CD, Jaffe DA (2018) US particulate matter air quality improves except in wildfire-prone areas. Proc Natl Acad Sci USA 115:7901–7906

McKenzie D, Littell JS (2017) Climate change and the eco-hydrology of fire: Will area burned increase in a warming western USA? Ecol Appl 27:26–36

Mendoza A, Garcia MR, Vela P et al (2005) Trace gases and particulate matter emissions from wildfires and agricultural burning in Northeastern Mexico during the 2000 fire season. J Air Waste Manag Assoc 55:1797–1808

National Interagency Fire Center (NIFC) 2021 Fire information statistics. https://www.nifc.gov/fireInfo/fireInfo_statistics.html. 4 Oct 2021

National Wildfire Coordinating Group (NWCG) (2020) NWCG Glossary of wildland fire (PMS 205). https://www.nwcg.gov/glossary/a-z/sort/p?combine=prescribed%20fire. 19 June 2020

Pechony O, Shindell DT (2010) Driving forces of global wildfires over the past millennium and the forthcoming century. Proc Natl Acad Sci USA 107:19167–19170

Prichard SJ, Larkin NK, Ottmar RD et al (2019) The fire and smoke model evaluation experiment—a plan for integrated, large fire–atmosphere field campaigns. Atmosphere 10:66

Pyne SJ (1997) Fire in America: a cultural history of wildland and rural fire. University of Washington Press, Seattle

Ryan KC, Knapp EE, Varner JM (2013) Prescribed fire in North American forests and woodlands: history, current practice, and challenges. Front Ecol Environ 11:e15–e24

Sommers WT, Loehman RA, Hardy CC (2014) Wildland fire emissions, carbon, and climate: science overview and knowledge needs. For Ecol Manage 317:1–8

Teakles AD, So R, Ainslie B et al (2017) Impacts of the July 2012 Siberian fire plume on air quality in the Pacific Northwest. Atmos Chem Phys 17:2593–2611

Vose JM, Peterson DL, Domke GM et al (2018) Forests. In: Reidmiller DR, Avery CW, Easterling DR et al (eds) Impacts, risks, and adaptation in the USA: fourth national climate assessment, vol II. U.S. Global Change Research Program, Washington, DC, pp 232–267

Williamson GJ, Bowman DMS, Price OF, et al (2016). A transdisciplinary approach to understanding the health effects of wildfire and prescribed fire smoke regimes. Environ Res Lett 11:125009

Wong JPS, Tsagkaraki M, Tsiodra I et al (2019) Effects of atmospheric processing on the oxidative potential of biomass burning organic aerosols. Environ Sci Technol 53:6747–6756

Open Access This chapter is licensed under the terms of the Creative Commons Attribution 4.0 International License (http://creativecommons.org/licenses/by/4.0/), which permits use, sharing, adaptation, distribution and reproduction in any medium or format, as long as you give appropriate credit to the original author(s) and the source, provide a link to the Creative Commons license and indicate if changes were made.

The images or other third party material in this chapter are included in the chapter's Creative Commons license, unless indicated otherwise in a credit line to the material. If material is not included in the chapter's Creative Commons license and your intended use is not permitted by statutory regulation or exceeds the permitted use, you will need to obtain permission directly from the copyright holder.

Chapter 2
Fuels and Consumption

Susan J. Prichard, Eric M. Rowell, Andrew T. Hudak, Robert E. Keane,
E. Louise Loudermilk, Duncan C. Lutes, Roger D. Ottmar,
Linda M. Chappell, John A. Hall, and Benjamin S. Hornsby

Abstract Wildland fuels, defined as the combustible biomass of live and dead vegetation, are foundational to fire behavior, ecological effects, and smoke modeling. Along with weather and topography, the composition, structure and condition of wildland fuels drive fire spread, consumption, heat release, plume production and smoke dispersion. To refine inputs to existing and next-generation smoke modeling tools, improved characterization of the spatial and temporal dynamics of wildland fuels is necessary. Computational fluid dynamics (CFD) models that resolve fire–atmosphere interactions offer a promising new approach to smoke prediction. CFD models rely on three-dimensional (3D) characterization of wildland fuelbeds (trees,

S. J. Prichard (✉)
School of Environmental and Forest Sciences, University of Washington, Seattle, WA, USA
e-mail: sprich@uw.edu

E. M. Rowell
Tall Timbers Research Station, Tallahassee, FL, USA
e-mail: erowell@talltimbers.org

A. T. Hudak
U.S. Forest Service, Rocky Mountain Research Station, Moscow, ID, USA
e-mail: andrew.hudak@usda.gov

R. E. Keane
U.S. Forest Service, Rocky Mountain Research Station, Missoula, MT, USA
e-mail: robert.keane@usda.gov

E. L. Loudermilk · B. S. Hornsby
U.S. Forest Service, Southern Research Station, Athens, GA, USA
e-mail: elloudermilk@fs.fed.us

B. S. Hornsby
e-mail: benjamin.hornsby@usda.gov

D. C. Lutes
U.S. Forest Service, Rocky Mountain Research Station, Missoula, MT, USA
e-mail: duncan.lutes@usda.gov

R. D. Ottmar
U.S. Forest Service, Pacific Northwest Research Station, Seattle, WA, USA
e-mail: roger.ottmar@usda.gov

This is a U.S. government work and not under copyright protection in the U.S.; foreign copyright protection may apply 2022
D. L. Peterson et al. (eds.), *Wildland Fire Smoke in the United States*,
https://doi.org/10.1007/978-3-030-87045-4_2

shrubs, herbs, downed wood and forest floor fuels). Advances in remote sensing technologies are leading to novel ways to measure wildland fuels and map them at submeter to multi-kilometer scales as inputs to next-generation fire and smoke models. In this chapter, we review traditional methods to characterize fuel, describe recent advances in the fields of fuel and consumption science to inform smoke science, and discuss emerging issues and challenges.

Keywords Fire behavior modeling · Fuel consumption · Measurement · Remote sensing · Vegetation dynamics · Wildland fuels

2.1 Introduction

Fuels, topography, and weather comprise the classic fire behavior triangle (Chap. 3). Fuels are the only one of the three variables that can be managed to influence fire behavior before an ignition occurs. In their most basic form, wildland fuels are the combustible biomass of live and dead vegetation. Because combustion of wildland fuels generates heat and emits pollutants, fuels science is a critical foundation of fire behavior and smoke modeling (Anderson 1976; Omi 2015; Keane 2019).

Along with weather and topography, characteristics of fuels that burn in a wildland fire event will drive fire spread, energy release, fuel consumption, and smoke production (Ottmar 2014). For example, a dry grassland with continuous cover can generate fast-moving fires with short-duration smoke production (Cook et al. 2016). In contrast, dense mixed-conifer forests with deep organic soils can support crown fires with large plume development followed by inefficient smoldering combustion in coarse wood and organic soil layers associated with long-duration smoke production (de Groot et al. 2007).

2.1.1 Understanding How Fuels Contribute to Smoke

A detailed accounting of how wildland fuels contribute to fire behavior and combustion is thus fundamental for smoke model predictions. Smoke emissions estimates are based on type and mass of fuel consumed, which is then used to determine smoke composition through emission factors for specific fuel categories (Urbanski 2014; Chap. 5). Each step of the smoke modeling process relies on source characterization of the composition and biomass of fuels and consumption in a wildland fire event (see

---

L. M. Chappell
U.S. Forest Service, Intermountain Region, Logan, UT, USA
e-mail: linda.chappell@usda.gov

J. A. Hall
Division of Research and Economic Development, Boise State University, Boise, ID, USA
e-mail: johnhall440@boisestate.edu

Fig. 5.7). Because fuels are dynamic over space and time, any effort to quantify fuels must be informed by the ecology of live and dead vegetation (Mitchell et al. 2009; Keane 2015). Of all variables involved in estimating smoke emissions, the amount of available fuel and proportion that is consumed are often the highest sources of uncertainty. Errors in estimates of available pre-burn fuels can create potentially large errors when estimating emissions due to fuel consumption (Peterson 1987). Reliable estimates of fuels also generally require more detailed site information than is provided by remotely sensed imagery and classified vegetation cover and type. For example, fuels that burn in a forest fire are often obscured by forest canopies and are strongly dictated by past disturbances or management activities (Keane 2015; Prichard et al. 2019a). Passive remote sensing imagery may provide operational maps of forest cover but cannot quantify the amount, structure, or condition of sub-canopy fuels that drive fire behavior and consumption (Keane et al. 2001).

Current geospatial datasets of wildland fuels, which are based on remote sensing, generally have a high degree of uncertainty (e.g., LANDFIRE; Keane et al. 2006; Reeves et al. 2006). The increased availability of remotely sensed datasets that enable 3D mapping of pre- and post-fire vegetation and fuels at multiple scales is contributing to a rapid evolution in the field of fuel characterization and consumption (Loudermilk et al. 2009; Wang and Glenn 2009; Hoff et al. 2019; Hudak et al. 2020). Next-generation fuel characterization will need to be at scales and resolutions appropriate for physics-based computational fluid dynamics (CFD) models that are capable of resolving fire–atmosphere interactions, heat release, and smoke production (Loudermilk et al. 2009; Rowell et al. 2016). Understanding the sources of uncertainty of aggregating fine-scale fuel characterization and consumption to the coarser scales used in smoke modeling and planning is an important area of study. For example, distribution of downed logs and stumps may vary at fine spatial scales (Brown 1974; Keane 2015), but reliable estimates of their consumption across burn units may be critical to anticipating long-term smoke impacts (Chaps. 3, 5 and 6).

Reliable fuel characterization is also needed to guide prescribed burn planning where fire managers need to take into account and mitigate potential smoke impacts to communities (Lavdas 1996). As timber harvest, mechanical fuel reduction, and prescribed burning modify fuels, fuel characterization after such treatments is critical for assessing effectiveness and how these activities influence fire behavior and smoke production (Reinhardt et al. 2008; Stephens et al. 2012).

Site-specific inventories of fuels and their predicted contribution to flaming and smoldering phases of fire inform forecasts used by fire managers during wildland fire events. If prescribed fire managers are aware of deep organic soil layers and large amounts of coarse wood that could contribute to long-term smoldering and low-buoyancy smoke production, they can model potential impacts and adjust burn prescriptions and mop-up procedures to mitigate associated impacts to air quality. The amount of consumption by combustion phase and duration of combustion (Ottmar 2014) directly influences smoke production, plume dynamics (Chap. 4), emissions (Chap. 5), carbon fluxes, tree mortality, soil heating, and other vegetation dynamics (Keane 2015). Furthermore, the amount and types of fuel consumed in flaming, smoldering and long-term smoldering (or glowing) phases of combustion are necessary for predicting emissions of specific pollutants (e.g., CO, $PM_{2.5}$) (Chaps. 5 and 6) and for anticipating smoke intrusions into communities (Peterson et al. 2018).

This chapter presents the current state of science for estimating the amount of wildland fuel and consumption as related to smoke management and future research needs. Topics covered include (1) an introduction to wildland fuels, (2) the current state of science on fuel characterization and consumption, (3) a vision for fuel and consumption science to inform smoke prediction, and (4) emerging issues and challenges in the field of fuel characterization and consumption research. Because source characterization of wildland fuels is critical to predicting smoke impacts, reviewing how to measure and map wildland fuel biomass and consumption provides useful context for fire and fuels managers, smoke scientists, and policy makers. We also review advances that are necessary for next-generation models of wildland fire behavior and smoke.

2.2 Wildland Fuels

Wildland fuels are often characterized as fuelbeds that are stratified by structure, continuity, and composition of biomass including tree canopies, snags, shrub stems and leaves, grass and herbaceous vegetation, sound and rotten wood, needle and leaf litter, and organic ground fuels (Ottmar et al. 2007). Numerous ecological processes influence wildland fuel dynamics, but four are particularly important in governing spatial and temporal distributions of wildland fuels (Keane 2015):

- Wildland fuels accumulate from the establishment, growth, phenology, and mortality of vegetation (*development*). The rate of biomass accumulation, or productivity of vegetation, is dictated by interactions of the plant species available to occupy a site and the physical environment (climate, soils, and topography).
- Over time, portions of living biomass shed or die and are *deposited* on the ground to become dead surface fuels, termed necromass.
- Below- and above-ground necromass is eventually *decomposed* by microbes and soil macrofauna.
- *Disturbances*, such as fire, insects, and disease, act on living and dead biomass to change the magnitude, trend, and direction of fuel accumulation in space and time.

These four processes interact to influence fuel development where the interactions depend on the ecosystem and corresponding climate and disturbance regimes. For example, live and dead vegetation characteristics often correlate to development and deposition, whereas climate drives decomposition and disturbance (Keane 2008). Vegetation is sometimes used as a surrogate for fuels (Keane et al. 1998; Menakis et al. 2000), but this assumption ignores the pivotal role of decomposition and disturbance on fuelbed development.

Wildland fuel properties and their distributions are a cumulative result of interactions of the four above processes across multiple spatial and temporal scales that create shifting mosaics of fuel conditions on fire-prone landscapes (Keane et al. 2012). The processes can also create heterogeneity in fuel loading and structure. For example, loading (biomass per unit area; kg m^{-2}) of fine woody debris can vary by 2–3 times its mean over a small (<10 ha) prescribed burn unit (Keane et al. 2012).

The spatial and temporal variability of wildland fuels can influence how fuel consumption influences smoke emissions (Anderson 1976) and, in turn, how fuel management influences fuel properties (Stephens et al. 2012). Because fuel dynamics are so heterogeneous, robust fuel classifications, sampling methods, and geospatial datasets are needed to improve predictions of fuel consumption and smoke production (Parsons et al. 2010; Keane 2015). Spatial configuration of fuel characteristics is needed for next-generation fire effects and behavior models that rely on 3D fuel inputs and represent fire with CFD modeling (Linn et al. 2002; Mell et al. 2007; King et al. 2008; Parsons et al. 2010). This variability, combined with uncertainty of fuel sampling techniques, makes estimating accurate fuel loadings for smoke prediction challenging.

2.2.1 Fuel Characteristics

The wildland fuelbed is generally divided into three vertical fuel *layers* including canopy, surface, and ground fuels (Keane 2015). *Canopy fuels* are the biomass above the surface fuel layer (>2 m high). *Surface fuels* generally include biomass within 2 m above the ground surface. *Ground fuels* are all organic matter below the ground line, where the ground line is usually just below the litter (Oi soil horizon, slightly decomposed) and include the Oe (moderately decomposed), and Oa (highly decomposed) soil horizons (collectively, "duff") (Soil Science Division Staff 2017).[1] Each fuelbed layer is composed of finer-scale elements called fuel strata and categories (Fig. 2.1).

Fuel strata describe the vertical profile of the wildland fuelbed, whereas *fuel categories* describe fuel types that are qualitatively and quantitatively defined for specific purposes or objectives, such as fire behavior prediction (Table 2.1). For example, the downed wood stratum often contains fuel categories including fine wood (<8 cm diameter), coarse wood (>8 cm diameter), stumps, and piles (Riccardi et al. 2007b). Fuels in the fine wood category are generally consumed during the flaming phase and drive fire spread, whereas coarse wood burns during the flaming phase of combustion but the majority of consumption is in smoldering combustion

---

[1] Ground fuels are defined as partially or fully decomposed soil organic matter. Organic soil horizons often consist of three vertical layers: the newly fallen leaf litter (Oi), partially decomposed material (Oe), and highly decomposed material (Oa). In the context of fuels, the Oi remains distinct from the Oe and the Oa, which are often combined into what is commonly called "duff" or ground fuels by fire and fuel managers. For this chapter, the Oi is referred to as leaf litter or litter, while the Oa and Oe horizons are combined and referred to as ground fuels.

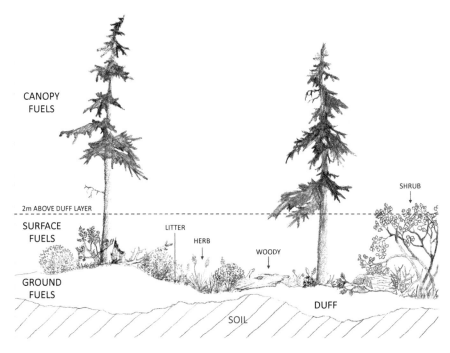

Fig. 2.1 Vertical fuel strata in a wildland fuelbed [Drawing by Ben Wilson, from Keane (2015)]

that occurs long after the passage of the flaming front, contributing to long-duration heat release and smoke (Albini 1976; Hyde et al. 2011).

Fuel strata and categories have specific physical and chemical properties, such as bulk density, loading (mass per area, kg m^{-2}), surface area (m^2), and heat content (J kg^{-1}), all of which are important inputs to fire behavior and effects models and descriptors of fuel characteristics (Chap. 3). The finest scale of fuel description is the fuel *particle,* which is a general term for a specific piece of fuel that is part of a fuel category. A fuel particle can be an intact or fragmented woody stick, grass blade, shrub leaf, or pine needle. Fuel particles have the widest diversity of properties, such as specific gravity (kg m^{-3}), heat content (J kg^{-1}), volume (m^3), and shape (unit or quality here). The properties of fuel categories, strata and fuelbeds, are often quantified from statistical summaries of properties of the particles that comprise them, thereby a source of uncertainty. For example, the means of quadratic mean diameter and surface area-to-volume ratio (m^{-1}) of all particles are often applied to size classes of wood particles (e.g., Brown 1974).

Within any given fuel strata, component or particle, wildland fuels are also defined as *dead* or *live*. Dead fuel is suspended or downed dead biomass (necromass), and live fuel is the biomass of living organisms including vascular plants (trees, shrubs, and herbs) and nonvascular plants such as mosses and ground lichens. The principal reason for distinguishing between live and dead fuels is the difference in fuel moisture dynamics that dictates the availability to burn, often called fuel condition. Both live

2 Fuels and Consumption

Table 2.1 Common canopy, surface fuel, and ground fuel categories used for fire and smoke modeling

Fuel stratum	Fuel category	Size	Description
Canopy fuels			
Canopy	Tree crowns		Fine branches (<6 mm diameter) and dead and live aerial foliage
	Snags		All burnable portions of dead trees including branches and stem wood
	Ladder fuels including vines, branches, tree regeneration		Any fuel that serves as a ladder between surface and canopy fuels
Surface fuels			
Shrub	Shrub crowns and stems	All shrubby material less than 5 cm diameter	All burnable shrubby biomass with branch diameters less than 5 cm
Herb	Grasses and forbs (non-woody vegetation)	All sizes	All live and dead grass, forb, and fern biomass
Downed wood	1-h wood (fine wood, twigs)	<0.6 cm diameter	Detached small wood fuel particles within 2 m of the ground
	10-h wood (fine wood, branches)	0.6–2.5 cm diameter	Detached small wood fuel particles within 2 m of the ground
	100-h wood (fine wood, branches)	2.5–8 cm diameter	Detached small wood fuel particles within 2 m of the ground
	1000-h wood (logs, coarse woody debris)	8 + cm diameter	Detached woody fuel particles within 2 m of the ground
Litter-lichen-moss	Litter	All	Freshly fallen non-woody material including leaves, cones, pollen cones
	Lichen	All	Lichen that grows on the ground surface (common in boreal forests)

(continued)

Table 2.1 (continued)

Fuel stratum	Fuel category	Size	Description
	Moss (bryophyte)	All	Moss that grows on the ground surface (common in boreal forests)
Ground fuels			
Organic soil horizons	Oe horizon Oa horizon	All	Partially decomposed and fully decomposed biomass, including decomposed litter and peat
Basal accumulations		All	Accumulated organic soil, bark slough, and litter around older trees

Fine woody debris (FWD) is a term often used for wood fuel particles <8 cm in diameter, and coarse woody debris (CWD) refers to woody fuel particles > 8 cm in diameter

and dead fuel properties are governed by antecedent weather, but live fuel moistures are primarily controlled by phenology, transpiration, evaporation, and soil water, which differ among taxa and across regional climate (Jolly et al. 2014). In contrast, dead fuel moisture is dictated by the physical properties of the fuel (e.g., size, density, surface area) and their interaction with local climate, short-term weather dynamics (wind, solar radiation and vapor pressure deficit), and available soil moisture (Fosberg et al. 1970; Viney 1991).

The 3D configuration of wildland fuels characterizes where fuels are and where they are not. Gaps in fuel structure influence fire spread, including whether a forest can support transitions from surface to crown fires (i.e., individual or group torching) and how readily fires can spread from tree crown to tree crown (crowning that is independent of surface fire dynamics) (Parsons et al. 2017; Ziegler et al. 2017). The spatial continuity of surface fuels also affects fire behavior. For example, although deserts and xeric rangelands may support vegetation that is dry enough to ignite, fire spread is unlikely due to sparse fuels and lack of continuity (Gill and Allan 2008; Swetnam et al. 2016).

2.2.2 Traditional Methods to Estimate Wildland Fuel Loadings

Numerous methods have been developed to estimate fuel loading (i.e., combustible biomass) to allow for flexibility in matching available resources with sampling objectives and constraints (Catchpole and Wheeler 1992). Keane (2015) reviewed traditional fuel sampling methods and the inherent challenges in measuring spatial and temporal variability of wildland fuels. Here, we summarize the main methods and review practical sampling limitations that are prompting evaluation of new technologies and methods.

Many traditional approaches to wildland fuel characterization rely on a variety of indirect methods to estimate loading and structure of wildland fuels. Methods such as photo series or mapping fuels based on major vegetation types rely on visual or associative techniques to relate fuel characteristics to available observations or datasets (Keane 2015). Associating fuel characteristics with remotely sensed products, such as Landsat Thematic Mapper, has limitations due to imagery resolution and forest and shrub canopies that obscure surface and ground fuels. In addition, high variability of fuel characteristics within a site or pixel may overwhelm unique fuelbed identification across sites (Keane et al. 2013; Prichard et al. 2019a). Another common method is to simplify fuel descriptions into fire behavior fuel models or broad vegetation types for fire simulations (Scott and Burgan 2005). Fuel models generally are too simplistic to represent the complexity of wildland fuels and ignore categories important to smoke and other fire effects such as coarse wood and organic soils (Sikkink and Keane 2008; Keane 2015).

Direct methods involve field sampling or measuring characteristics of fuel particles in situ or in the lab to calculate loading and usually involve direct contact with the fuel (e.g., measuring dimensions and weight of particles). Within fixed-area plots, mass is often measured using *destructive sampling*, which involves physically clipping and collecting the fuel, then drying the material and weighing it (Mueller-Dombois and Ellenberg 1974; Sokal and Rohlf 1981). Methods for sampling litter and ground fuel loading have remained virtually unchanged over the last four decades (Brown et al. 1985; DeBano et al. 1998) and include destructive sampling and estimations based on depth measurements.

Some ecosystems may have patchy soil organic matter coverage (e.g., deserts, woodlands, sagebrush, grasslands), making sampling difficult and often requiring a field measurement of ground fuel and litter cover. Several factors affect the accuracy and precision of estimates for monitoring and calculations of ground fuel consumption. First, the spatial variability of litter often requires a high number of measurements. Depth measurements are challenging because the interface between the duff and litter layers can be diffuse. Sampling also disrupts the ground fuel layer and can compromise pre- and post-fire measurements. Discontinuities in some litter and ground fuels are also challenging to quantify, including animal scat, mineral content, tree cones, and basal accumulations (Ottmar et al. 2007). Finally, reliable bulk density values are lacking for many fuelbeds in North America, and accurate characterization of litter and ground fuel loading require destructive sampling to include depth and bulk density measurements.

Due to the high spatial variability of wildland fuels and lack of correlation between fuel strata and categories, estimations based on traditional fuel measurement techniques often result in high variance and lack of precision (Keane 2013). For example, planar intersect sampling of woody fuel loadings incorporates only one dimension (Brown 1971). Given that fine and coarse wood can vary differently across space, linear sampling may not capture spatial variability of fine and coarse wood (Keane and Gray 2013). Other conventional fuel inventory techniques, such as photo series, may also be inappropriate because fuels vary at spatial scales that might be different from the scales represented by the photo (Keane and Gray 2013).

Many fuel assessments involve sampling fuels before and after treatment, especially when estimating fuel consumption. Making consistent measurements is challenging because accurate fuel sampling involves direct manipulation of the fuelbed. For example, destructive sampling removes fuel from a fixed-area plot, rendering the plot unusable for post-fire monitoring. High variability of fuels may preclude paired sampling (i.e., plots outside the treatment area used to quantify pre-treatment conditions) or quantifying pre-burn conditions using classification, mapping, and modeling. Accurate and consistent sampling methods are needed to sample fuels for the same sampling frame throughout the monitoring period. Some have used the photoload method (Keane and Dickinson 2007) as a way to sample fuels within a sample frame without disturbance with mixed results (Tinkham et al. 2016).

2.2.3 Emerging Technologies and Methods

Advances in remote sensing offer a number of promising methods to characterize wildland fuels including airborne and *ground-based light detection and ranging (Lidar)* and *structure-from-motion photogrammetry* (SfM) (Loudermilk et al. 2009; Hudak et al. 2016; Cooper et al. 2017) that allow for synoptic, 3D characterization of many wildland fuels.

Ground-based *Lidar*, also known as *terrestrial laser scanning* (TLS), is used to estimate the loading and structure of surface and sub-canopy fuels (Loudermilk et al. 2009; Seielstad et al. 2011). Mounted on a tripod or vehicle, TLS units obtain scan distances at sub-cm scales from the instrument location to vegetation, surface fuel, and other object surfaces and can penetrate through foliage layers. The Lidar signal, which amounts to a 3D cloud of X, Y, and Z points, can then be related to fuel loading by constructing statistical models where destructively sampled loadings for various categories are correlated to statistical metrics derived from the Lidar point cloud data (Fig. 2.2). It can be difficult to differentiate between fuel categories using TLS in heterogeneous fuelbeds, and integration with multispectral imagery is sometimes necessary for image interpretation. The cost of TLS instruments and image processing generally relegates their use to research.

Airborne Lidar scanning (ALS) is used operationally for precision forest inventory of tree stems and crowns. Its coarser resolution (9–12 returns per m^2) as well as the influence of overstory objects and noise limits its ability to adequately characterize understory and surface fuels, especially through an overstory forest canopy (Hudak et al. 2016a, b). Active Lidar remote sensing adds a vertical dimension to other remotely sensed datasets, because it can penetrate vegetation biomass and characterize pre- and post-burn vegetation structure, biomass, and fuel consumption (Lefsky et al. 2001; Hyde et al. 2007; Sexton et al. 2009). Lidar offers advances in forest biomass mapping, because physical measures of canopy height and density

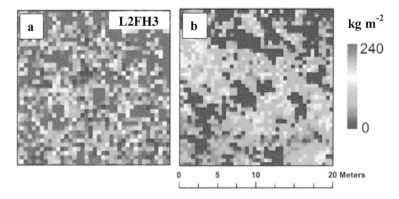

Fig. 2.2 Example of pre-fire TLS-derived fuel mass for (**a**) managed forest plots (Rowell 2017) and **b** post-fire residual fuels for the same site. This dataset demonstrates variability of fuels consumption for prescribed fire, where 3D structure, ignition pattern, fuel moisture and fluid flow of air affect how fire consumes fuels

can be extracted from point cloud data and reduce the uncertainties in biomass (i.e., fuel load) estimation. Neither Lidar nor other remote sensing systems can penetrate the forest floor to measure litter and ground fuel depth, although recent work suggests that robust estimates of the litter-layer fuel mass are possible (Rowell et al. 2020).

SfM technology uses photogrammetry of high-resolution images, often collected from cameras mounted on an unmanned aerial system (UAS) to create 3D multi-spectral images of vegetation and fuels (Zarco-Tejada et al. 2014). Although photogrammetric points have inferior vegetation penetration compared to Lidar, the multi-spectral capabilities of digital cameras make assignment of plant functional type or live/dead status more feasible than from the single near-infrared or green channel data in most Lidar sensors (Bright et al. 2016; Hudak et al. 2020). Integrating short-range SfM using digital cameras, mobile phones, or high definition (4K) digital video allows for fine-scale, 3D representations of wildland fuels in true color or multispectral images (Wallace et al. 2019). Once calibrated with field-based measurements, these datasets can provide 3D mapping of live and dead canopy and surface fuel loading and structure with applications for biomass mapping, fire behavior modeling, and fuel consumption measurements (Figs. 2.3 and 2.4).

Highly resolved spatial data from TLS and SfM expand sampling beyond the domains of traditional destructive plots and planar intersect fuel surveys. As data from TLS and SfM images can be sampled at high resolution, they can be merged into 3D point clouds for fine-scale mapping and quantification of live and dead surface and canopy fuels. TLS excels at capturing detailed pre- and post-fire 3D data that represent continuous changes in estimates of bulk density at fine scales (Rowell et al. 2016; Hudak et al. 2020). Such spatially explicit fuels consumption data provide linkages between fire behavior and smoke production by describing interactions that produce smoke from a range of fire types and behavior (Moran et al. 2019).

Fig. 2.3 Structure from motion point cloud generated for a mixed conifer site (roughly 500 m by 500 m in size) at the Lubrecht Experimental Forest, Montana

2.3 Fuel Consumption

Fuels are consumed in a complex set of combustion phases that differ with each wildland fire (Ottmar 2014). Because different fuel categories (i.e., tree crowns, shrubs, herbs, downed wood, litter, and ground fuels) have different propensities to burn, consumption varies across time and space (Weise and Wright 2014). Fuel type and condition, moisture content, arrangement, and ignition patterns affect the amount of biomass consumed.

Fuel consumption is the amount of fuel that is consumed during all combustion phases. During combustion, vegetative matter is decomposed through a thermal/chemical reaction where plant organic material is rapidly oxidized producing carbon dioxide, water, and heat (Byram 1959; Johnson and Miyanishi 2001). During the pre-ignition phase, pyrolysis occurs first and is the heat-absorbing reaction that removes moisture and converts fuel elements such as cellulose into char, carbon dioxide, carbon monoxide, water vapor, combustible vapors and gases, and particulate matter (Kilzer and Broido 1965). Flaming combustion follows as the escaping organic hydrocarbon vapors released from the surface of the fuels burn (Williams

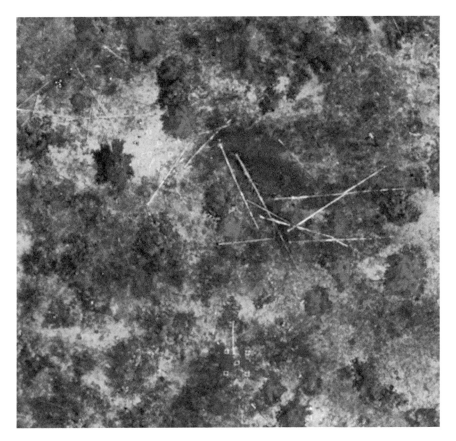

Fig. 2.4 Multi-spectral orthophoto mosaic, approximately 100 × 100 m in size, generated from unmanned aerial system imagery collected at the Lubrecht Experimental Forest, Montana, demonstrating potential discrimination between live fuels (shown as red tree crowns and surface vegetation) and downed dead wood (linear blue objects)

2018) (Fig. 2.5). Combustion efficiency is usually high if volatile emissions remain near the flames.

During the smoldering phase, emissions of combustible gases and vapors above the fuel are insufficient to support a flame (Ohlemiller 1986; Johnson and Miyanishi 2001) (Fig. 2.5). Gases and vapors condense, appearing as visible smoke as they escape into the atmosphere; smoke consists mostly of particles <1.0 μm diameter. The amount of particulate emissions generated per mass of fuel consumed during the smoldering phase, generally expressed as an emission factor (Chap. 5), is more than double that of the flaming phase. Smoldering combustion is more common in densely packed and highly lignified fuel types (e.g., organic soils and decayed logs) due to the lack of oxygen necessary to support flaming combustion. For example, deep ground fuel, such as peatland soils, can smolder for weeks, contributing greatly to smoke emissions (Rappold et al. 2011). In boreal ecosystems, approximately

Fig. 2.5 Representative photos of **a** flaming and smoldering of surface fuels (flaming dominates), **b** flaming and smoldering of large log and surrounding grass and litter (smoldering dominates) and **c** short- and long-term smoldering (glowing) phases of combustion in a large log (long-term smoldering dominates) (Photos by Roger Ottmar)

90% of emissions can be attributed to burning of deep ground fuel characterizing peatland soils. Given these impacts, methods of quantifying depth of burn and its spatial variability are critical (van der Werf et al. 2010; Thompson and Waddington 2014).

Because heat generated from smoldering is seldom sufficient to sustain an active convection column, smoke often concentrates in nearby drainages and valley bottoms, compounding the effect of the fire on local air quality (Chap. 5). Smoldering combustion is less prevalent in fuels with high surface-area-to-volume ratios (e.g., grasses, shrubs, small-diameter woody fuels) (Sandberg and Dost 1990). Near the end of the smoldering phase, pyrolysis nearly ceases, leaving unconsumed fuel as black char. This is often referred to as the glowing or residual smoldering phase (DeBano et al. 1998).

Combustion phases occur both sequentially and simultaneously as a fire front moves across the landscape. Combustion efficiency is rarely constant, resulting in a different set of chemical compounds being released at different rates into the atmosphere during each combustion phase (Fig. 2.6) (Ferguson and Hardy 1994). The flaming stage has a high combustion efficiency and generally emits the least amount of $PM_{2.5}$ emissions relative to fuel mass consumed. The smoldering phase has a lower combustion efficiency, producing more $PM_{2.5}$ relative to fuel mass consumed.

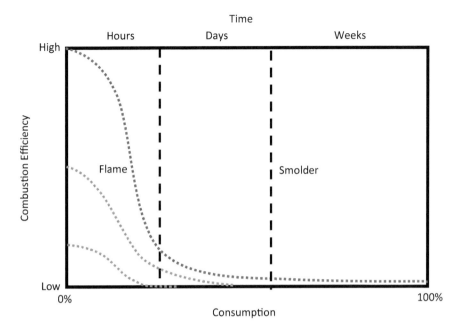

Fig. 2.6 Conceptual diagram of combustion efficiency over time and combustion phase. The red dotted line represents a fire event with a large burned area; the orange dotted line represents a small fire that is constrained by local inversions and has minimal combustion efficiency; the gray dotted line represents a low-intensity prescribed fire

The surface-area-to-volume ratio of fuels also influences the amount of fuel consumed. Smaller particles (e.g., grass and small twigs) require less heat to ignite and combust compared to larger fuel particles (e.g., large logs). Small particles generally burn during the flaming stage, and larger fuels often burn during the smoldering stage. Fuel geometry also determines moisture uptake and release from individual particles. For example, particles with high surface-area-to-volume ratios such as grass can absorb and release moisture quickly compared to fuels with low surface-to-volume ratios.

The compactness of fuel particles in fuelbeds can enhance or diminish fuel consumption and affect smoke emissions. Packing ratio—the fraction of the fuelbed volume occupied by organic material—is a measure of fuelbed compactness. A loosely packed fuelbed (low packing ratio), such as a sparse grassland or shrubland, has ample oxygen for combustion but may inefficiently transfer heat between burning and adjacent unburned fuel particles. Alternatively, a dense fuelbed (high packing ratio), such as decayed soil organic matter, can efficiently transfer heat between the particles, but low availability of oxygen reduces consumption and combustion efficiency.

Fuel continuity also affects fuel consumption. Sustained ignition and combustion continue only if fuel particles are close enough that heat can be transferred between particles, allowing combustion to occur. For example, piles of branches and leaves are often optimally packed with particles close enough for adequate heat transfer with large enough spaces between particles for oxygen availability. As a result, pile burning, when appropriately executed, often results in nearly complete combustion (Hardy 1996).

Canopy fuels exemplify the importance of particle size and surface-to-volume ratio in determining fuel consumption. Severe crown fires burn tree crowns and generally leave boles and large branches behind. Even under extreme fire conditions, live tree boles and large branches are not generally available to burn due to their low surface area and high moisture. In fire behavior modeling, canopy bulk density is used to quantify available canopy fuel. The diffuse distribution of canopy bulk density makes it difficult to measure with traditional methods. However, Lidar and other 3D point-cloud data offer promising approaches for characterizing pre- and post-burn canopy fuel (Skowronski et al. 2011, 2020).

2.3.1 Indirect Estimates of Fuel Consumption

Consumption of wildland fuels can be measured directly by measuring pre and post-fire loadings (Ottmar 2014), but because of time and labor constraints, it is typically estimated from indirect, or non destructive, measurements that use remote sensing to map consumption in 2D or 3D. To reduce uncertainties in estimated consumption for smoke modeling, pre- and post-fire fuel measurements ideally would be co-located rather than selecting proxy sites to represent pre-burn fuels.

Predictive models are commonly used to estimate fuel consumption based on pre-burn fuel loadings. CONSUME (Prichard et al. 2007) and the First Order Fire Effects Model (FOFEM; Reinhardt et al. 1997) are used operationally for prescribed burn planning to predict fuel consumption, heat release, and emissions. They can also estimate fuel consumption based on remotely sensed maps of area burned and pre-burn fuel loadings. For example, the Fuel Characteristic Classification System (FCCS) (Ottmar et al. 2007; Riccardi et al. 2007a, 2017b) supports fuelbed datasets that are available as a map layer within LANDFIRE, based on crosswalks to existing vegetation type (https://www.landfire.gov/evt.php). Fuelbed data from FCCS can be used as inputs to CONSUME or FOFEM to estimate fuel consumption for a burned area or planning unit. Model predictions can be improved with field-based observations to refine fuelbed assignments or pre-burn fuel loading values.

Consumption can also be estimated using a satellite-derived estimate of biomass burned (M, g) from pre- and post-burn imagery in the classic equation (Seiler and Crutzen 1980; Kaufman et al. 1989; Wooster et al. 2005):

$$M = A \times B \times \beta \tag{2.1}$$

where A is burned area (m^2) measured from imagery, B is biomass (fuel load) per unit area (g m^{-2}) estimated from pre- and post-burn imagery, and ß is the burning efficiency or combustion factor (fraction of fuel burned) (Vermote et al. 2009).

Burning efficiency, the amount of fuel that burns, is coupled to intrinsic fuel conditions (type, physical arrangement, chemical composition, and fuel moisture) and extrinsic abiotic factors, such as weather conditions (temperature, relative humidity, and wind), that vary at daily and seasonal time-scales. These factors must be measured or modeled on site close to the time of burning, then inputted into consumption models to constrain the efficiency of simulated combustion to conditions at the time of burning (Ottmar 2014).

Burned area (A) can be estimated from airborne or satellite imagery, although estimations will differ depending on the scene, the type of imagery used (van der Werf et al. 2006), and the algorithms applied (Roy et al. 2005). Multispectral satellite imagery is commonly used for burned area mapping (Lentile et al. 2006; Hudak et al. 2007). With the many satellites in orbit today, errors in burn area estimation can be reduced by using post-fire imagery with higher spatial resolution (250 m or better) and shorter latency (daily or sub-daily) after fire.

Biomass (B) can also be estimated from optical imagery but with less certainty (Tucker 1977; Sellers 1985; Gitelson and Merzlyak 1997; Thenkabail et al. 2000). In multilayered forest canopies with high leaf area index (leaf area per unit ground area), passive optical sensors saturate and lose sensitivity, reducing the utility of spectral indices such as normalized difference vegetation index (NDVI) or normalized burn ratio (NBR) (Goel and Qin 1994; Haboudane et al. 2004; Hudak et al. 2007).

Because canopy biomass is often correlated to canopy height, statistical metrics calculated from the distribution of height measures provided by airborne Lidar can be used to estimate biomass and other forest structure attributes such as stem density, basal area, and volume (Lefsky et al. 1999, 2002; Hudak et al. 2008; Dubayah et al. 2010; Silva et al. 2016, 2017).

Canopy height and density information based on Lidar-based 3D point cloud data can be converted to 2D raster maps (with height and density attributes) that are more easily manipulated and processed with geospatial analysis. Fuel biomass density can be estimated from airborne Lidar resampled to 30-m resolution bins, commensurate with LANDFIRE fuel maps (Hudak et al. 2016b), or as fine as 5-m resolution (Hudak et al. 2016a). Ground-based TLS can be used at scales down to 10 cm. At this fine grain size, it is feasible to differentiate fuel components that are a heterogeneous mixture of materials (or species), each with their own emission factor (EF) (Chap. 5). For finer scales, Eq. 2.1, which predicts the amount of consumed biomass (M, g) at the level of individual fuel components (or species) x (Seiler and Crutzen 1980; Brönnimann et al. 2009), can be revised to

$$M_x = A \times B \times \beta \times EF_x \quad (2.2)$$

In the fine-scale 3D domain, fuel volume (V, m^{-3}) can be substituted for area A (m^{-2}), and fuel bulk density (BD, g m^{-3}) can be substituted for biomass (B) density (g m^{-2}), traditionally characterized in 2D, to estimate M (g), the mass of emissions due to consumption of fuel component (or species) x.

Terrestrial Lidar has also been used to estimate shrub consumption. Hudak et al. (2020) demonstrated that 3D estimates of shrub volume, combined with co-located field measures of bulk density, can provide spatially explicit estimates of vegetation bulk density. Comparison of pre- and post-fire 3D fuel maps can provide 3D maps of consumption, although at slightly coarser resolution, given errors in co-registration between pre- and post-fire maps.

2.3.2 Direct Measures of Fuel Consumption

Direct measurements of heat flux using thermal imagery can be calibrated to estimate consumption rates and to map consumption which are important for smoke prediction. The rate of biomass loss (i.e., consumption) is linearly related to the rate of heat flux from an active fire (Wooster et al. 2005; Freeborn et al. 2008; Smith et al. 2013). Heat flux can be measured remotely from the thermal infrared radiation emitted by the fire, which amounts to 10–20% of the total heat flux (Byram 1959). Temperatures of heat sources, as measured by calibrated thermal infrared sensors, can be converted to fire radiative power (FRP, W), which equates to Joules per second (J s^{-1}). Continuous measurements of FRP over the duration of the fire can be integrated with respect to time(s) to estimate total heat flux, also known as fire radiative energy (FRE) in J (Fig. 2.7). The integral of the FRP time series can be approximated (Boschetti and Roy 2009) as

$$\text{FRE} = \sum_{i}^{n} 0.5(\text{FRP}_i + \text{FRP}_{i-1})(t_i - t_{i-1}) \qquad (2.3)$$

where time t is the time in seconds (s) for each FRP observation i in the time series (Wooster et al. 2013). This integration can be applied to every pixel in a multi-temporal stack of FRP observations to produce an FRE image that estimates total consumption (Hudak et al. 2016a; Klauberg et al. 2018).

Comparisons between the (direct) FRE approach to estimating fuel consumption and the (indirect) approach to consumption estimates derived from remotely sensed burn area (A) and pre-fire fuel biomass (B) measurements by Eqs. 2.1 and 2.2 are reasonably linear (Roberts et al. 2009; Wooster et al. 2013). The relationship scales because it is linear, permitting a simplification of Eq. 2.2:

$$M_x = \text{FRE} \times C \times \text{EF}_x \qquad (2.4)$$

where C is a "combustion factor" (g kJ^{-1}) for a given vegetation fuel type (x).

Accuracy of FRE-derived estimates of consumption depends on the frequency of FRP observations and whether they span the full duration of the fire, including the

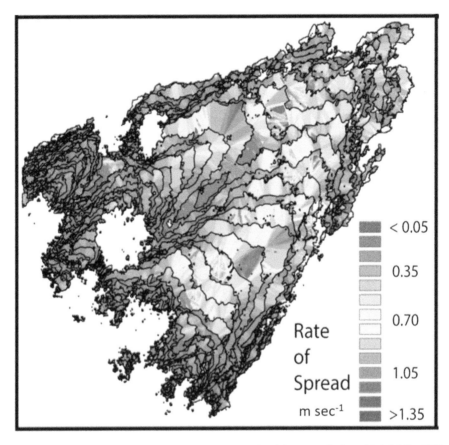

Fig. 2.7 Using digital thermography from an unmanned aerial system platform, high fidelity FRE and rate of spread can be extracted from these data. Moran et al. (2019) demonstrate the utility of these platforms, describing the points of head, flanking and backing fire (Image used with permission from the author)

flaming and smoldering phases of combustion. Thermal sensors mounted on fixed-wing aircraft can image a given site for only a few seconds, separated by several minutes needed to turn the aircraft around and re-image the same location on the fire (Hudak et al. 2016a; Klauberg et al. 2018). Visible and near-infrared (NIR) sensors can capture flame location and geometry and distinguish flaming combustion from residual smoldering combustion. The dual-band technique, using both mid-wave infrared (MWIR) and longwave infrared (LWIR) wavelengths, provides for more robust FRP estimation than using MWIR or LWIR alone (Dozier 1981).

Current measurement technologies are unable to partition the FRP signal between different fuel components burning simultaneously within the same pixel space. For surface fires beneath forest canopies, the FRP signal may be attenuated from overstory canopy occlusion, which may differ with canopy cover (Mathews et al. 2016).

Correcting for canopy occlusion may be possible through Lidar-derived canopy structure (Hudak et al. 2016a).

2.4 Gaps in Wildland Fuels Characterization

Until recently, a major gap in our understanding of wildland fuels has been a lack of spatial dimensionality in fuel characterization, which is necessary to reduce uncertainty and increase precision of inputs to fire behavior, fuel consumption and smoke models (Chaps. 3 and 4). Advances in remote sensing techniques offer promising approaches to 3D fuel characterization for fine-scale inputs of CFD models of fire behavior to landscape fire spread, fuel consumption, and smoke models. These methods are currently under development (Rowell et al. 2020), employing a hierarchical sampling method from fine-scale characterization to coarse-scale mapping applications (Fig. 2.8).

Broad-scale mapping and modeling applications present an additional challenge to quantifying fuel characteristics and represent them hierarchically across spatial scales. Field and remote sensing measurements may be taken at similar scales, but they are inherently difficult to integrate due to the complexity of fuels and challenges in co-locating and coordinating field and remote sensing measurements. For example, a new approach to 3D field sampling (Hawley et al. 2018) was designed specifically to link 3D fuel types and fuel mass, collected within 1000-cm^3 cubes to the same resolution of volume TLS point clouds of vegetation structure, with 1 cm^3 precision

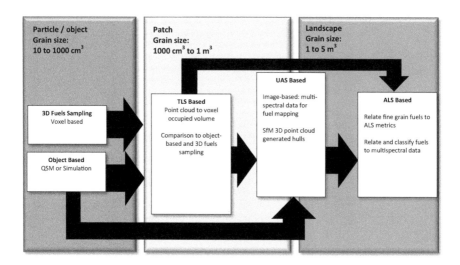

Fig. 2.8 Conceptual diagram of multi-scaled estimates of 3D fuels characterized using a hierarchical sampling method from individual fuel particles or objects to patch and landscape extents and corresponding sampling resolutions (grain size)

Fig. 2.9 Voxel sampling frame, vertical view showing the 10 × 10 × 10 cm sample voxel grid of a mixed shrub, herb, grass, ad litter fuelbed (Photo by Susan Prichard)

(Fig. 2.9). In 3D imagery, volumetric pixels are termed *voxels*, and the 1000-cm^3 cubes are also referred to as voxels within the field sampling frame.

Calibrated with voxel field datasets, TLS is a novel and scalable advancement in fuel characterization with highly resolved bulk density estimates for known volumes (Rowell et al. 2020). Robust coupling involves co-locating techniques between individual 3D field plots and TLS point clouds. However, this approach has limitations. First, voxel sampling provides explicit representation of fuel types and fuel mass, but the 1000 cm^3 space of each voxel is assumed fully occupied due to lack of measurements at finer spatial scales. Second, the TLS is limited by occlusion near to the ground where most fine and consumable fuels occur. Additional work is needed to create machine-learning algorithms to classify 3D point cloud datasets generated from TLS and/or photogrammetry into objects and apply rule-based assignments of metrics such as bulk density, surface-area-to-volume ratios, and fuel moisture content to each classified object or volume.

2.4.1 Scaling from Fine-Scale to Coarse-Scale Fuel Characterization

The structure and condition of fuels influence their availability to burn and how much exogenous work must be applied to release their energy. For example, coarse-scale grid cells (e.g., $5 \times 5 \times 5$ m) may be sufficient to represent crown fuels during extreme fire spread events, where fire weather and topography dominate fire behavior and smoke production patterns. In contrast, fine-scale fuel heterogeneity measurements are often critical for accurate fire behavior predictions in a low-intensity surface fire such as a prescribed burn. A forest that has been recently thinned and burned contains combustible fuels but in a structure that is less available to burn in a subsequent fire. However, column-driven fire spread combined with strong winds could exceed the burning threshold for that site. Similarly, sites with high live fuel moisture in grass and shrub fuels may present barriers to fire spread under normal fire weather conditions, but burning thresholds can be exceeded by exceptional fire weather.

At present, no established method exists to scale 3D fuels data from fine-scale field measurements to the larger spatial scales (e.g., burn units or watersheds) useful for decision making. Before such mapping applications can be developed, modelers need to identify how fuel metrics (e.g., loading, bulk density, heat content) and characteristics (e.g., fuel type and live/dead) can be assigned from sampled values to large spatial scales and across fire types (e.g., prescribed fire, wildfire, surface fire versus canopy fire).

Fire atmosphere interactions that contribute to fire behavior, plume dynamics, and smoke production are beginning to be resolved in models such as WRF-SFIRE (Mandel et al. 2011), FIRETEC (Linn et al. 2002), and Wildland-Urban-Interface Fire Dynamics Simulator (WFDS; Mell et al. 2007). However, evaluation datasets are needed to determine how the scale and precision of fuel inputs influence model predictions of fire behavior, heat release, and smoke production.

Large-scale studies such as the Fire and Smoke Model Evaluation Experiment (FASMEE; Prichard et al. 2019b) and the FIREX-AQ Western wildfires campaign (Werneke et al. 2018) include synchronized and coordinated measurements of source characterization, fire behavior, plume dynamics, and smoke production. Investments in these coordinated measurement campaigns are necessary to improve our understanding of fire atmosphere interactions and inform future model evaluation and development (Liu et al. 2019, Chap. 4).

2.4.2 Challenges in Forest Floor Characterization

Organic soil layers, including litter and ground fuels, can be a substantial portion of total fuel loading and contribute disproportionately to smoke emissions including long-term smoldering events. However, methods for characterizing peatland and forest floor layers have not advanced much in recent decades. Remote sensing techniques, such as TLS, can be used for litter characterization but are unable to penetrate organic soils and cannot resolve their density or depth. Models of organic soil accumulation, decomposition, and changing moisture characteristics are needed to complement 3D fuel measurement techniques.

No models exist that provide accurate representations of ground fuel consumption as it relates to forest structure, climate, weather, leaf chemistry, and time since last fire, all of which are dynamic through space and time. For example, depending on fire intensity and soil moisture, wildland fires rarely consume entire organic soil layers. Variability in ground fuel consumption and smoldering patterns adds further complexity to smoke production. Recent research on spatial distributions of ground fuel depth, biomass, and other characteristics in long-unburned forests of the southeastern USA emphasizes fine-scale spatial and temporal variability in ground fuels and the potential challenges of sampling across forest stands or burn units (Kreye et al. 2014). In boreal ecosystems, where the majority of biomass is stored in peatland soils, Chasmer et al. (2017) showed that variations in forest floor depth could be quantified by comparing ground surface elevation models derived from separate pre- and post-fire Lidar collections. However, in most fuelbeds, ground fuel layers are too shallow relative to the vertical precision of airborne Lidar to detect changes in depth as a result of consumption.

2.4.3 Modeling Spatial and Temporal Dynamics of Wildland Fuels

The biggest gap in our knowledge of wildland fuels is creating up-to-date and accurate models of fuel dynamics to inform smoke modeling. This challenge has been termed the "ecology of fuels" (Mitchell et al. 2009), requiring an understanding of the entire life cycle of wildland fuels, including vegetative reproduction, growth, senescence, deposition of fine and coarse debris, decay, mortality and connections to weather, climate, soils, and nutrient cycling (Agee and Huff 1987; Harmon et al. 2000). The 3D spatial complexity of fuels and their dynamics over time, translate to similar complexity and variability in the availability of fuels to burn and their contribution to fire behavior and effects. However, the life cycle of fuels as it relates to vegetation dynamics and feedbacks with fire has not been fully defined. Furthermore, the temporal dynamics of fuels can be distinct between fine-scale changes in fine-fuel moisture and coarse-scale changes (e.g., vegetation structure, productivity and climate).

Limited understanding of live and dead fuel moisture dynamics also constrains our ability to model fire behavior, fuel consumption, and smoke production. Fuel moisture varies across ecosystems, seasons, and fuel components, and moisture dynamics often exhibit large sub-daily changes on local scales (Viney 1991; Banwell et al. 2013; Kreye et al. 2018). Fuel moisture often dictates the availability of fuels for ignition and consumption, with pronounced differences across arid, semi-arid, and humid climates. Summer climate in western North America is generally characterized by a long period of drying, making coarse wood and organic soils generally available to burn during the peak of wildfire season (Estes et al. 2012). In contrast, the southeastern USA has a humid, subtropical climate; downed wood decays quickly, and where coarse wood exists it can act as a fuel break during low-intensity fire spread. Live and dead fine-fuel dynamics determine if fuels are available to burn, either promoting or inhibiting fire spread. For example, across ecosystems with grass-dominated fuelbeds, spring green-up is generally considered a barrier to fire spread. Differences in fuel moisture and the corresponding availability of fuels to burn over hours to months are well known among practitioners, but these fundamentals are not explicitly represented in predictive fire behavior, fuel consumption, and smoke models.

2.5 Vision for Improving Fuel Science in Support of Smoke Science

Fuel characterization and mapping to support smoke science will need to rely on a range of methods. Because some fuels, including forest floor and peatland soils, cannot be remotely sensed, future approaches to fuel characterization will involve a combination of traditional methods and new technologies. Rather than describing fuel characteristics as modeled estimates across raster maps, the ranges and variations of fuel distributions will be required, particularly for CFD models that rely on gridded, 3D inputs of fuels, terrain, and atmospheric turbulence. Fuel inventory and modeling methods also need to be developed to capture the nested spatial variability of wildland fuels and dynamics of wildland fuelbeds over time (Keane 2015).

As more work is devoted to 3D fuel characterization for CFD models, we envision a library of 3D fuels, mapping tools, and parameters for customization of fuelbeds for specific applications and fine- to coarse-scale mapping of pre- and post-burn canopy and surface fuels (Fig. 2.10). To date, CFD models such as FIRETEC and WFDS are used only for research due to their complex input and computational requirements. However, progress is being made to advance real-time models of fire spread and smoke production that can be used operationally for prescribed burn planning and wildfire monitoring (e.g., QUIC-Fire; Linn et al. 2020).

For CFD models to move into operational use, applications will be needed to translate 3D fuel characteristics into model inputs at appropriate scales for smoke management applications (e.g., prescribed burn planning, wildfire smoke modeling). CFD modeling requirements mean that next-generation fuel mapping will need synthetic, gridded fuelbeds from remotely sensed data, machine-learning algorithms to identify objects within 3D point clouds, and assigned fuel properties for each identified object or fuel complex (based on statistical models and known probability distributions) (Fig. 2.11). User-friendly technology and analytical tools will be required

Fig. 2.10 Remotely sensed datasets can be used to characterize and quantify patterns of bulk density, consumption and fire effects. For example, Plots **a** and **b** represent pre-fire and post-fire short-range, photogrammetry-based 3D point clouds for an individual plot that can be calibrated with field data to estimate fuel consumption. Estimated consumption can then be scaled to prescribed burn units using synoptic pre- and post-burn TLS imagery (**c**) where bright yellow on the ground is burned and blue hues are unburned

to guide smoke managers in novel but practical approaches to improve 3D fuel characterization and mapping.

Better characterization of sources of smoldering consumption can also improve estimates of the severity and duration of smoke impacts to communities, especially from prescribed burning (Hyde et al. 2011). Advances with SfM from both UAS

Fig. 2.11 Synthetic 3D broadleaf and long-needled pine litter fuelbeds developed from object-based scanning and statistical models of leaf litter composition and depth [From Rowell et al. (2016)]

platforms and short-range photogrammetry offer access to fine grain data that can be used to map fuels that contribute most to smoldering combustion and long-duration smoke production (Wallace et al. 2012; Cooper et al. 2017). SfM photogrammetry can complement ALS imagery by providing true color or multispectral images that allow for delineation of live and dead fuels and fuel classification refinement (Fig. 2.10). For example, integration of SfM imagery can assist in object-based classification of large coarse woody debris, and these objects can then be attributed with mass estimates to improve modeling of flaming and smoldering emissions (Fig. 2.4).

TLS-based estimates can be used to refine coarser-scale estimates of surface and canopy fuels (García et al. 2011; Seielstad et al. 2011; Rowell et al. 2016, 2017). Fuel libraries from TLS tied empirically or probabilistically to large-scale ALS or passive remote sensing datasets will be a significant step toward broad-scale 3D mapping applications. A limitation of ALS and TLS has been cost, efficiency, and time since acquisition. There are a growing number of ALS datasets nationally, but these snapshots in time do not encompass disturbances that could alter fuel loading and distribution or expected fire behavior. Forest growth models, such as the Forest Vegetation Simulator, can use ALS data and their derivatives to calculate estimates of growth and biomass accumulation in forest canopies.

Maintaining reliable, up-to-date maps of wildland fuels will require linkages between remotely sensed datasets and ecological process models. High deposition of vegetation, coupled with severe disturbance effects, may alter fuelbed characteristics and render fuel maps outdated (Keane et al. 2001). It may be especially important to capture fuel dynamics in frequently burned or actively managed ecosystems. Ecosystem models typically fall short in simulating realistic fuel characteristics needed by existing fire models (Thornton et al. 2002). Ecological models that simulate development, deposition, decomposition, and disturbance (Sect. 2.2) can capture multi-scale fuel dynamics and translate them to fire behavior and smoke modeling at relevant spatial scales (Hatten and Zabowski 2009; Dunn and Bailey 2015). Linking fuel characteristics with ecological processes can inform fire behavior and smoke dynamics. Improved representation of fuel dynamics within ecological models will also refine how they simulate wildfires, insects, disease, fuel treatment and ecological restoration activities, and climate change.

2.6 Science Delivery to Managers

Over the past two decades, several fire effects and smoke models have been used by managers to characterize fuels and inform fire and smoke management decisions. Table 2.2 presents examples of models used to predict smoke production and, in some cases, dispersion. To appropriately apply their products to smoke management decisions or ensemble predictions, it is important to understand the error, bias, assumptions and limitations of the models. The BlueSky Smoke Modeling Framework (Larkin et al. 2010) is an operational smoke prediction tool that uses ensemble modeling to estimate available fuel, consumption, emissions, and smoke dispersion. BlueSky estimates fuel loadings from a 1-km fuelbed map of the USA or user inputs and models fuel consumption with CONSUME as a first step to smoke production and dispersion modeling (Larkin et al. 2010).

The Interagency Fuel Treatment Decision Support System (IFTDSS, https://iftdss.firenet.gov) was designed to provide a Web-based system to assist managers in fire, fuel, and smoke planning; reduce the number of tools for which access is needed; and reduce error propagation caused by using multiple, ensemble models. IFTDSS is working to incorporate CONSUME and FOFEM modules that use mapped fuel loadings values from LANDFIRE (Rollins 2009) or user inputs. CONSUME and FOFEM rely on a combination of empirical, semi-empirical, and physical process-based models of consumption. Command-line versions of calculators for both models are available for smoke modeling applications.

Every approach to modeling smoke emissions has limitations. Point-based models such as CONSUME and FOFEM use many empirical equations for estimating fuel consumption and smoke emissions. However, most equations were developed with data collected from a limited number of ecosystems and fuelbeds, and under a limited range of fire and fuel conditions. The physics-based process model in FOFEM simplifies many complex processes and was calibrated using relatively few lab and field burns (Albini et al. 1995; Albini and Reinhardt 1995). Although point models provide smoke estimates based on published research and expert opinion, model precision is limited by the high variability inherent in the production of smoke (Larkin et al. 2012). For example, Prichard et al. (2014) used CONSUME and FOFEM to compare predicted and actual fuel consumption in the southeastern USA, finding that predicted

Table 2.2 Selected smoke models, ranging from relatively simple to complex, including general benefits and drawbacks

Model scale	Example	Simulation area	Benefits	Drawbacks
Point	CONSUME[a]	Project	Relatively easy to install and use. Fast execution time	Limitations of simple empirical equations. Unstudied fire, fuel, and consumption relationships
Point	FOFEM[b]	Project	Relatively easy to install and use. Fast execution time	Generalizations in woody consumption model. Limitations of empirical equations. Unstudied fire, fuel, and consumption relationships. Poor correlation of default and actual fuel loadings
Landscape	Emissions Estimation System (EES)[c]	State	Provides statewide smoke estimates	Drawbacks of the FOFEM module. Untested fuel moisture assumptions. Uses daily fire perimeters; no predictive capability
Landscape	BlueSky[d]	Regional, state, national	Provides variable-scale smoke forecasts	Drawbacks of the CONSUME module. Poor correlation of fuel maps and actual fuels
Landscape	Weather Research and Forecasting (WRF)-Sfire WRF-Chem[e]	Regional, state, national	Provides variable-scale smoke forecasts. Real-time atmospheric boundary layer and weather forecast component. Coupled weather-fire modeling	Errors associated with fire spread model and unburned areas inside the estimated fire perimeter. Limited number of fuelbeds. Fuelbeds developed for fire behavior estimation, not smoke. Computationally intensive
Landscape	High-Resolution Rapid Refresh—Smoke (HRRR-Smoke)[f]	Regional, state, national	Provides variable-scale smoke forecasts. Based on WRF and WRF-Chem. Radiative power is remotely sensed; no fuel inputs required	Fire detection is at a relatively coarse spatial resolution (~3 km) and variable temporal resolution. Interpolation of fire spread rate between satellite passes. In development

(continued)

2 Fuels and Consumption

Table 2.2 (continued)

Model scale	Example	Simulation area	Benefits	Drawbacks
Landscape	HIGRAD FIRETEC[g]	Up to large projects	Coupled 3D, physics-based models of combustion and atmospheric processes	Very computationally intensive. Not real time. Currently for research only

[a] Prichard et al. (2007)
[b] Reinhardt and Crookston (2003)
[c] Clinton et al. (2003)
[d] Larkin et al. (2010)
[e] Mandel et al. (2011), Grell et al. (2005)
[f] Ahmadov et al. (2019)
[g] Linn et al. (2002)

fuel consumption had high uncertainty in some cases, particularly with high pre-burn fuel loading.

For smoke model applications to be useful for managers, models must be updated to include recent research. A formal process is needed to provide periodic version updates to ensure that smoke modeling applications include the "best available science" for estimating smoke emissions. This is of particular concern as existing point models are integrated or merged into spatial modeling frameworks.

There are relatively few training options for the wide variety of available smoke models and products. The Introduction to Fire Effects (RX-310) and Smoke Management Techniques (RX-410) classes developed by the National Wildfire Coordinating Group provide limited training using CONSUME and FOFEM and an introduction to BlueSky. The annual Air Resource Advisor training class (administered by the US Forest Service) focuses on large-scale (wildfire) smoke impacts and primarily uses BlueSky for simulations. Students in this class are members of fire Incident Management Teams and use air quality modeling to assess smoke risks to fire personnel and local communities. The limited options for smoke model training can lead to misinterpretation of model results or overreliance on model estimates without understanding underlying limitations and assumptions.

2.7 Research Needs

For fuel and consumption research related to smoke management, scientific challenges can be summarized in six categories as follows:

- *Consistent methodologies to address sampling of wildland fuels*—Although field sampling is needed to represent a fuelbed from ground to canopy, the required sampling methods do not easily overlap (e.g., planar intersect for downed wood, depths and bulk density for litter, ground fuels), and most traditional fuel sampling methods have low repeatability and high uncertainty. Because fuel categories are not necessarily well correlated, predicting one component based on available sampling of another is unrealistic. Hierarchical sampling methods that employ a range of remotely sensed and field-based datasets (Fig. 2.5) are needed to integrate fuels data and support characterization at the scale of prescribed burn units and wildland fire events.

- *Better understanding of the role of sampling scale in error propagation in fuel characterization and mapping*—The appropriate sampling area and intensity may differ by fuel component (e.g., bulk density and biomass of litter and ground fuels). Scale considerations are important for coordinated sampling design and to inform applications that apply fine-scale fuel characterization to coarser-scale mapping applications. CFD models can be integrated with smoke simulations to evaluate sensitivity of smoke prediction to fuelbed heterogeneity and spatial scales of fuel inputs. More work is needed to evaluate the sensitivity of current CFD models (e.g., FIRETEC, WFDS) to spatial scales of fuel characterization across different vegetation types.
- *Improved methods for characterizing fuels that are major sources of smoke, including coarse wood, peatland soils, and other ground fuels*—Although TLS and SfM offer promising advances in characterizing wildland fuels, these techniques cannot quantify deep organic soil layers. Intensive field sampling is needed to characterize variability in peatland soils and other ground fuels and to contribute to predictive models of ground fuels, potentially paired with innovations in remote sensing techniques or soil mapping. In contrast, TLS and photogrammetry may aid in more accurate surveys and characterization of coarse wood. However, more work is needed to understand and characterize fuel moisture, decay class, and contribution of coarse wood to fuel consumption and emissions.
- *Improvements to 3D fuel characterization using ALS, TLS, and SfM photogrammetry*—Remote sensing techniques, including integrated ALS, TLS, and SfM datasets, have advanced fuel characterization, but research is needed to inform image interpretation and quantification of wildland fuel loadings and structure. Some of the remaining challenges with these methods include:
    - Resolutions of available remotely sensed imagery may not match (e.g., Landsat TM vs. Lidar vs. photogrammetry) and may not fit the spatial scale or match the temporal dynamics of the component of interest (i.e., downed wood vs. stand structure).
    - Wildland fuels are inherently variable in 3D space, and correlations are often weak between canopy fuels and surface or ground fuels, which are obscured by forest canopies.
    - Fuel moisture dynamics are critical for fire behavior and smoke production but are difficult to measure with remote sensing.
- *Use of 3D fuels mapping for improved estimates of fuel consumption*—As methods to map fuels in 3D become more widely available, improved maps of fuel consumption based on pre- and post-burn imagery will be possible. Field validation will be required to inform fuel consumption mapping that can improve

emission estimates for flaming front fires and post-flaming front smoldering combustion.
- *Improved models of fuel dynamics*—More research is needed on modeling vegetation and fuel dynamics over time and space, with emphasis on climate change effects on vegetation and consequences for fuel properties. Live fuel moisture is particularly dynamic and a critical aspect of fire behavior and effects (e.g., Jolly et al. 2014). Spatiotemporal dynamics of fuels has implications for fire, climate, and carbon modeling at local to regional scales. Research is needed to refine existing ecological process models and potentially develop new ones to project vegetation and fuel dynamics, tailoring projections to next-generation fire behavior and smoke models.

2.8 Conclusions

Fuels are foundational to smoke prediction, often being the largest source of potential uncertainty and error in the chain of biophysical components involved in combustion and smoke production from ground to atmosphere. Until recently, fuels and fuel consumption have been studied using traditional methods to estimate the cover, height, and biomass of wildland fuels across dominant ecosystems of North America, providing a good knowledge base in both the scientific and management communities. Over the past decade, significant progress has been made in describing and quantifying fuels more accurately; new technologies have improved 3D characterization and quantification across large spatial scales.

Despite this progress, improved smoke modeling will require coordinated advances in fuel characterization, consumption by combustion phase and fire atmosphere interactions associated with fire behavior, and plume dynamics modeling. One of the biggest challenges in characterizing fuels is the high spatial and temporal variability that is present in wildland fuels in nearly all types of ecosystems. Quantifying fuel loadings across large landscapes continues to be a major issue, for both technical and practical reasons. In addition, up-to-date fuel inventories are relatively rare, with measurement scale and mapping applications often being a barrier for agencies that manage vegetation and fuels.

Although most fire and fuel managers are generally well informed about traditional methods for characterizing fuels, greater emphasis is often placed on fire behavior than smoke production. Potential smoke impacts on human health and other activities (Chap. 7) provide an important context for smoke science and for applications of scientific tools and concepts in managing both prescribed fire and wildfire (Engel 2013; Ryan et al. 2013; Long et al. 2018). Improved linkages, both technically and logistically, are needed to inform estimates of smoke production that may exceed National Ambient Air Quality Standards, as well as phenomena such as long-term smoldering events and nighttime inversions. Although some targeted work has been conducted on coarse wood and ground fuel consumption (Brown et al. 1985; Varner et al. 2007; Prichard et al. 2017), the sample size and range of fire weather and fuel moisture conditions are currently inadequate to improve existing fuel consumption models.

Most fuels managers do not have routine access to high-tech tools or high-resolution data to estimate smoke production (e.g., 3D characterization of fuels). Therefore, practical approaches are needed to improve field-based fuel characterization, fire behavior modeling, and consumption modeling, which will in turn elucidate the potential contribution of specific fuels (coarse wood, rotten stumps, basal accumulations, and deep organic soil layers) to fire emissions and smoke. Given the spatial and temporal complexity of wildland fuel dynamics, a better understanding is needed on the ecology of vegetation and fuels—concurrently, not as separate topics. In future decades, we anticipate that climate change will drive substantial changes in vegetation and fire dynamics, with concomitant changes in fuelbeds and their contribution to fuel consumption and emissions. Developing or revising ecological process models to ensure compatibility with next-generation fire behavior and smoke models will improve characterization of wildland fuel dynamics as well as smoke predictions.

References

Agee JK, Huff MH (1987) Fuel succession in a western hemlock/Douglas-fir forest. Can J for Res 17:697–704

Ahmadov R, James E, Grell G et al (2019) Forecasting smoke, visibility and smoke-weather interactions using a coupled meteorology-chemistry modeling system: rapid refresh and high-resolution rapid refresh coupled with smoke (RAP/HRRR-Smoke). Geophysical Research Abstracts, EGU2019, 2118605A

Albini FA, Reinhardt ED (1995) Modeling ignition and burning rate of large woody natural fuels. Int J Wildland Fire 5:81–91

Albini FA, Brown JA, Reinhardt ED, Ottmar RD (1995) Calibration of a large fuel burnout model. Int J Wildland Fire 5:173–192

Albini FA (1976) Estimating wildfire behavior and effects (General Technical Report INT-GTR-30). Ogden: U.S. Forest Service, Intermountain Forest and Range Research Station

Anderson HE (1976) Fuels - the source of the matter. Air quality and smoke from urban and forest fires. The National Academies Press, Washington, DC, pp 318–321

Banwell FM, Varner JM, Knapp EE, Van Kirk RW (2013) Spatial, seasonal, and diel forest floor moisture dynamics in Jeffery pine-white fir forests of the Lake Tahoe Basin, USA. For Ecol Manage 305:11–20

Boschetti L and Roy DP (2009) Strategies for the fusion of satellite fire radiative power with burned area data for fire radiative energy derivation. J Geophys Res 114:D20302

Bright BC, Loudermilk EL, Pokswinski SM et al (2016) Introducing close-range photogrammetry for characterizing forest understory plant diversity and surface fuel structure at fine scales. Can J Remote Sens 42:460–472

Brönnimann S, Volken E, Lehmann K, Wooster M (2009) Biomass burning aerosols and climate–a 19th century perspective. Meteorol Z 18:349–353

Brown JK (1971) A planar intersect method for sampling fuel volume and surface area. Forest Sci 17:96–102

Brown JK, Marsden MA, Ryan KC, Reinhardt ED (1985) Predicting duff and woody fuel consumed by prescribed fire in the Northern Rocky Mountains (General Technical Report INT-GTR-337). Ogden: U.S. Forest Service, Intermountain Forest Experiment Station

Brown JK (1974) Handbook for inventorying downed woody material (General Technical Report INT-GTR-16). Ogden: U.S. Forest Service, Intermountain Forest Experiment Station

Byram GM (1959) Combustion of forest fuels. In: Brown KP (ed) Forest fire: control and use. McGraw-Hill, New York

Catchpole WR, Wheeler CJ (1992) Estimating plant biomass: a review of techniques. Aust J Ecol 17:121–131

Chasmer LE, Hopkinson CD, Petrone RM, Sitar M (2017) Using multitemporal and multispectral airborne Lidar to assess depth of peat loss and correspondence with a new active normalized burn ratio for wildfires. Geophys Res Lett 44:11851–11859

Clinton N, Scarborough J, Tian Y, Gong P (2003) A GIS based emissions estimation system for wildfire and prescribed burning. In: Proceedings of the EPA 12th annual emission inventory conference. San Diego, CA. Available at: https://www3.epa.gov/ttn/chief/conference/ei12/part/clinton.pdf. 10 July 2020

Cook GD, Meyer CP, Muepu M, Liedloff AC (2016) Dead organic matter and the dynamics of carbon and greenhouse gas emissions in frequently burnt savannas. Int J Wildland Fire 25:1252–1263

Cooper SD, Roy DP, Schaaf CB, Paynter I (2017) Examination of the potential of terrestrial laser scanning and structure-from-motion photogrammetry for rapid nondestructive field measurement of grass biomass. Remote Sens 9:531

de Groot WJ, Landry R, Kurz WA et al (2007) Estimating direct carbon emissions from Canadian wildland fires. Int J Wildland Fire 16:593–606

DeBano LF, Neary DG, Ffolliott PF (1998) Fire's effect on ecosystems. Wiley, New York

Dozier J (1981) A method for satellite identification of surface temperature fields of subpixel resolution. Remote Sens Environ 11:221–229

Dubayah RO, Sheldon SL, Clark DB et al (2010) Estimation of tropical forest height and biomass dynamics using lidar remote sensing at La Selva, Costa Rica. J Geophys Res Biogeosci 115:G2

Dunn CJ, Bailey JD (2015) Modeling the direct effects of salvage logging on long-term temporal fuel dynamics in dry-mixed conifer forests. For Ecol Manage 341:93–109

Engel KH (2013) Perverse incentives: the case of wildfire smoke regulation. Ecol Law Quart 40:623–672

Estes BL, Knapp EE, Skinner CN, Uzoh FCC (2012) Seasonal variation in surface fuel moisture between unthinned and thinned mixed conifer forests, northern California, USA. Int J Wildland Fire 21:428–435

Ferguson SA, Hardy CC (1994) Modeling smoldering emissions from prescribed broadcast burns in the Pacific-Northwest. Int J Wildland Fire 4:135–142

Fosberg MA, Lancaster JW, Schroeder MJ (1970) Fuel moisture response-drying relationships under standard and field conditions. Forest Sci 16:121–128

Freeborn PH, Wooster MJ, Hao WM et al (2008) Relationships between energy release, fuel mass loss, and trace gas and aerosol emissions during laboratory biomass fires. J Geophys Res: Atmosp 113:D01301

García M, Danson FM, Riano D et al (2011) Terrestrial laser scanning to estimate plot-level forest canopy fuel properties. Int J Appl Earth Obs Geoinf 13:636–645

Gill AM, Allan G (2008) Large fires, fire effects and the fire-regime concept. Int J Wildland Fire 17:688–695

Gitelson AA, Merzlyak MN (1997) Remote estimation of chlorophyll content in higher plant leaves. Int J Remote Sens 18:2691–2697

Goel NS, Qin W (1994) Influences of canopy architecture on relationships between various vegetation indices and LAI and FPAR: A computer simulation. Remote Sens Rev 10:309–347

Grell GA, Peckham SE, Schmitz R et al (2005) Fully coupled "online" chemistry within the WRF model. Atmos Environ 39:6957–6975

Haboudane D, Miller JR, Pattey E et al (2004) Hyperspectral vegetation indices and novel algorithms for predicting green LAI of crop canopies: modeling and validation in the context of precision agriculture. Remote Sens Environ 90:337–352

Hardy C (1996) Guidelines for estimating volume, biomass, and smoke production for piled slash (General Technical Report PNW-GTR-364). U.S. Forest Service, Pacific Northwest Research Station, Portland

Harmon ME, Krankina ON, Sexton J (2000) Decomposition vectors: a new approach to estimating woody detritus decomposition dynamics. Can J for Res 30:76–84

Hatten JA, Zabowski D (2009) Changes in soil organic matter pools and carbon mineralization as influenced by fire severity. Soil Sci Soc Am J 73:262–273

Hawley CM, Loudermilk EL, Rowell EM, Pokswinski S (2018) A novel approach to fuel biomass sampling for 3D fuel characterization. MethodsX 5:1597–1604

Hoff V, Rowell E, Teske C, Queen L, Wallace T (2019) Assessing the relationship between forest structure and fire severity on the North Rim of the Grand Canyon. Fire 2:10

Hudak AT, Morgan P, Bobbitt MJ et al (2007) The relationship of multispectral satellite imagery to immediate fire effects. Fire Ecol 3:64–74

Hudak AT, Crookston NL, Evans JS et al (2008) Nearest neighbor imputation of species-level, plot-scale forest structure attributes from LiDAR data. Remote Sens Environ 112:2232–2245

Hudak AT, Dickinson MB, Bright BC et al (2016a) Measurements relating fire radiative energy density and surface fuel consumption–RxCADRE 2011 and 2012. Int J Wildland Fire 25:25–37

Hudak AT, Bright BC, Pokswindki S et al (2016b) Mapping forest structure and composition from low-density LiDAR for informed forest, fuel, and fire management at Eglin Air Force Base, Florida, USA. Can J Remote Sens 42:411–427

Hudak AT, Kato A, Bright BC et al (2020) Towards spatially explicit quantification of pre-and postfire fuels and fuel consumption from traditional and point cloud measurements. Forest Sci 66:428–442

Hyde P, Nelson R, Kimes D, Levine E (2007) Exploring LiDAR–RaDAR synergy—predicting aboveground biomass in a southwestern ponderosa pine forest using LiDAR, SAR and InSAR. Remote Sens Environ 106:28–38

Hyde JC, Smith AMS, Ottmar RD et al (2011) The combustion of sound and rotten coarse woody debris: a review. Int J Wildland Fire 20:163–174

Johnson EA, Miyanishi K (eds) (2001) Forest fires: behavior and ecological effects. Elsevier, Amsterdam

Jolly WM, Hadlow AM, Huguet K (2014) De-coupling seasonal changes in water content and dry matter to predict live conifer foliar moisture content. Int J Wildland Fire 23:480–489

Kaufman YJ, Tucker CJ, Fung IY (1989) Remote sensing of biomass burning in the tropics. Adv Space Res 9:265–268

Keane RE (2008) Biophysical controls on surface fuel litterfall and decomposition in the northern Rocky Mountains, USA. Can J for Res 38:1431–1445

Keane RE (2013) Describing wildland surface fuel loading for fire management: a review of approaches, methods and systems. Int J Wildland Fire 22:51–62

Keane RE (2015) Wildland fuel fundamentals and applications. Springer, New York

Keane RE, Gray K (2013) Comparing three sampling techniques for estimating fine woody down dead biomass. Int J Wildland Fire 22:1093–1107

Keane RE, Burgan RE, van Wagtendonk JV (2001) Mapping wildland fuels for fire management across multiple scales: integrating remote sensing, GIS, and biophysical modeling. Int J Wildland Fire 10:301–319

Keane R, Gray K, Bacciu V, Leirfallom S (2012) Spatial scaling of wildland fuels for six forest and rangeland ecosystems of the northern Rocky Mountains, USA. Landscape Ecol 27:1213–1234

Keane RE, Herynk JM, Toney C et al (2013) Evaluating the performance and mapping of three fuel classification systems using forest inventory and analysis surface fuel measurements. For Ecol Manage 305:248–263

Keane RE, Dickinson LJ (2007) The Photoload sampling technique: estimating surface fuel loadings using downward looking photographs (General Technical Report RMRS-GTR-190). Fort Collins: U.S. Forest Service, Rocky Mountain Research Station

Keane RE, Garner JL, Schmidt KM et al (1998) Development of input spatial data layers for the FARSITE fire growth model for the Selway-Bitterroot Wilderness complex, USA (General Technical Report RMRS-GTR-3). Ogden: U.S. Forest Service, Rocky Mountain Research Station

Keane RE, Frescino TL, Reeves MC, Long J (2006) Mapping wildland fuels across large regions for the LANDFIRE prototype project. In: Rollins MG, Frame C (eds) The LANDFIRE prototype project: nationally consistent and locally relevant geospatial data for wildland fire management, 2006 (General Technical Report RMRS-GTR 175, pp. 367–396). Fort Collins: U.S. Forest Service, Rocky Mountain Research Station

Keane RE (2019) Natural fuels. In Manzello SL (ed) Encyclopedia of wildfires and Wildland-Urban Interface (WUI) fires. Springer International Publishing, Cham

Kilzer FJ, Broido A (1965) Speculations on the nature of cellulose pyrolysis. Pyrodynamics 2:151–163

King KJ, Bradstock RA, Cary GJ et al (2008) The relative importance of fine-scale fuel mosaics on reducing fire risk in south-west Tasmania, Australia. Int J Wildland Fire 17:421–430

Klauberg C, Hudak A, Bright B et al (2018) Use of ordinary kriging and Gaussian conditional simulation to interpolate airborne fire radiative energy density estimates. Int J Wildland Fire 27:228–240

Kreye JK, Varner JM, Dugaw CJ (2014) Spatial and temporal variability of forest floor duff characteristics in long-unburned *Pinus palustris* forests. Can J for Res 44:1477–1486

Kreye JK, Hiers JK, Varner JM et al (2018) Effects of solar heating on the moisture dynamics of forest floor litter in humid environments: composition, structure, and position matter. Can J for Res 48:1331–1342

Larkin NK, O'Neill SM, Solomon R et al (2010) The BlueSky smoke modeling framework. Int J Wildland Fire 18:906–920

Larkin NK, Strand TM, Drury SA et al (2012) Phase I of the smoke and emissions model intercomparison project (SEMIP): creation of SEMIP and evaluation of current models (Final report 42). Joint Fire Science Program, Boise

Lavdas LG (1996) Improving control of smoke from prescribed fire using low visibility occurrence risk index. South J Appl for 20:10–14

Lefsky MA, Cohen W, Acker S et al (1999) Lidar remote sensing of the canopy structure and biophysical properties of Douglas-fir western hemlock forests. Remote Sens Environ 70:339–361

Lefsky MA, Cohen WB, Parker GG, Harding DJ (2002) Lidar remote sensing for ecosystem studies. Bioscience 52:19–30

Lefsky M, Cohen W, Spies T (2001) An evaluation of alternate remote sensing products for forest inventory, monitoring, and mapping of Douglas-fir forests in western Oregon. Canadian J Forest Res 31:78–87

Lentile LB, Holden AZ, Smith AMS et al (2006) Remote sensing techniques to assess active fire characteristics and post-fire effects. Int J Wildland Fire 15:319–345

Linn R, Reisner J, Colman JJ, Winterkamp J (2002) Studying wildfire behavior using FIRETEC. Int J Wildland Fire 11:233–246

Linn RR, Goodrick S, Brambilla S et al (2020) QUIC-fire: A fast-running simulation tool for prescribed fire planning. Environ Model Software 125;104616

Liu Y, Kochanski A, Baker KR et al (2019) Fire behaviour and smoke modelling: model improvement and measurement needs for next-generation smoke research and forecasting systems. Int J Wildland Fire 28:570–588

Long JW, Tarnay LW, North MP (2018) Aligning smoke management with ecological and public health goals. J Forest 116:76–86

Loudermilk EL, Hiers JK, O'Brien JJ et al (2009) Ground-based LIDAR: a novel approach to quantify fine-scale fuelbed characteristics. Int J Wildland Fire 18:676–685

Mandel J, Beezley JD, Kochanski AK (2011) Coupled atmosphere-wildland fire modeling with WRF-fire. Geosci Model Development 4:591–610

Mathews BJ, Strand EK, Smith AM et al (2016) Laboratory experiments to estimate intersection of infrared radiation by tree canopies. Int J Wildland Fire 25:1009–1014

Mell W, Jenkins MA, Gould J, Cheney P (2007) A physics-based approach to modelling grassland fires. Int J Wildland Fire 16:1–22

Menakis JP, Keane RE, Long DG (2000) Mapping ecological attributes using an integrated vegetation classification system approach. J Sustain for 11:245–265

Mitchell RJ, Hiers JK, O'Brien J, Starr G (2009) Ecological forestry in the Southeast: Understanding the ecology of fuels. J Forest 107:391–397

Moran CJ, Seielstad CA, Cunningham MR et al (2019) Deriving fire behavior metrics from UAS imagery. Fire 2:36

Mueller-Dombois D, Ellenberg H (1974) Aims and methods of vegetation ecology. Wiley, New York

Ohlemiller TJ (1986) Smoldering combustion (NBSIR 85-3294). U.S. Department of Commerce National Bureau of Standards, Gaithersburg

Omi PN (2015) Theory and practice of wildland fuels management. Current Forestry Reports 1:100–117

Ottmar RD (2014) Wildland fire emissions, carbon, and climate: modeling fuel consumption. For Ecol Manage 317:41–50

Ottmar RD, Sandberg DV, Riccardi CL, Prichard SJ (2007) An overview of the fuel characteristic classification system – quantifying, classifying, and creating fuelbeds for resource planning. Can J for Res 37:2383–2393

Parsons RA, Mell WE, McCauley P (2010) Linking 3D spatial models of fuels and fire: effects of spatial heterogeneity on fire behavior. Ecol Model 222:679–691

Parsons RA, Linn RR, Pimont F et al (2017) Numerical investigation of aggregated fuel spatial pattern impacts on fire behavior. Land 6:43

Peterson JL (1987) Analysis and reduction of the errors of predicting prescribed burn emissions. Master's thesis. University of Washington, Seattle

Peterson J, Lahm P, Fitch M et al (eds) (2018) NWCG Smoke management guide for prescribed fire (PMS 420-2, NFES 1279). National Wildfire Coordinating Group, Boise

Prichard S, Ottmar R, Anderson G (2007) CONSUME user's guide and scientific documentation. U.S. Forest Service, Pacific Northwest Research Station, Seattle. https://www.fs.fed.us/pnw/fera/research/smoke/consume/consume30_users_guide.pdf. 10 July 2020

Prichard SJ, Karau EC, Ottmar RD et al (2014) Evaluation of the CONSUME and FOFEM fuel consumption models in pine and mixed hardwood forests of the eastern United States. Can J for Res 44:784–795

Prichard S, Kennedy M, Wright C et al (2017) Predicting forest floor and woody fuel consumption from prescribed burns in southern and western pine ecosystems of the United States. For Ecol Manage 405:328–338

Prichard S, Larkin NS, Ottmar R et al (2019a) The fire and smoke model evaluation experiment—a plan for integrated, large fire-atmosphere field campaigns. Atmosphere 10:66

Prichard SJ, Kennedy MC, Andreu AG et al (2019b) Next-generation biomass mapping for regional emissions and carbon inventories: incorporating uncertainty in wildland fuel characterization. J Geophys Res Biogeosci 124:3699–3716

Rappold AG, Stone SL, Cascio WE et al (2011) Peat bog wildfire smoke exposure in rural North Carolina is associated with cardiopulmonary emergency department visits assessed through syndromic surveillance. Environ Health Perspect 119:1415–1420

Reeves MC, Kost JR, Ryan KC (2006) Fuels products of the LANDFIRE project. In: Andrews PL, Butler BW (eds) Fuels management—how to measure success (Proceedings RMRS-P-41, pp. 239–249). U.S. Forest Service, Rocky Mountain Research Station, Fort Collins

Reinhardt ED, Keane RE, Calkin DE, Cohen JD (2008) Objectives and considerations for wildland fuel treatment in forested ecosystems of the interior western United States. For Ecol Manage 256:1997–2006

Reinhardt E, Crookston NL (2003) The fire and fuels extension to the forest vegetation simulator. (General Technical Report RMRS-GTR-116). U.S. Forest Service, Rocky Mountain Research Station, Ogden

Reinhardt E, Keane RE, Brown JK (1997) First order fire effects model: FOFEM 4.0 user's guide (General Technical Report INT-GTR-344). U.S. Forest Service, Intermountain Research Station, Ogden

Riccardi CL, Prichard SJ, Sandberg DV, Ottmar RD (2007a) Quantifying physical characteristics of wildland fuels using the fuel characteristic classification system. Can J for Res 37:2413–2420

Riccardi CL, Ottmar RD, Sandberg DV et al (2007b) The fuelbed: a key element of the fuel characteristic classification system. Can J for Res 37:2394–2412

Roberts G, Wooster M, Lagoudakis E (2009) Annual and diurnal African biomass burning temporal dynamics. Biogeosciences 6:849–866

Rollins MG (2009) LANDFIRE: A nationally consistent vegetation, wildland fire, and fuel assessment. Int J Wildland Fire 18:235–249

Rowell E, Loudermilk EL, Seielstad C, O'Brien JJ (2016) Using simulated 3D surface fuelbeds and terrestrial laser scan data to develop inputs to fire behavior models. Can J Remote Sens 42:443–459

Rowell E, Loudermilk EL, Hawley C et al (2020) Coupling terrestrial laser scanning with 3D fuel biomass sampling for advancing wildland fuels characterization. Forest Ecol Manage 462:117945

Rowell EM (2017) Virtualization of fuelbeds: building the next generation of fuels data for multiple-scale fire modeling and ecological analysis. Ph.D. thesis. University of Montana, Missoula

Roy DP, Jin Y, Lewis P, Justice C (2005) Prototyping a global algorithm for systematic fire-affected area mapping using MODIS time series data. Remote Sens Environ 97:137–162

Ryan KC, Knapp EE, Varner JM (2013) Prescribed fire in North American forests and woodlands: history, current practice, and challenges. Front Ecol Environ 11:e15–e24

Sandberg DV, Dost FN (1990) Effects of prescribed fire on air quality and human health. In: Wasltad JW, Radosevich SR, Sandberg DV (eds) Natural and prescribed fire in Pacific Northwest forests. Oregon State University Press, Corvallis, pp 191–218

Scott JH, Burgan RE (2005) Standard fire behavior fuel models: a comprehensive set for use with Rothermel's surface fire spread model (General Technical Report RMRS-GTR-153). U.S. Forest Service, Rocky Mountain Research Station, Fort Collins

Seielstad C, Stonesifer C, Rowell E, Queen L (2011) Deriving fuel mass by size class in Douglas-fir (*Pseudotsuga menziesii*) using terrestrial laser scanning. Remote Sens 3:1691–1709

Seiler W, Crutzen PJ (1980) Estimates of gross and net fluxes of carbon between the biosphere and atmosphere from biomass burning. Clim Change 2:207–247

Sellers PJ (1985) Canopy reflectance, photosynthesis and transpiration. Int J Remote Sens 6:1335–1372

Sexton JO, Bax T, Siqueira P et al (2009) A comparison of lidar, radar, and field measurements of canopy height in pine and hardwood forests of southeastern North America. For Ecol Manage 257:1136–1147

Sikkink PG, Keane RE (2008) A comparison of five sampling techniques to estimate surface fuel loading in montane forests. Int J Wildland Fire 17:363–379

Silva CA, Hudak AT, Vierling LA et al (2016) Imputation of individual longleaf pine (*Pinus palustris* Mill.) tree attributes from field and LiDAR data. Can J Remote Sens 42:554–573

Silva CA, Hudak AT, Vierling LA et al (2017) Impacts of airborne lidar pulse density on estimating biomass stocks and changes in a selectively logged tropical forest. Remote Sens 9:1068

Skowronski NS, Clark KL, Duveneck M, Hom J (2011) Three-dimensional canopy fuel loading predicted using upward and downward sensing LIDAR systems. Remote Sens Environ 115:703–714

Skowronski NS, Gallagher MR, Warner TA (2020) Decomposing the interactions between fire severity and canopy fuel structure using multi-temporal, active, and passive remote sensing approaches. Fire 3:7

Smith AM, Tinkham WT, Roy DP et al (2013) Quantification of fuel moisture effects on biomass consumed derived from fire radiative energy retrievals. Geophys Res Lett 40:6298–6302

Sokal RR, Rohlf FJ (1981) Biometry. W.H. Freeman and Company, San Francisco

Soil Science Division Staff (2017) Soil survey manual. In: Ditzler C, Scheffe K, Monger HC (eds) USDA Handbook 18. Government Printing Office, Washington, D.C.

Stephens SL, McIver JD, Boerner REJ et al (2012) The effects of forest fuel-reduction treatments in the United States. Bioscience 62:549–560

Swetnam TW, Farella J, Roos CI et al (2016) Multiscale perspectives of fire, climate and humans in western North America and the Jemez Mountains, USA. Philos Trans R Soc b: Biol Sci 371:20150168

Thenkabail PS, Smith RB, De Pauw E (2000) Hyperspectral vegetation indices and their relationships with agricultural crop characteristics. Remote Sens Environ 71:158–182

Thompson DK, Waddington JM (2014) A Markov Chain method for simulating bulk density profiles in boreal peatlands. Geoderma 232–234:123–129

Thornton PE, Law BE, Gholz HL et al (2002) Modeling and measuring the effects of disturbance history and climate on carbon and water budgets in evergreen needleleaf forests. Agric for Meteorol 113:185–222

Tinkham WT, Hoffman CM, Canfield JM et al (2016) Using the photoload technique with double sampling to improve surface fuel loading estimates. Int J Wildland Fire 25:224–228

Tucker CJ (1977) Asymptotic nature of grass canopy spectral reflectance. Appl Opt 16:1151–1156

Urbanski S (2014) Wildland fire emissions, carbon, and climate: emission factors. For Ecol Manage 317:51–60

van der Werf GR, Randerson JT, Giglio L et al (2006) Interannual variability of global biomass burning emissions from 1997 to 2004. Atmos Chem Phys 6:3423–3441

van der Werf GR, Randerson JT, Giglio L et al (2010) Fire emissions and the contribution of deforestation, savanna, forest, agricultural and peat fires (1997–2009). Atmos Chem Phys 10:11707–11735

Varner JM, Hiers JK, Ottmar RD et al (2007) Tree mortality resulting from re-introducing fire to long-ununburned longleaf pine ecosystems: the importance of duff moisture. Can J for Res 37:1349–1358

Vermote E, Ellicott E, Dubovik O et al (2009) An approach to estimate global biomass burning emissions of organic and black carbon from MODIS fire radiative power. J Geophys Res: Atmosp 114:D18205

Viney NR (1991) A review of fine fuel moisture modelling. Int J Wildland Fire 1:215–234

Wallace L, Lucieer A, Watson C, Turner D (2012) Development of a UAV-LiDAR system with application to forest inventory. Remote Sens 4:1519–1543

Wallace L, Saldias DS, Reinke K et al (2019) Using orthoimages generated from oblique terrestrial photography to estimate and monitor vegetation cover. Ecol Ind 101:91–101

Wang C, Glenn NF (2009) Estimation of fire severity using pre- and post-fire LiDAR data in sagebrush steppe rangelands. Int J Wildland Fire 18:848–856

Warneke C, Schwarz JP, Ryerson T et al (2018) Fire influence on regional global environments and air quality (FIREX-AQ): A NOAA/NASA interagency intensive study of North American fires. https://www.esrl.noaa.gov/csd/projects/firex/whitepaper.pdf. 13 July 2020

Weise DR, Wright CS (2014) Wildland fire emissions, carbon and climate: characterizing wildland fuels. For Ecol Manage 317:26–40

Williams FA (2018) Combustion theory, 2nd edn. CRC Press, Taylor and Francis Group, Boca Raton

Wooster MJ, Roberts G, Perry G, Kaufman Y (2005) Retrieval of biomass combustion rates and totals from fire radiative power observations: FRP derivation and calibration relationships between biomass consumption and fire radiative energy release. J Geophys Res: Atmosp 110:D24311

Wooster MJ, Roberts G, Smith AM et al (2013) Thermal remote sensing of active vegetation fires and biomass burning events. In: Kuenzer C, Dech S (eds) Thermal infrared remote sensing: sensors, methods, applications. Springer, Dordrecht, pp 347–390

Zarco-Tejada PJ, Diaz-Varela R, Angileri V, Loudjani P (2014) Tree height quantification using very high resolution imagery acquired from an unmanned aerial vehicle (UAV) and automatic 3D photo-reconstruction methods. Eur J Agron 55:89–99

Ziegler JP, Hoffman C, Battaglia M, Mell W (2017) Spatially explicit measurements of forest structure and fire behavior following restoration treatments in dry forests. For Ecol Manage 386:1–12

Open Access This chapter is licensed under the terms of the Creative Commons Attribution 4.0 International License (http://creativecommons.org/licenses/by/4.0/), which permits use, sharing, adaptation, distribution and reproduction in any medium or format, as long as you give appropriate credit to the original author(s) and the source, provide a link to the Creative Commons license and indicate if changes were made.

The images or other third party material in this chapter are included in the chapter's Creative Commons license, unless indicated otherwise in a credit line to the material. If material is not included in the chapter's Creative Commons license and your intended use is not permitted by statutory regulation or exceeds the permitted use, you will need to obtain permission directly from the copyright holder.

Chapter 3
Fire Behavior and Heat Release as Source Conditions for Smoke Modeling

Scott L. Goodrick, Leland W. Tarnay, Bret A. Anderson, Janice L. Coen, James H. Furman, Rodman R. Linn, Philip J. Riggan, and Christopher C. Schmidt

Abstract Modeling smoke dispersion from wildland fires is a complex problem. Heat and emissions are released from a fire front as well as from post-frontal combustion, and both are continuously evolving in space and time, providing an emission source that is unlike the industrial sources for which most dispersion models were originally designed. Convective motions driven by the fire's heat release strongly couple the fire to the atmosphere, influencing the development and dynamics of the smoke plume. This chapter examines how fire events are described in the smoke

---

S. L. Goodrick (✉)
U.S. Forest Service, Southern Research Station, Athens, GA, USA
e-mail: scott.l.goodrick@usda.gov

L. W. Tarnay
U.S. Forest Service, Pacific Southwest Region, McClellan, Truckee, CA, USA
e-mail: eleland.tarnay@usda.gov

B. A. Anderson
U.S. Forest Service, National Headquarters, Fort Collins, CO, USA
e-mail: bret.a.anderson@usda.gov

J. L. Coen
National Center for Atmospheric Research, Boulder, CO, USA
e-mail: janicec@ucar.edu

J. H. Furman
U.S. Forest Service, Eglin AFB, Niceville, FL, USA
e-mail: james.h.furman@usda.gov

R. R. Linn
Los Alamos National Laboratory, Los Alamos, NM, USA
e-mail: rrl@lanl.gov

P. J. Riggan
U.S. Forest Service, Pacific Southwest Research Station, Riverside, CA, USA
e-mail: philip.riggan@usda.gov

C. C. Schmidt
University of Wisconsin, Space Sciences and Engineering, Madison, WI, USA
e-mail: chris.schmidt@ssec.wisc.edu

This is a U.S. government work and not under copyright protection in the U.S.; foreign copyright protection may apply 2022
D. L. Peterson et al. (eds.), *Wildland Fire Smoke in the United States*,
https://doi.org/10.1007/978-3-030-87045-4_3

modeling process and explores new research tools that may offer potential improvements to these descriptions and can reduce uncertainty in smoke model inputs. Remote sensing will help transition these research tools to operations by providing a safe and reliable means of measuring the fire environment at the space and time scales relevant to fire behavior.

Keywords Atmosphere models · Coupled fire · Energy release · Fire behavior · Fire progression · Remote sensing

3.1 Introduction

Many tools used to simulate smoke impacts from wildland fires evolved from tools used in the air quality community for assessing anthropogenic pollution impacts. As such, it has been necessary to describe a wildland fire event in terms common to these anthropogenic pollutant sources—often characterized as point, line, and area sources. Descriptions of a fire event, or of an individual burn period of interest, are often reduced to simply an amount of fuel consumed at a specified location during a period of time, perhaps with diurnal variability. As the sources of the emissions and energy that drive plume dynamics (Chaps. 4 and 5), fire behavior and associated heat release (this chapter) are critical links between fuels (Chap. 2) and downwind impacts (Chaps. 6 and 7).

Fire–atmosphere interactions are tied to the energy released by the combustion process that heats the surrounding air. This heating drives a convective circulation whereby the heated air expands, decreases in density, and is forced upwards by denser ambient air. The drawing in of ambient air to replace the buoyant updraft is referred to as entrainment and is determined by the conservation laws of mass and momentum. The spatial pattern of entrainment is governed by fireline shape, ambient winds, topography, and drag induced by vegetation structure. A sustained release of heat, such as from a wildland fire, induces a feedback that allows the scale of these convective circulations to grow and interact throughout the deepening layer of the atmosphere and form a plume. This chapter focuses on how to better capture the spatial evolution of this heat source in the description of fire events used in the smoke modeling process, and how these descriptions can be improved to reduce the error associated with forcing the fire emissions modeling process to conform to an overly idealized anthropogenic emissions source.

The description of a fire consists of both temporal and spatial components. Accurately describing the evolution of a fire through time connects the release of emissions to varying atmospheric conditions such as wind direction and atmospheric stability that can greatly affect transport and dispersion. The spatial component of a fire description is more complex: The atmosphere varies temporally and spatially, requiring that the fire location and the time component are correct. But a fire is much more than a passive emitter of pollutants to the atmosphere. Distribution of heat across the landscape creates feedbacks between the fire and atmosphere, altering flow patterns and affecting downwind plume characteristics (Chap. 4). For a simple idealized source, Cunningham et al. (2005) illustrate how a buoyant plume interacts with a surface shear layer to yield variations in plume spread and depth (Fig. 3.1). For a larger fire, such processes interact across different scales to produce more complex plumes (Fig. 3.2), and that multiscale interaction can influence fire spread as well.

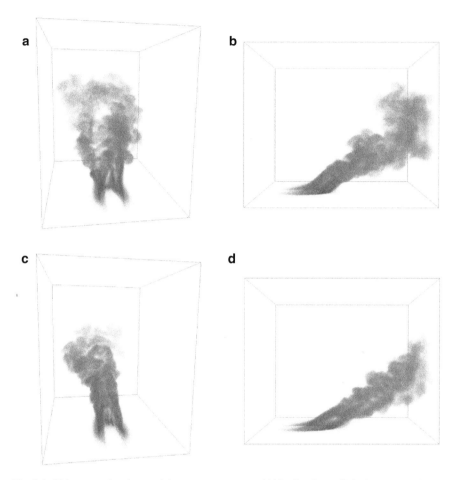

Fig. 3.1 Volume-rendered potential temperature at $t = 1100$ s for the realistic heat source large-eddy simulation. The upper two images are for the deep shear layer ($z_0 = 150$ m) case with views from **a** the inflow boundary and **b** lateral boundary. The lower two images are for the shallow shear layer ($z_0 = 50$ m) case with views in (**c**) and (**d**) identical to those in (**a**) and (**b**) respectively. Darker shades represent higher values of potential temperature. From Cunningham et al. (2005)

Fig. 3.2 Panorama image of the smoke plume above a prescribed burn at Magazine Mountain, Arizona, on February 27, 2004, revealing a complex structure of merging multiple updraft "cores" when the plume is viewed from the ground. From Achtemeier et al. (2011)

3.2 Current State of Science

3.2.1 Representing Fire in Smoke Models

Smoke models are numerical tools that provide information on the spatial distribution of pollutant species through time, such that the ecological, human health, economic, and societal effects of wildland fires can be simulated and assessed. Box models, Gaussian plume models, and Lagrangian particle and puff models, among others, are based on atmospheric transport and dispersion theory and may include complex chemical mechanisms for describing the generation of ozone and secondary organic carbon (Goodrick et al. 2012). Each tool must include a description of their emissions source, based on a representation of a wildland fire that includes fire behavior and heat release. For this assessment, we examine two smoke modeling tools commonly used on an operational basis in the USA and explore some additional models used within the research community.

3.2.1.1 Operational Tools

Operational tools are those models used for real-time decision making and planning for wildland fire management. With the exception of some planning applications, operational tools must make calculations faster than real time and be able to tailor outputs in order to effectively support decision making; for operational smoke models, the output is focused on surface pollutant concentrations rather than the full three-dimensional (3D) distribution. In contrast, research tools are focused more on advancing our understanding of a phenomenon and thus operate without the faster than real-time constraint and provide a broader range of outputs that are useful to scientists but of little practical value to land managers.

VSMOKE

VSMOKE (Lavdas 1996) is a Gaussian plume model designed for estimating smoke impacts from prescribed fires in the southeastern USA. It is best suited for simulating the effects of a single fire within periods of constant or slowly changing fire behavior, emissions, and weather conditions, during which the smoke can be adequately depicted within a steady-state framework. The fire is treated as either a point source or a specific fire area that releases emissions and heat at a constant rate. The atmosphere is described by a mixing height, transport wind, and a stability class which is treated as steady state and spatially homogeneous. The stability class and heat release affect dispersion calculations through the determination of the plume rise as determined by the commonly used Briggs equations (Briggs 1982).

Because VSMOKE is designed for prescribed fires, the model accommodates a wide range of fire behaviors, such as fires dominated by combinations of backing and flanking fire rather than conditions dominated by head fire. This is accomplished through a parameter controlling the fraction of smoke released from the surface versus that released at the plume-rise height, or uniformly distributed between the surface and the plume-rise height to achieve a range of possible plume behaviors (Lavdas 1996) (Fig. 3.3). The user can assign the parameter, although a default value based on unpublished observations of prescribed fires is provided by Lavdas. Unfortunately, there is little work connecting variations of this fraction of emissions subject to plume rise to proportions of head, flank, and backing fire or other descriptions of firing method used.

BlueSky

BlueSky is a modular smoke modeling framework which links a series of processing steps containing datasets or individual component models to estimate smoke emissions and transport for smoke forecasts and decision support (Larkin et al. 2009). Figure 3.4 shows the array of models that can be incorporated into the BlueSky framework. The minimum fire information input data required by BlueSky are fire location and daily fire growth. This fire information is transformed into dispersion model inputs by identifying appropriate fuel loads for the location, applying a consumption model to estimate daily fuel consumption, and then constructing a time profile of heat and emissions release.

By default, wildfires use the Western Regional Air Partnership wildfire profile (Air Resources Inc. 2005) that allocates 68% of the emissions to an afternoon active

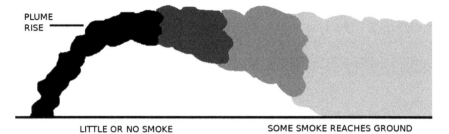
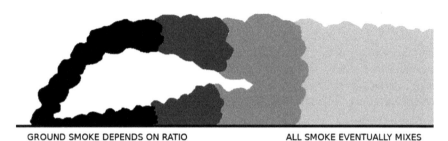
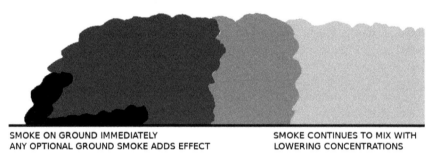

Fig. 3.3 Effects of plume rise options on ground-level smoke concentrations for VSMOKE. From Lavdas (1996)

burning period (1300–1700 h local time) along with a nocturnal smoldering component. Prescribed fires default to a time profile generated by the Fire Emissions Production System, which is based on simple rise and decay curves initially derived for estimating emissions from coniferous logging slash in the Pacific Northwest (Sandberg and Peterson 1984).

Validation efforts for older versions of the BlueSky framework found a tendency to underestimate near-field surface smoke concentrations while potentially overestimating far-field surface smoke concentrations (Riebau et al. 2006). Sensitivity studies found that predictions of surface smoke concentrations could be improved

3 Fire Behavior and Heat Release as Source Conditions … 57

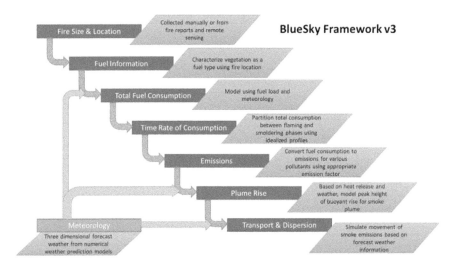

Fig. 3.4 Overview of BlueSky smoke modeling framework. Adapted from Goodrick et al. (2012)

by splitting fires into multiple emissions sources, effectively mimicking the concept of multiple updraft core plumes (Solomon 2007). A subsequent study of the 2008 northern California wildfires found that BlueSky predictions of $PM_{2.5}$ were in closer agreement with observations in both the near- and far-field (Strand et al. 2012). This simple application of the core plume concept with multiple updrafts is an example of a research tool transitioning to operations.

3.2.1.2 Research Tools

Although a wide range of research tools could be discussed in this section, our focus is on those tools that can provide insight into improving the representation of wildland fires within smoke dispersion models. This is not an exhaustive list, but a sampling of tools that are advancing our knowledge of the linkage between the fire and smoke dispersion processes.

DaySmoke

DaySmoke is a hybrid plume particle model that consists of four sub-models: an entraining turret model, a detraining particle model, a model of large-eddy parameterization for the mixed boundary layer, and a relative emissions model that describes the emission history of the prescribed burn (Achtemeier et al. 2011). The entraining turret model handles the convective lift phase of plume development and represents the updraft within a buoyant plume. This updraft is not constrained to remain within the mixed layer. A burn in DaySmoke may have multiple, simultaneous updraft cores. In comparison with single-core updrafts, multiple-core updrafts have smaller updraft velocities, are smaller in diameter, are more affected by entrainment, and are therefore less efficient in the vertical transport of smoke.

The importance of multiple-core updraft plumes was demonstrated with the Brush Creek prescribed burn in eastern Tennessee on March 18, 2006, where visual observations identified between 1 and 5 cores throughout the duration of the fire (Jackson et al. 2007; Liu et al. 2010). DaySmoke simulations with 1 to 10 updraft cores produced estimates of hourly $PM_{2.5}$ concentrations in Asheville, North Carolina, ranging from 45 mg m^{-3} (single updraft core) to 240 mg m^{-3} (10 updraft cores). The simulation with 4 updraft cores produced an hourly peak $PM_{2.5}$ concentration of 140 mg m^{-3}, which agreed well with observations at the air quality monitor location in Asheville.

In applying the Fourier amplitude sensitivity test (FAST) to DaySmoke simulations of prescribed burning in the southeastern USA, the most important parameters for determining plume rise were the entrainment coefficient and number of updraft cores (Liu et al. 2010). Both of these parameters relate to the distribution of heat across the landscape as temperature gradients enhance turbulent mixing and therefore entrainment. Areas of elevated fire intensity indicate enhanced buoyancy and therefore stronger updrafts.

Although DaySmoke can represent multiple core updrafts, it has no method for determining the appropriate number of cores to include. Achtemeier et al. (2012) determined the number of updraft cores by linking DaySmoke to a cellular automata fire model, tested on an aerial ignition prescribed burn conducted at Eglin AFB on February 6, 2011, as part of the Prescribed Fire Combustion and Atmospheric Dynamics Research (RxCADRE) collaborative research project. Originally described by Achtemeier (2013), the fire model incorporates a two-dimensional wind flow model to represent coupled fire–atmosphere circulations and provides DaySmoke with the following input information: (1) 2-m winds for calculating indraft velocities and estimates for calculating initial plume updraft velocities, (2) location and number of updraft cores, (3) approximate initial plume diameter, and (4) relative emissions production. During the simulation of the RxCADRE burn, pressure anomalies were as low as -1.4 mb, and the number of updraft cores ranged from 1 to 6 but typically was 4. Figure 3.5 shows the distribution of fire and associated pressure anomalies.

The coupled DaySmoke simulation produced a strong vertical plume that extended 1000 m above the mixing height and resulted in the majority of the flaming phase emissions being injected above the mixed layer. The observed plume heights measured with a ceilometer verified the model results, as did minimal ground-level smoke concentrations measured by a small network of downwind particulate samplers (Achtemeier et al. 2012). Linking a fire model with the dispersion model allowed the simulated plume to provide burn managers with more accurate information for their ignition planning. Linking a fire model with a smoke plume model also improved descriptions of the fire as input into the smoke model (Achtemeier et al. 2012). Tools that more strongly couple the fire and atmosphere promise further benefits.

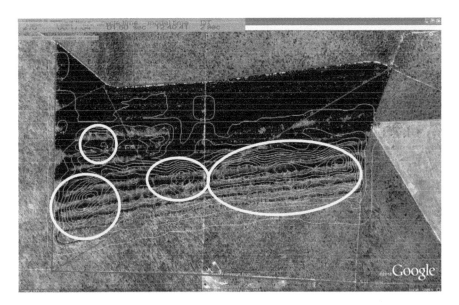

Fig. 3.5 Pressure anomalies (white lines) generated by the Achtemeier et al. (2012) fire model at 12:40:41. The yellow ellipses highlight centers that might correspond to updraft cores. From Achtemeier et al. (2012)

WRF-SFIRE

Kochanski et al. (2016) proposed an integrated system for fire, smoke, and air quality simulations by coupling WRF-SFIRE with WRF-Chem to construct an integrated forecast system for wildfire behavior and smoke prediction (Fig. 3.6). The Weather Research and Forecast (WRF) model is a mesoscale numerical weather prediction system designed for both atmospheric research and operational forecasting applications (Skamarock et al. 2008). WRF is designed to allow for incorporation of new functionalities: WRF-SFIRE and WRF-Chem are two extensions to the WRF model. WRF-SFIRE is a two-way, coupled fire–atmosphere model that estimates fire spread based on local meteorological conditions, taking into account feedback between the fire and atmosphere (Mandel et al. 2011, 2014). WRF provides a multi-scale domain, with fine scales for modeling fire behavior nested inside coarser scales for resolving the larger-scale synoptic flow.

WRF-Chem is a chemical transport model used to investigate regional-scale air quality by simulating the emission, transport, mixing, and chemical transformation of trace gases and aerosols simultaneously with meteorology (Grell et al. 2005). In current operational modeling frameworks, prescribed fire activity and fire emissions are simplified to a single plume whose vertical extent is estimated by a simple plume-rise model. However, with the coupled WRF-SFIRE-Chem system, pyroplume development, smoke dispersion, and air quality impacts are comprehensively

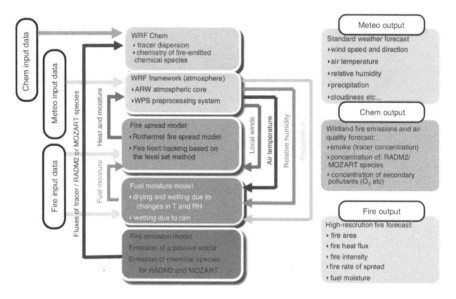

Fig. 3.6 Diagram of WRF-SFIRE coupled with fuel moisture model and WRF-Chem. From Kochanski et al. (2016)

modeled by one system that includes fire spread, heat release, fire emissions, fire plume rise, and smoke transport and dispersion with associated plume chemistry.

Application of the WRF-SFIRE-Chem system on two California fires, the 2007 Witch/Guejito fires and the 2012 Barker Canyon fire, yielded promising results (Kochanski et al. 2016). For the Witch/Guejito fire, simulated and observed local- and long-range fire spread and smoke transport agreed well, but ozone, $PM_{2.5}$, and NO concentrations were generally underestimated in the simulations. Simulated plume-top heights exhibited considerable variation throughout the day, with the standard deviation of time-averaged plume heights as high as 600 m. The simulations clearly exhibited multiple plume-rise peaks associated with multiple core updrafts and reinforced that a single Gaussian-shaped plume and injection height provides an unrealistic representation of a wildfire plume.

Simulations of several large 2015 wildfires in northern California highlighted the ability of the WRF-SFIRE-Chem system to capture feedback effects between smoke and weather (Kochanski et al. 2019). Smoke from the wildfires induced a positive feedback loop in which aerosols aloft in the smoke plume absorbed incoming solar radiation, warming the top of the plume. Less solar radiation was received at the ground, resulting in surface cooling. This warming aloft and cooling below develops a local smoke-enhanced inversion that inhibits the growth of the planetary boundary layer and reduces surface winds, resulting in smoke accumulation that further reduces near-surface temperatures. Such results are possible only in a system that fully integrates fire and smoke processes within the weather model.

MesoNH-ForeFire

Similar to WRF-SFIRE, the European MesoNH-ForeFire modeling system combines a fire area simulator and a mesoscale meteorological model to simulate fire–atmosphere interactions (Filippi et al. 2011). The fire area simulator is based on the spread model of Balbi et al. (2009) and describes the mean propagation velocity of the fire front as a function of slope, surface wind speed, and fuel properties. Initial application of MesoNH-ForeFire to real-case scenarios in predominantly Mediterranean Maquis shrublands yielded plume structures that agreed qualitatively with photographs of the plume, with distinct updrafts developing over each fire flank that merge over the head of the fire. Although the model produced some of the observed plume structures, the 50-m grid used in the atmospheric simulation limited the model's ability to reproduce finer-scale structures. In a more recent study using a finer grid resolution, MesoNH-ForeFire plumes compared well with Lidar-based plume observations (Leroy-Cancellieri et al. 2014). The more refined model grid improved the representation of the fire in space and time, resulting in improved forcing of the atmospheric processes governing plume behavior.

CAWFE

The Coupled Atmosphere Wildland Fire Environment (CAWFE) is an alternative system that employs a numerical weather prediction model designed specifically for simulating small-scale weather processes in complex terrain (Clark and Hall 1991; Clark et al. 1996, 1997; Coen 2013). Coupling of numerical weather prediction models to fire-spread simulations provides many benefits, as the coupling allows dynamic interaction among the components of the fire environment.

However, fire presents an interesting problem to many weather models, depending on their formulation and inherent assumptions. Models such as WRF are designed to simulate a broad range of weather phenomena ranging from hundreds of meters to tens of kilometers. This flexibility and scalability do not come without a cost. Models designed for these scales tend to dissipate energy at fine scales due to choices in numerical schemes used in its solver and grid refinement methodology. Thus, the model tends to dissipate energy at the scales that the fire is trying to add energy.

The fire component of CAWFE is a front-tracking approach similar to that of WRF-SFIRE. CAWFE simulations helped Coen et al. (2018) evaluate the relative roles of climate, fuel accumulation, and forest structure changes tied to fire exclusion, and nonlinear effects tied to dynamic coupling of fire environment components on the 2014 King fire (California). The CAWFE atmospheric formulation has shown promise in reproducing significant features of major wildfire events, such as a 25-km up-canyon run on the King fire (Coen et al. 2018), but its less dissipative nature may be more applicable for lower-intensity prescribed fires.

FIRETEC and WFDS-PB

FIRETEC (Linn et al. 2003) and the Wildland Urban Interface Fire Dynamics Simulator (WFDS-PB; Mell et al. 2009) are physics-based fire models that use a finite-volume, large-eddy simulation approach to model the atmosphere. FIRETEC (Linn et al. 2003) is a 3D model designed to simulate the constantly evolving relationships between wildland fire and its environment. FIRETEC describes a range of processes that drive fire behavior and how these processes interact with the overlying atmosphere (Fig. 3.9). Vegetation is described as a highly porous 3D medium characterized by bulk quantities (e.g., surface area-to-volume ratio, moisture content, bulk density) of the thermally thin components of the vegetation. FIRETEC and WFDS-PB have been used to simulate crown fires (Linn et al. 2012; Hoffman et al. 2016), bark beetle effects on fire behavior (Hoffman et al. 2013; Linn et al. 2013), and fireline interactions (Morvan et al. 2011, 2013). The primary drawback of such physics-based models is the very high computation requirements inherent in the fine resolution of the computational grid (1–2 m).

The WFDS model builds on the Fire Dynamics Simulator, which was developed by the National Institute of Standards and Technology to model structural fire. WFDS is intended to help understand fine-scale fire behavior within wildland fires and between wildland and developed areas. WFDS uses computational fluid dynamics to represent buoyant flow, heat transfer, combustion, and thermal degradation of vegetative fuels. This approach uses large-eddy simulation to solve the gas-phase equations on computational grids that are too coarse to directly resolve the detailed physical phenomena.

Computational costs can be lowered by implementing a "level-set method" to propagate the fireline (e.g., WFDS-LS; Bova et al. 2016). This numerical technique tracks the evolution of an interface between two locations (e.g., burned and unburned fuels), thus simplifying issues from merging and splitting fronts that are difficult to track (Mallet et al. 2009). Explicitly resolving gas-phase combustion is not necessary for smoke plume simulations of this scale if the heat release per unit area is known (Liu et al. 2019). Models such as WFDS can be used to inform our ability to design communities to withstand an approaching wildfire.

3.2.2 Remote Sensing

Although models provide one means of developing a more complete description of a fire for input into smoke models, empirical observations are also a vital source of information about individual fires and for fire model verification. Wildland fires are difficult to measure due to high temperatures, but many remote sensing techniques have emerged over the last 20 years that are capable of observing wildland fires across a broad range of spatial and temporal scales. This approach is often capable of deriving spatial and temporal distributions of heat release as inputs for smoke models.

3.2.2.1 Fire Area

Measurements of area burned are critical for estimating fire emissions, which are one of the largest sources of potential error in modeling (Soja et al. 2009). As described in Chap. 2, satellites provide a consistent means for estimating burned area. Two satellite platforms commonly used for estimating burned area are the Geostationary Operational Environmental Satellite (GOES) and Moderate Resolution Imaging Spectroradiometer (MODIS). The GOES product is geosynchronous with temporal resolutions of 5 min for the Fire Detection and Characterization Algorithm (FDCA) at the cost of relatively low resolution of 2 km. Despite geostationary satellites tending to have coarser resolution than polar orbiters, the more frequent observations have shown utility beyond just detection. For example, Liu et al. (2018) were able to extract near real-time rate-of-spread estimates for fires in Western Australia using data from the Himawari-8 satellite.

The polar-orbiting MODIS instruments provide improved spatial resolution of 500 m but at lower temporal resolution of 4 overpasses per day. The GOES and MODIS products capture the inherent tradeoff between spatial and temporal resolution which limits their current utility for describing the evolution of fire events. Satellite products successfully capture large wildfires that account for the majority of emissions (Soja et al. 2009), but are less useful for prescribed fires due to a low detection rate as most prescribed fires are of lower intensity and shorter duration (Nowell et al. 2018). Although satellite instruments and algorithms will continue to advance, alternative instrument platforms, such as aircraft and unmanned aircraft systems (UASs), provide better spatial and temporal resolution for select events.

3.2.2.2 Energy Release

Knowing the fire location is the first step in describing a fire for use in smoke models. The next piece is knowing the rate and amount of heat released. Measurements of Fire Radiative Power (FRP) detect the rate of radiant heat output from a fire, and FRP integrated over time provides an estimate of the fire radiative energy (FRE), which is proportional to the total mass of fuel biomass consumed (Chap. 2). In their review of fire meteorology, Kremens et al. (2010) outlined several methods for estimating the energy radiated by the combustion of fuels within each fire-affected pixel (Kaufman et al. 1996; Butler et al. 2004; Riggan et al. 2004; Ichoku and Kaufman 2005; Smith and Wooster 2005). The FRP and (by time integration) FRE are calculated by combining two infrared bands to estimate the mean radiant fire temperature and emissivity-area product for an individual pixel (Dozier 1981; Matson and Dozier 1981; Riggan et al. 2004).

In an examination of the 2013 Rim Fire (California), Peterson et al. (2015) employed FRP estimates from the GOES-14 satellite to study extreme fire spread and pyroconvection. Peaks in FRP during the Rim Fire likely coincided with the most intense burning (Fig. 3.7). Although diurnal variability in FRP is evident, it is equally evident that variation in FRP does not follow a simple diurnal distribution as

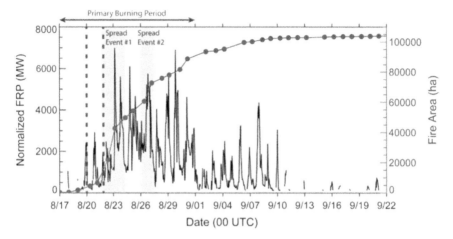

Fig. 3.7 Time series of normalized hourly FRP from GOES-West (black) and cumulative fire area derived from National Infrared Operations observations (red). Spread events 1 and 2 are highlighted with yellow shading, and the pyroCb events of 19 August and 21 August are denoted by dashed brown vertical lines. From Peterson et al. (2015)

described for the Western Regional Air Partnership and Western Governors Association (Air Resources Inc. 2005). The co-occurrence of high FRP on days with weaker atmospheric stability is likely tied to a greater vertical extent of the smoke plumes and an enhanced probability of smoke injection into the free troposphere (Val Martin et al. 2010; Peterson et al. 2014).

Satellites are not the only remote sensing platforms from which fire information can be derived. One example of an aircraft-based platform, the FireMapper thermal-imaging radiometer, allows quantitative measurements of fire-spread rates, fire temperatures, radiant-energy flux, residence time, and fireline geometry (Riggan et al. 2010). Figure 3.8 is a FireMapper thermal image of the Esperanza Fire (California) depicting thermal anomalies indicative of biomass burning on October 26, 2006, between 14:07 and 14:17 PDT (Fig. 3.9).

Coen and Riggan (2014) examined the Esperanza Fire to test the CAWFE model and examined how dynamic interactions of the atmosphere with large-scale fire spread and energy release affect observed patterns of fire behavior as mapped by FireMapper. This is a case of FireMapper being used to verify a model projection of fire behavior. The CAWFE simulation correctly depicted the fire location at the time of an early-morning incident involving firefighter fatalities. Periods of deep plume growth were also well captured by the model and verified by FireMapper, highlighting the importance of fire–atmosphere coupling in reproducing the evolution of a fire.

Fig. 3.8 FireMapper thermal image of the Esperanza Fire (southern California), showing thermal anomalies indicative of biomass burning on October 26, 2006, between 14:07 and 14:17 PDT. Higher fire intensity is indicated by orange and yellow pixels. From Riggan et al. (2010)

3.2.3 Effects of Management Actions

3.2.3.1 Prescribed Fire

A shortcoming of many fire behavior tools is their inability to consider interactions between multiple lines of fire (e.g., counter-firing operations); operational tools do not account for convective heating or interactions between multiple heat sources and are typically limited to describing fire behavior for a point ignition spreading in a homogeneous environment (Furman et al. 2019; Hiers et al. 2020). Fire operations are planned to accomplish multiple, specific objectives. Prescribed fire objectives often include maintaining fire within a limited range of intensities (e.g., rates of spread, flame lengths) to minimize damage to the resource but supply enough heat to aid in smoke transport and dispersion. A key part of the burn plan is developing sufficient heat to generate a plume that rises above the mixed layer such that surface impacts to nearby communities are minimized (Achtemeier et al. 2012).

The widely used Briggs plume-rise schemes used in air quality forecasting assume the plume rises through a passive environment that does not consider the complex ways a fire and the environment interact (Moisseeva and Stull 2020). Neglecting such interactions can lead to overestimation of plume rise and underestimation of surface smoke concentration for "highly tilted" plumes characterized by weak buoyancy and strong winds (Achtemeier et al. 2011), or underestimation of plume rise

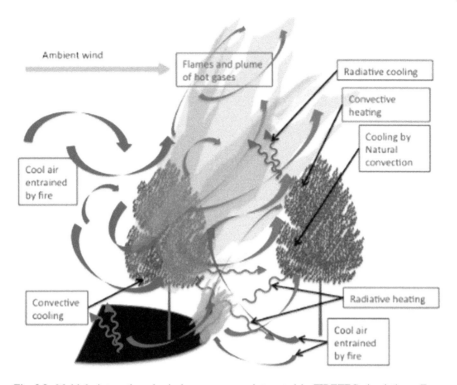

Fig. 3.9 Multiple interactive physical processes are integrated in FIRETEC simulations. From Furman et al. (2019)

and subsequent overestimation of surface smoke concentration for strongly buoyant plumes (Achtemeier et al. 2012).

Using FIRETEC, Furman et al. (2019) examined whether a coupled fire–atmosphere model could reproduce a range of fire phenomena common to prescribed fires. They examined questions that current operational tools are ill-suited to answer, including:

- How does distance between lines of fires and multiple ignition points affect fire intensity and plume lofting?
- How does spot ignition moderate fire intensity compared to line ignition?
- How does unit boundary ignition affect fire behavior and fire effects in the interior of the burned area?
- How do mid-story vegetation and other forest structure variables influence wind fields and resulting fire behavior?

Furman et al. (2019) evaluated different ignition patterns for prescribed fires in longleaf pine (*Pinus palustris*) forest fuels. Figure 3.10 illustrates FIRETEC results that depict general fire phenomenology associated with multiple ignition lines ignited

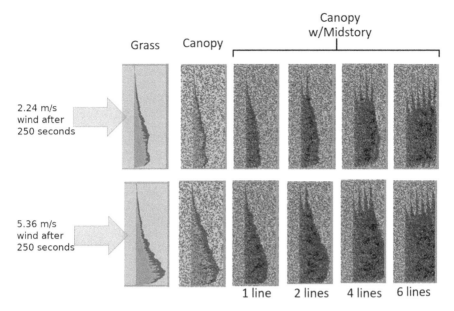

Fig. 3.10 Baseline fire scenarios modeled with FIRETEC. The images are bounded by the fuel breaks and therefore do not show the entire computation domain. From Furman et al. (2019)

by all-terrain vehicles (ATVs). Higher fire behavior and mid-story/canopy consumption in response to an increased number of simultaneous ignition lines (Fig. 3.10) is common knowledge among experienced prescribed fire managers. However, the effects of line spacing on convective lift and subsequent plume lofting were not as well known (Figs. 3.11 and 3.12). These FIRETEC simulations revealed that the ATV-ignition strip-head fires reached greater plume height and volume than the plastic sphere dispenser (or "ping pong ball") aerial ignition, as the small individual ignitions of the aerial ignition were widely dispersed and burned together more slowly than solid lines ignited by the ATVs.

A new simulation tool called QUIC-Fire (Linn et al. 2020) is designed to rapidly simulate fire–atmosphere feedbacks by coupling the 3D rapid wind solver QUIC-URB to a physics-based cellular automata fire-spread model (Fire-CA). QUIC-Fire uses 3D fuels inputs similar to those used by the CFD-based FIRETEC model, allowing this tool to simulate the effects of fuel structure on local winds and fire behavior. Preliminary comparisons between QUIC-Fire and FIRETEC show that the model outputs agree well. QUIC-Fire is the first tool intended to provide an opportunity for prescribed fire planners to compare, evaluate, and design burn plans, including complex ignition patterns and coupled fire-atmospheric feedback. Additional work to incorporate process-based emissions production into QUIC-Fire has also shown promise (Josephson et al. 2020).

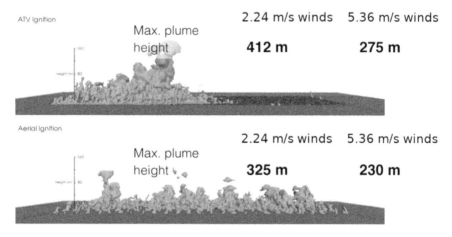

Fig. 3.11 View from upwind of burn unit illustrating differences in modeled maximum plume heights between 5-line, 16-ha ATV, and aerial ignitions for two surface wind speeds. From Furman et al. (2019)

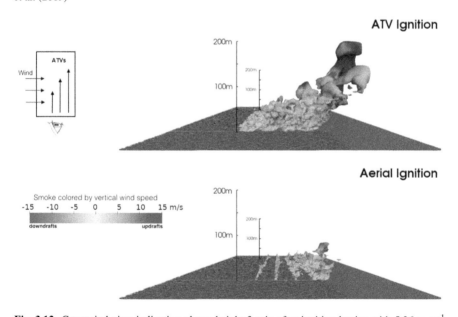

Fig. 3.12 Crosswind view indicating plume height 3 min after ignition begins with 5.36 m s^{-1} wind. "Trees" were removed for visual clarity. Plume color denotes vertical wind speed of heated gasses. From Furman et al. (2019)

3.3 Gaps in Understanding the Link Between Fire Behavior and Plume Dynamics

Many current smoke modeling tools have a number of limitations that are largely linked to the fire event not being an explicit part of the simulation (Liu et al. 2019). By excluding the fire event from the simulation, these tools are unable to incorporate detailed and rapidly varying spatial distributions of heat release across the landscape, which links the fire source to the atmosphere, often leading to the development of multiple plume cores. In addition, emissions must be estimated with a method such as climatological diurnal trends as in the Global Fire Emissions Database (Randerson et al. 2017) or the Smoke Emissions Reference Application database (Prichard et al. 2020).

Advancing our modeling capability beyond these empirically derived methods and toward more process-based methods is critical for predicting emissions in the no-analog climate expected in future decades. Making this shift to process-based models requires an improved understanding of fire and smoke processes, as well as collecting data tailored to rigorous testing, evaluation, and validation of model performance under real-world conditions (Liu et al. 2019).

Many currently available observational datasets are not suitable for evaluating coupled fire–atmosphere models, because these tools require integrated datasets that comprehensively characterize fuels, energy released, local micrometeorology, plume dynamics, and smoke chemistry (Alexander and Cruz 2013; Cruz and Alexander 2013). To fill such data gaps, several field campaigns have been conducted or are planned in the USA. In 2012, the RxCADRE field campaign collected integrated data on fuels, fire behavior, fire effects, and smoke on large prescribed fires at Eglin Air Force Base (Florida) for the specific purpose of evaluating fire and smoke models (Ottmar et al. 2016). The RxCADRE data are currently being used to evaluate coupled fire–atmosphere modeling systems.

Moisseeva and Stull (2020) examined plume rise from an experimental burn of the RcCADRE campaign, using WRF-SFIRE by taking advantage of the combined fire behavior and plume measurements collected by the project. Their model results capture the timing, rise, and dispersion of the fire plume reasonably well compared with observations of emissions and dispersion data collected from an airborne platform during the experiment. Although the plume observations available in RxCADRE were limited, other efforts are working to increase the amount and quality of plume observations, including the Fire Influence on Regional and Global Environments Experiment (FIREX-AQ) (Warneke et al. 2018), Western Wildfire Experiment for Cloud Chemistry, Aerosol Absorption and Nitrogen (WE-CAN) project (https://www.eol.ucar.edu/field_projects/we-can), and Fire and Smoke Model Evaluation Experiment (FASMEE) (Prichard et al. 2019). Liu et al (2019) has outlined specific information needed to advance our knowledge of fire–atmosphere coupling and its ties to plume dynamics (Chap. 4) (Fig. 3.13).

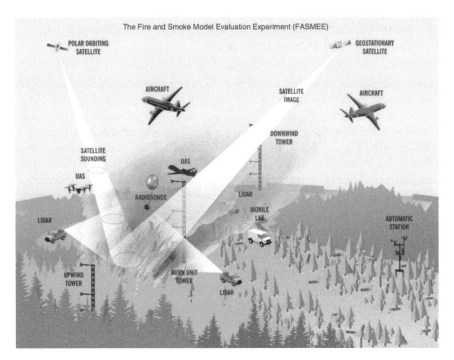

Fig. 3.13 Schematic representation of the Fire and Smoke Model Evaluation Experiment (FASMEE) project measurement platforms. From Liu et al. (2019)

3.3.1 Heat Release

Measurements of fire-base depth, spread rate, and total mass consumption during flaming can be used to calculate a first-order estimate of heat release per unit area for fire behavior model validation and as inputs for smoke models. Note that a single-point measurement can be misleading, because firelines are not uniform. For this reason, a more complete set of measurements to support model testing would provide the fire-base depth, spread rate, and total mass consumption along the fire perimeter. Furthermore, surface heat is vertically distributed over the first few grid-cell layers in some fire–atmosphere coupled models (e.g., WRF-SFIRE), which means the appropriate vertical decay scale (extinction depth) needs to be assessed. Also, fire heat varies in both space and time, leading to complex dynamical structures of smoke plumes. Dynamical structure is an important factor for the formation of separate smoke plume cores. Measurements of the structures together with smoke dynamics are needed to understand the relations of smoke dynamics to horizontal and vertical heat fluxes (radiative and convective) during fires.

3.3.2 Fire Spread

Fire spread is important for determining fuel consumption and spatial and temporal variation of heat release, burned area, and burn duration. Lateral fire progression, spread perpendicular to the predominant wind direction, is particularly affected by atmospheric turbulence. In models such as WRF-SFIRE, the lateral rate of spread is parameterized using (1) local wind perturbations normal to the flank, and (2) the Rothermel formula (Rothermel 1972) for head-fire rate of spread. Some of these normal wind perturbations can be created by fine-scale differences in topography, fuels, and pressure gradients that are dampened or smoothed; these differences are not neglected in the WRF scheme and are partially preserved in the CAWFE modeling scheme. Characterization of lateral fire spread and atmospheric turbulence, in concert with variation in fuel and topography, is needed to validate and improve this approach (Bebieva et al. 2020). We must better understand where WRF-SFIRE type simplifications are "good enough," or when fine-scale modeling of fuels and topography to produce wind perturbations are necessary (Coen 2018).

3.3.3 Plume Cores

Individual plume cores within a smoke plume are highly dynamic, often forming as a result of local fuel accumulations and ignition processes. Once formed, they can instantly affect heat fluxes, exit velocity, and temperature, which are important for smoke plume rise and vertical profile simulation. Despite their importance, the number of plume cores is rarely noted for prescribed burns. Because the dynamic nature of plume cores makes them difficult to define and track, observational and modeling evidence is needed to understand the roles of sub-plumes.

3.4 Vision for Improving Smoke Science

A scale-appropriate abstraction of fire is needed to supply heat and emissions to smoke models. The typical current level of abstraction—representing a fire as a simple point-source with either a constant emission rate or diurnal emissions profile—may be appropriate for coarse-scale continental assessments. However, smoke models are often used to address a range of scales for local visibility and air quality concerns where a more detailed description of a fire in space and time may be required for accurate results. At the local scale, the modeling approach of Kochanski et al (2019) captures the coupling between fire and atmosphere to provide a detailed abstraction of the smoke source. This coupled approach is also ideal for prescribed fire applications, because it allows for complex ignition patterns common in prescribed fire operations.

Between these two extremes in scale lies the regional scale where the simple point-source is inadequate; as at finer scales, we must begin to account for variations in fuels and fire geometry that influence plume organization and lead to near-field underestimation and far-field overestimation of surface smoke concentrations (Riebau et al. 2006). The fully coupled approach of the local scale may become too computationally intensive at the regional scale to be useful for forecasting as the number of fires in a region increases. At regional scales, a more flexible abstraction of the fire event is required that differs in level of detail between the two extremes.

Coupled models such as FIRETEC, WFDS, CAWFE, QUIC-Fire, and WRF-SFIRE could be the central tools in developing such an abstraction. Running a series of simulations with these models with known fire behavior and plume behavior at a range of spatial scales would allow for calculating plume rise and fine-scale wind perturbations along ignition perimeters. This would include common ignition methods for prescribed fire, such as aerial ignitions, strip-head fires, and spot ignitions. These relationships would connect some basic information about the ignition (e.g., line spacing) and return appropriate inputs of heat and emissions through time, scaled to the spatial and temporal elements of the burn unit. For coupled models in these scenarios, developing multiple-core updrafts would be explicitly simulated as the atmosphere responds to the distribution of heat across the landscape, effectively building this fine-scale process into quantitative relationships. An important outcome would be improved estimates of plume rise for use in dispersion models (Chap. 4).

Advances in modeling heat release from wildland fires must be accompanied by advances in our ability to observe fires. Technological advances and improved affordability of both sensors and sensor platforms are revolutionizing our ability to collect information on wildland fires. Sensor systems which were previously cost prohibitive for widespread use in fire research, such as hyperspectral cameras, image intensifiers, and thermal cameras, are now less expensive and are being more commonly used. Allison et al. (2016) provide a review of the application of a number of these technologies to wildfire detection and monitoring. The technology is advancing toward an integrated hierarchical system of sensors that combine continuous monitoring for early detection with field-deployable small sensing platforms to provide detailed data for specific fire incidents.

Challenges identified by Allison et al. (2016) include developing robust automatic detection algorithms, integrating sensors of varying capabilities and modalities, and developing best practices for introducing new sensor platforms (e.g., small UASs) in a safe and effective manner within a fire perimeter. Image processing techniques are advancing rapidly due to increased computing power and the emergence of machine learning tools. Moran et al. (2019) describe a hybrid threshold gradient method for detecting areas of flaming combustion that combines the use of a temperature threshold value such as the Draper point (525 °C) with a gradient-based edge detection algorithm. This combined approach yields solutions that maintain observed variability while maximizing indifference to sensor resolution and spectral band differences.

Zhao et al. (2018) demonstrated the effectiveness of using saliency detection combined with a deep convolutional neural network to segment wildfire images into regions of smoke, flaming combustion, and burned areas. Deep convolutional neural networks are a state-of-the-art machine learning method for image recognition (Lecun et al. 2015), and saliency detection extracts core objects from a complicated scene (Itti et al. 1998). Combining these techniques provides a more robust solution, as individual images are broken down into a set of core objects which are then compared by the neural network.

Adding to the value of new imaging techniques and platforms is the ability to integrate information across sensors and platforms to provide an enhanced view of the environment. Jimenez et al. (2018) describe an experimental design and preliminary results for linking highly resolved ground-based fire measurements collocated with in situ and thermography remotely sensed by UASs. Linkage of the in situ and UAS thermography offers an opportunity to link the combustion environment with post-fire processes and wildland fire modeling efforts across a broader spatial scale.

Fassnacht et al. (2021) combined satellite-based differenced Normalized Burn Ratio (dNBR) information with high-resolution orthoimages from a UAS to identify sources of variability in satellite data related to pre- and post-fire vegetation structure. Their results suggest that the fraction of consumed canopy cover, along with shadows of snags and standing dead trees with remaining crown structure, influenced what the satellite detects, providing an underestimate of dNBR. Improving our ability to examine the fire environment across scales will improve our understanding of variability in the data and inform modeling of fire processes.

3.5 Emerging Issues and Challenges

3.5.1 Magnitude of Fire and Smoke Impacts

In recent years, prominent smoke impacts have been observed in many locations in the USA. In California, long-duration smoke events—termed "smoke waves" by Liu et al. (2016)—are now emitting enough $PM_{2.5}$ to become the primary source in the annual emissions inventory. The modeling work of Koman et al. (2019) found that 97.4% of California residents lived in a county with at least one smoke wave during the 2007–2013 study period, and 24.7% of the population lived in a county averaging at least one smoke wave per year. Based on data from the California Air Resources Board (CARB 2020), for the period 2014–2019, the average annual area burned was 34% higher and the $PM_{2.5}$ emissions from wildfires 43% higher than during the study period of Koman et al. (2019), indicating that smoke-wave events were more common in recent years.

3.5.2 Managing Fuels to Minimize Air Quality Impacts

At the root of increasing impacts on regional air quality is accumulation of fuels on the landscape, exacerbated by a warming climate that is creating increased likelihood and opportunities for fuels to drive extreme fire behavior, thus leading to an extended duration of smoke waves (Chap. 2). These impacts are not necessarily uniform, predictable, or even consistent from one year to the next, one fire to the next, or one day to the next for a given fire. Fuel continuity as affected by fire exclusion, previous wildfire and other disturbance footprints, and previous fuel treatments interact with weather (and climate) to create the conditions for large-fire growth and smoke waves.

Koontz et al. (2020) showed that fire severity (damage to natural resources, typically mortality in overstory trees) in dry forests of the Sierra Nevada (California) was higher in forests with higher homogeneity of fuels at all scales above 90 m (the smallest scale tested). The resilience of these forests, which may have been reduced by structural homogenization caused by several decades of fire exclusion, could be restored with management that targets increased forest structural variability (Chap. 8). A smoke modeling framework that links changes in forest structure and associated changes in fire behavior at fine scale (<90 m, and probably <30 m) with plume development and smoke dispersion could evaluate the potential for fuel reduction treatments to limit smoke-wave impacts. From the perspective of operations, planning, and State Implementation Plans this modeling ability could provide a foundation for strategic application of fuel reductions, thus informing prioritization of treatments that would minimize smoke impacts.

3.5.3 Need for Dispersion Climatologies

Simultaneous, Monte Carlo modeling of fuels, fire, smoke emissions, and meteorology from micro- to mesoscales is beyond current computational capabilities, thus complicating assessment of model sensitivities (Bakhshaii and Johnson 2019). Kochanski et al. (2019) provided an example of potential sensitivity, showing that spatial resolutions of 1.3 km or finer were required to resolve canyon winds and smoke-enhanced inversions in the complex terrain of the Klamath Mountains (California), an area characterized by steep elevations and relatively fine-scale "corrugation" of the landscape.

Integrating smoke climatologies (Kaulfus et al. 2017), high-resolution modeling (Kochanski et al. 2019; Kiefer et al. 2019), and reanalysis climatologies may provide a tool for linking regional weather patterns with dispersion and spread parameter sensitivities, as the plume climatologies provide constraints against which to test models. Such climatologies would allow us to develop scenario-based, high-resolution, and atmospherically coupled modeling exercises on these "modes" of transport; developing a library of likely dispersion scenarios that could be used operationally for wildfires and planning prescribed fires across large landscapes.

Table 3.1 Atmosphere-fire models cited in the text, including appropriate spatial and temporal scales for their use

Model	Coupling method	Spatial scale	Domain area	Temporal scale	CPU usage
FIRETEC	One way	1–5 m	1–5 km^2	1 s–2 h	No real time
WFDS-PB	One way	1 mm–10 m	1–5 km^2	1 s–4 h	No real time
CAWFE	Two way	10 m–30 m	1–10 km^2	1 s–h	No real time
WRF-SFIRE	Two way	10 m–10 km	1–km^2	1 s–days	Real time
MesoNH-ForeFire	Two way	10 m–1 km	1–100 km^2	1 s–h	Approx. real time
ARPS/DEVS-FIRE	Two way	10 m–1 km	1–100 km^2	1 s–h	No real time

Adapted from Bakhshaii and Johnson (2019)

3.5.4 When and Where is Coupled Fire–Atmosphere Modeling Needed?

The coupled fire–atmosphere models discussed in this chapter would seem ideal for providing time-varying inputs of emissions and heat release for use with smoke dispersion tools, but these models should not be viewed as a universal solution, because each model has characteristics best suited to different spatial and temporal scales (Table 3.1). Among the coupled models using numerical weather prediction models, differences in semiempirical assumptions inherent in model formulation affect model performance, depending on the degree of topographic complexity and synoptic conditions (Coen 2018).

When large-scale synoptic conditions are the dominant driver of fire spread, the simplifying assumptions used by WRF-SFIRE (Kochanski et al. 2013) show promise and allow for real-time forecasting with modern 100-plus processor computing clusters (Kochanski et al. 2019). As topographic slopes exceed 40°, and fine-scale fuel consumption and fire-induced winds start driving fire behavior, the WRF-SFIRE approach may be limited due to the aggressive dissipation of fine-scale motions and gradients that occur under these conditions (Coen 2018).

Finer-scale models such as CAWFE may improve the resolution of fire-induced flows but lack the ability to do so operationally until faster computers or more computationally efficient approaches are available. However, these retrospective approaches have the potential to help in planning and prioritizing areas of the landscape where fuels, weather, and topography might create fire–atmosphere coupling characteristic of some large wildfires [e.g., the previously mentioned Rim fire (Peterson et al. 2015) and King fire (Coen et al. 2018)].

3.6 Conclusions

This chapter focuses on how fire events are described in the smoke modeling process and how these descriptions can reduce the error associated with forcing a fire to conform to an idealized anthropogenic emissions source, such as a point source (e.g., industrial stack emissions). Liu et al. (2019) highlighted several needs for next-generation smoke research and forecasting systems:

- Acquire dynamic and high-resolution fire energy and emissions information for smoke modeling of large burns.
- Improve the capability to describe multiple sub-plumes and understand mechanisms governing their evolution.
- Understand feedbacks between atmospheric disturbances induced by fire and smoke processes.
- Link combustion processes to speciation of fire emissions across fuel types and combustion conditions.

Achieving such model improvements requires extensive and detailed observations of the spatial and temporal evolution of heat released by wildland fires, as heat release connects the fire to the overlying smoke plume. Remote sensing provides a safe and reliable means of collecting such observations. The emergence of new observing platforms such as small UASs and thermal sensors, combined with new processing techniques allowing integration of multiple data streams, will help improve the temporal and spatial resolution of heat release measurements at scales appropriate for model development. Integrated field campaigns such as FASMEE (Prichard et al. 2019) that seek to measure the fire environment as thoroughly as possible will help facilitate the transition of new modeling tools from research to operations by providing a testbed for developing the data necessary for model advancements.

3.7 Key Findings

- Wildland fires are poorly described as emission sources in current operational smoke modeling tools. Fires are complex emission sources that evolve through time and are not well represented as traditional point, line, or area sources used in air quality models.
- A number of newer research tools have improved descriptions of wildland fires as heat and emission sources. Transitioning such tools to operations will require further data collection for understanding model performance and uncertainty but will enhance our ability to cope with evolving environmental conditions during wildland fires.
- Remote sensing provides a robust platform for a wide range of fire measurements such as energy release and fire area. Technological advances are expected to improve sensor abilities in the future.

3.8 Key Information Needs

In addition to the fuel information described in Chap. 2, key fire behavior and energy release measurements include:

- Quantitative fire radiation from satellite, airborne, and tower-based platforms.
- Flame front dimensions and spread rates that include forward, lateral, and backing spread.
- Combustion efficiency, and emissions partitioning between flaming and smoldering combustion.
- Convective fluxes are needed in space and through time as inputs to plume models.

References

Achtemeier GL (2013) Field validation of a free-agent cellular automata model of fire spread with fire-atmosphere coupling. Int J Wildland Fire 22:148–156

Achtemeier GL, Goodrick SA, Liu Y et al (2011) Modeling smoke plume-rise and dispersion from southern united states prescribed burns with daysmoke. Atmosphere 2:358–388

Achtemeier GL, Goodrick SA, Liu Y (2012) Modeling multiple-core updraft plume rise for an aerial ignition prescribed burn by coupling Daysmoke with a cellular automata fire model. Atmosphere 3:352–376

Air Resources Inc. (2005) 2002 Fire emissions inventory for the WRAP Region—Phase II (Report, Project 178–6). Prepared for the Western Governors Association/Western Regional Air Partnership. http://www.wrapair.org/forums/fejf/tasks/FEJFtask7PhaseII.html. 7 July 2020

Alexander ME, Cruz MG (2013) Limitations on the accuracy of model predictions of wildland fire behaviour: a state-of-the-knowledge overview. For Chron 89:370–381

Allison RS, Johnston JM, Craig G et al (2016) Airborne optical and thermal remote sensing for wildfire detection and monitoring. Sensors 16:1310

Bakhshaii A, Johnson EA (2019) A review of a new generation of wildfire–atmosphere modeling. Can J for Res 49:565–574

Balbi JH, Morandini F, Silvani X et al (2009) A physical model for wildland fires. Combust Flame 156:2217–2230

Bebieva Y, Oliveto J, Quaife B et al (2020) Role of horizontal eddy diffusivity within the canopy on fire spread. Atmosphere 11:672–689

Bova AS, Mell WE, Hoffman CM (2016) A comparison of level set and marker methods for the simulation of wildland fire front propagation. Int J Wildland Fire 25:229–241

Briggs GA (1982) Plume rise predictions. In: Haugen DA (ed) Lectures on air pollution and environmental impact analyses. American Meteorological Society, Boston, pp 59–111

Butler BW, Cohen J, Latham D et al (2004) Measurements of radiant emissive power and temperatures in crown fires. Can J for Res 34:1577–1587

California Air Resources Board (CARB) (2020) California wildfire burn acreage and preliminary emissions estimates. https://ww3.arb.ca.gov/cc/inventory/pubs/ca_wildfire_co2_emissions_estimates.pdf. 19 Mar 2021

Clark TL, Hall WD (1991) Multi-domain simulations of the time-dependent Navier-Stokes equations: Benchmark error analysis of nesting procedures. J Comput Phys 92:456–481

Clark TL, Hall WD, Coen JL (1996) Source code documentation for the Clark-Hill cloud scale model: code version G3CH01. (Technical Note NCAR/TN-426+STR). University Corporation for Atmospheric Research, Boulder

Clark TL, Keller T, Coen JL et al (1997) Terrain induced turbulence over Lantau Island: 7 June 1994 tropical storm Russ case study. J Atmos Sci 54:1795–1814

Coen J (2018) Some requirements for simulating wildland fire behavior using insight from coupled weather-wildland fire models. Fire 1:6

Coen JL (2013) Modeling wildland fires: a description of the Coupled Atmosphere-Wildland Fire Environment model (CAWFE). (Technical Note. NCAR/TN-500+STR). University Corporation for Atmospheric Research, Boulder

Coen JL, Riggan PJ (2014) Simulation and thermal imaging of the 2006 Esperanza Wildfire in southern California: application of a coupled weather-wildland fire model. Int J Wildland Fire 23:755–770

Coen JL, Stavros EN, Fites-Kaufman JA (2018) Deconstructing the King megafire. *Ecological Applications, 28*, 1565–1580.

Cruz MG, Alexander ME (2013) Uncertainty associated with model predictions of surface and crown fire rates of spread. Environ Model Softw 47:16–28

Cunningham P, Goodrick SL, Hussaini MY, Linn RR (2005) Coherent vortical structures in numerical simulations of buoyant plumes from wildland fires. Int J Wildland Fire 14:61–75

Dozier J (1981) A method for satellite identification of surface temperature fields of sub-pixel resolution. Remote Sens Environ 11:221–229

Fassnacht F, Schmidt-Riese E, Kattenborn T, Hernández J (2021) Explaining Sentinel 2-based dNBR and RdNBR variability with reference data from the bird's eye (UAS) perspective. Int J Appl Earth Observ Geoinf 95:102262

Filippi JB, Bosseur F, Pialat X et al (2011) Simulation of coupled fire/atmosphere interactions with the MesoNH-ForeFire models. J Combust 2011:540390

Furman JH, Linn RR, Williams BW, Winterkamp J (2019) Using a computational fluid dynamics model to guide wildland fire management (Final Report, Project RC-2013-03). Department of Defense, Environmental Security Technology Certification Program, Washington, DC

Goodrick SL, Achtemeier GL, Larkin NK et al (2012) Modelling smoke transport from wildland fires: a review. Int J Wildland Fire 22:83

Grell GA, Peckham SE, Schmitz R et al (2005) Fully coupled 'online' chemistry in the WRF model. Atmos Environ 39:6957–6976

Hiers JK, O'Brien JJ, Varner JM et al (2020) Prescribed fire science: the case for a refined research agenda. Fire Ecol 16:1–5

Hoffman CM, Morgan P, Mell W et al (2013) Surface fire intensity influences simulated crown fire behavior in lodgepole pine forests with recent mountain pine beetle-caused tree mortality. Forest Sci 59:390–399

Hoffman CM, Canfield J, Linn RR et al (2016) Evaluating crown fire rrate of spread predictions from physics-based models. Fire Technol 52:221–237

Ichoku C, Kaufman YJ (2005) A method to derive smoke emission rates from MODIS fire radiative energy measurements. IEEE Trans Geosci Remote Sens 43:2636–2649

Itti L, Koch C, Niebur E (1998) A model of saliency-based visual attention for rapid scene analysis. IEEE Trans Pattern Anal Mach Intell 20:1254–1259

Jackson WA, Achtemeier GL, Goodrick SL (2007) A technical evaluation of smoke dispersion from the brush creek prescribed fire and the impacts on Asheville, North Carolina (White paper). National Interagency Fire Center, Boise. http://www.nifc.gov/smoke/documents/Smoke_Incident_Impacts_Asheville_NC.pdf. 7 July 2020

Jimenez D, Butler B, Queen L et al (2018) A comparison of in-situ fire energy measurements to remote sensed thermography using Unmanned Aerial Systems (UAS). In: Viegas DX (ed) Advances in forest fire research 2018. Universidade de Coimbra, Coimbra, pp 1244–1248

Josephson AJ, Holland TM, Brambilla S et al (2020) Predicted emission source terms in a reduced-order fire spread model—part 1: particle emissions. Fire 3:4

Kaufman YJ, Remer LA, Ottmar R et al (1996) Relationship between remotely sensed fire intensity and rate of emission of smoke: SCAR-C experiment. In: Levine JS (ed) Global biomass burning: Atmosphere, climate, and biospheric implications. MIT Press, Cambridge, pp 685–696

Kaulfus AS, Nair U, Jaffe D et al (2017) Biomass burning smoke climatology of the United States: implications for particulate matter air quality. Environ Sci Technol 51:11731–11741

Kiefer MT, Charney JJ, Zhong S et al (2019) Evaluations of the ventilation index in complex terrain: a dispersion modeling study. J Appl Meteorol Climatol 58:551–568

Kochanski AK, Jenkins MA, Mandel J et al (2013) Evaluation of WRF-SFIRE performance with field observations from the FireFlux experiment. Geoscientific Model Dev 6:1109–1126

Kochanski AK, Jenkins MA, Yedinak K et al (2016) Toward an integrated system for fire, smoke and air quality simulation. Int J Wildland Fire 25:534–546

Kochanski AK, Mallia DV, Fearon MG et al (2019) Modeling wildfire smoke feedback mechanisms using a coupled fire-atmosphere model with a radiatively active aerosol scheme. J Geophys Res Atmos 124:9099–9116

Koman PD, Billmire M, Baker KR et al (2019) Mapping modeled exposure of wildland fire smoke for human health studies in California. Atmosphere 10:308

Koontz MJ, North MP, Werner CM et al (2020) Local forest structure variability increases resilience to wildfire in dry western U.S. coniferous forests. Ecol Lett 23:483–494

Kremens RJ, Smith AMS, Dickinson MB (2010) Fire meteorology: current and future directions in physics-based measurements. Fire Ecol 6:13–35

Larkin NK, O'Neill S, Solomon R et al (2009) The BlueSky smoke modeling framework. Int J Wildland Fire 18:906–920

Lavdas LG (1996) Program VSMOKE—User's manual (General Technical Report SRS-GTR-6). U.S. Forest Service, Southeastern Forest Experiment Station, Macon

Lecun Y, Bengio Y, Hinton G (2015) Deep learning. Nature 527:436–444

Leroy-Cancellieri V, Augustin P, Filippi JB et al (2014) Evaluation of wildland fire smoke plume dynamics and aerosol load using UV scanning lidar and fire-atmosphere modelling during the Mediterranean Letia 2010 experiment. Nat Hazard 14:509–523

Linn R, Anderson K, Winterkamp J et al (2012) Incorporating field wind data into FIRETEC simulations of the international crown fire modeling experiment (ICFME): preliminary lessons learned. Can J for Res 42:879–898

Linn RR, Goodrick SL, Brambilla S et al (2020) QUIC-fire: a fast-running simulation tool for prescribed fire planning. Environ Model Softw 125:104616

Linn RR, Reisner JM, Edminster CB et al (2003) FIRETEC—A physics-based wildfire model (2003 R&D 100 joint entry). Los Alamos National Laboratory, and U.S. Forest Service, Rocky Mountain Research Station, Los Alamos

Linn RR, Sieg CH, Hoffman CM et al (2013) Modeling wind fields and fire propagation following bark beetle outbreaks in spatially-heterogeneous pinyon-juniper woodland fuel complexes. Agric for Meteorol 173:139–153

Liu JC, Mickley LJ, Sulprizio MP et al (2016) Particulate air pollution from wildfires in the Western US under climate change. Clim Change 138:655–666

Liu X, He B, Quan X et al (2018) Near real-time extracting wildfire spread rate form Himawari-8 satellite data. Remote Sens 10:1654

Liu Y, Achtemeier GL, Goodrick SL, Jackson W (2010) Important parameters for smoke plume rise simulation with Daysmoke. Atmos Pollut Res 1:250–259

Liu Y, Kochanski A, Baker KR et al (2019) Fire behaviour and smoke modelling: model improvement and measurement needs for next-generation smoke research and forecasting systems. Int J Wildland Fire 28:570–588

Mallet V, Keyes DE, Fendell FE (2009) Modeling wildland fire propagation with level set methods. Comput Math Appl 57:1089–1101

Mandel J, Beezley JD, Kochanski AK (2011) Coupled atmosphere-wildland fire modeling with WRF 3.3 and SFIRE 2011. Geoscientific Model Dev 4:591–610

Mandel J, Amram S, Beezley JD et al (2014) Recent advances and applications of WRF–SFIRE. Nat Hazard 14:2829–2845

Matson M, Dozier J (1981) Identification of sub-resolution high temperature sources using a thermal IR sensor. Photogramm Eng Remote Sens 47:1311–1318

Mell W, Maranghides A, McDermott R, Manzello SL (2009) Numerical simulation and experiments of burning Douglas-fir trees. Combust Flame 156:2023–2041

Moisseeva N, Stull R (2020) Capturing plume rise and dispersion with a coupled large-eddy simulation: case study of a prescribed burn. Atmosphere 10:579

Moran CJ, Seielstad CA, Cunningham MR et al (2019) Deriving fire behavior metrics from UAS imagery. Fire 2:36

Morvan D, Hoffman C, Rego F, Mell W (2011) Numerical simulation of the interaction between two fire fronts in grassland and shrubland. Fire Saf J 46:469–479

Morvan D, Meradji S, Mell W (2013) Interaction between head fire and backfire in grasslands. Fire Saf J 58:195–203

Nowell HK, Holmes CD, Robertson K et al (2018) A new picture of fire extent, variability, and drought interaction in prescribed fire landscapes: Insights from Florida government records. Geophys Res Lett 45:7874–7884

Ottmar RD, Hiers JK, Butler BW et al (2016) Measurements, datasets and preliminary results from the RxCADRE project-2008, 2011 and 2012. Int J Wildland Fire 25:1–9

Peterson DA, Hyer EJ, Campbell JR et al (2015) The 2013 Rim fire: implications for predicting extreme fire spread, pyroconvection, smoke emissions. Bull Am Meteor Soc 96:229–247

Peterson DA, Hyer E, Wang J (2014) Quantifying the potential for high-altitude smoke injection in the North American boreal forest using the standard MODIS fire products and subpixel-based methods. J Geophys Res Atmos 119:3401–3419

Prichard S, Larkin N, Ottmar R et al (2019) The fire and smoke model evaluation experiment—a plan for integrated, large fire–atmosphere field campaigns. Atmosphere 10:66

Prichard SJ, O'Neill SM, Eagle P et al (2020) Wildland fire emission factors in North America: Synthesis of existing data, measurement needs and management applications. Int J Wildland Fire 29:132–147

Randerson JT, van der Werf GR, Giglio L et al (2017) Global fire emissions database, version 4.1 (GFEDv4). Oak Ridge National Laboratory, Oak Ridge. Distributed Active Archive Center. https://doi.org/10.3334/ORNLDAAC/1293. 7 July 2020

Riebau A, Larkin N, Pace T et al (2006) BlueSkyRAINS West (BSRW) demonstration project (Final report). www.airfire.org/pubs/BlueSkyRAINS_West_November_2006.pdf. 14 Mar 2021

Riggan PJ, Tissell RG, Lockwood RN et al (2004) Remote measurement of energy and carbon flux from wildfires in Brazil. Ecol Appl 14:855–872

Riggan PJ, Wolden LG, Tissell RG et al (2010) Remote sensing fire and fuels in southern California. In: Wade DD (ed) Proceedings of 3rd fire behavior and fuels conference. International Association of Wildland Fire, Birmingham. https://www.fs.usda.gov/treesearch/pubs/38813. 7 July 2020

Rothermel RC (1972) A mathematical model for predicting fire spread in wildland fuels (Research Paper INT-115). U.S. Forest Service, Intermountain Forest and Range Experiment Station, Ogden

Sandberg DV, Peterson J (1984) A source-strength model for prescribed fires in coniferous logging slash. In: Proceedings of the 21st annual meeting of the Air Pollution Control Association, Pacific Northwest International Section. Air Pollution Control Association, Pittsburgh

Skamarock WC, Klemp JB, Dudhia J et al (2008) A description of the advanced research WRF version 3 (Technical Report NCAR/TN-475+STR). University Corporation for Atmospheric Research, Boulder

Smith AMS, Wooster MJ (2005) Remote classification of head and backfire types from MODIS fire radiative power observations. Int J Wildland Fire 14:249–254

Soja AJ, Al-Saadi JA, Giglio L et al (2009) Assessing satellite-based fire data for use in the national emissions inventory. J Appl Remote Sens 3:031504

Solomon R (2007) An automated system for evaluating Bluesky predictions of smoke impacts on community health and ecosystems (Final report, JFSP 03-1-3-09). U.S. Forest Service, Pacific Northwest Research Station, Seattle. http://digitalcommons.unl.edu/jfspresearch/98. 1 July 2020

Strand TM, Larkin NK, Craig KJ et al (2012) Analyses of BlueSky gateway PM2.5 predictions during the 2007 southern and 2008 northern California fires. J Geophys Res 117:D17301

Val Martin M, Logan JA, Kahn RA et al (2010) Smoke injection heights from fires in North America: analysis of 5 years of satellite observations. Atmos Chem Phys 10:1491–1510

Warneke C, Schwarz JP, Ryerson T et al (2018) Fire influence on regional global environments and air quality (FIREX-AQ): A NOAA/NASA interagency intensive study of North American fires. https://www.esrl.noaa.gov/csd/projects/firex/whitepaper.pdf. 6 July 2020

Zhao Y, Ma J, Li X, Zhang J (2018) Saliency detection and deep learning-based wildfire identification in UAV imagery. Sensors 18:712

Open Access This chapter is licensed under the terms of the Creative Commons Attribution 4.0 International License (http://creativecommons.org/licenses/by/4.0/), which permits use, sharing, adaptation, distribution and reproduction in any medium or format, as long as you give appropriate credit to the original author(s) and the source, provide a link to the Creative Commons license and indicate if changes were made.

The images or other third party material in this chapter are included in the chapter's Creative Commons license, unless indicated otherwise in a credit line to the material. If material is not included in the chapter's Creative Commons license and your intended use is not permitted by statutory regulation or exceeds the permitted use, you will need to obtain permission directly from the copyright holder.

Chapter 4
Smoke Plume Dynamics

Yongqiang Liu, Warren E. Heilman, Brian E. Potter, Craig B. Clements,
William A. Jackson, Nancy H. F. French, Scott L. Goodrick,
Adam K. Kochanski, Narasimhan K. Larkin, Peter W. Lahm,
Timothy J. Brown, Joshua P. Schwarz, Sara M. Strachan, and Fengjun Zhao

Abstract Smoke plume dynamic science focuses on understanding the various smoke processes that control the movement and mixing of smoke. A current challenge facing this research is providing timely and accurate smoke information for

Y. Liu (✉) · S. L. Goodrick · F. Zhao
U.S. Forest Service, Southern Research Station, Athens, GA, USA
e-mail: yongqiang.liu@usda.gov

S. L. Goodrick
e-mail: scott.l.goodrick@usda.gov

F. Zhao
e-mail: zhaofengjun1219@163.com

W. E. Heilman
U.S. Forest Service, Northern Research Station, Lansing, MI, USA
e-mail: warren.heilman@usda.gov

B. E. Potter · N. K. Larkin
U.S. Forest Service, Pacific Northwest Research Station, Seattle, WA, USA
e-mail: brian.potter@usda.gov

N. K. Larkin
e-mail: sim.larkin@usda.gov

C. B. Clements · A. K. Kochanski
San Jose State University, San Jose, CA, USA
e-mail: craig.clements@sjsu.edu

A. K. Kochanski
e-mail: adam.kochanski@sjsu.edu

W. A. Jackson
U.S. Forest Service (Retired), Asheville, NC, USA

N. H. F. French
Michigan Tech University, Ann Arbor, MI, USA
e-mail: nancy.french@mtu.edu

P. W. Lahm
Fire and Aviation Management, U.S. Forest Service, Washington, DC, USA
e-mail: peter.lahm@usda.gov

This is a U.S. government work and not under copyright protection in the U.S.; foreign copyright protection may apply 2022
D. L. Peterson et al. (eds.), *Wildland Fire Smoke in the United States*,
https://doi.org/10.1007/978-3-030-87045-4_4

the increasing area burned by wildfires in the western USA. This chapter synthesizes smoke plume research from the past decade to evaluate the current state of science and identify future research needs. Major advances have been achieved in measurements and modeling of smoke plume rise, dispersion, transport, and superfog; interactions with fire, atmosphere, and canopy; and applications to smoke management. The biggest remaining gaps are the lack of high-resolution coupled fire, smoke, and atmospheric modeling systems, and simultaneous measurements of these components. The science of smoke plume dynamics is likely to improve through development and implementation of: improved observational capabilities and computational power; new approaches and tools for data integration; varied levels of observations, partnerships, and projects focused on field campaigns and operational management; and new efforts to implement fire and stewardship strategies and transition research on smoke dynamics into operational tools. Recent research on a number of key smoke plume dynamics has improved our understanding of coupled smoke modeling systems, modeling tools that use field campaign data, real-time smoke modeling and prediction, and smoke from duff burning. This new research will lead to better predictions of smoke production and transport, including the influence of a warmer climate on smoke.

Keywords Measurement · Modeling · Management · Plume rise · Smoke impacts · Smoke plume · Transport and dispersion

4.1 Introduction

4.1.1 Scientific Significance

The fate of smoke from a wildland fire depends in large part on the airflow carrying it away from the fire which can involve a complex interaction of eddies that may occur near the ground or expand beyond the atmospheric mixed layer. The buoyancy of smoke and ambient atmospheric conditions determines how high and how quickly the smoke rises, and thus where it travels. Understanding the processes that control the movement and mixing of smoke is essential to any endeavor to predict (and

---

T. J. Brown
Desert Research Institute, Reno, NV, USA
e-mail: tim.brown@dri.edu

J. P. Schwarz
Earth System Research Laboratory, National Oceanic and Atmospheric Administration, Boulder, CO, USA
e-mail: joshua.p.schwarz@noaa.gov

S. M. Strachan
Idaho Department of Environmental Quality, Boise, ID, USA
e-mail: sara.strachan@deq.idaho.gov

manage) the impacts of smoke on public health and safety, including air quality, visibility, aviation, and climate. These processes are commonly referred to as "plume dynamics".

An improved scientific understanding of smoke plume dynamics will allow for more accurate assessments and predictions of the fate and impacts of wildland fire emissions. These emissions are a major source of particulates, black carbon, and organic carbon, and can contribute substantially to the formation of aerosols and gases that comprise air pollution and haze (Baars et al. 2011; Strand et al. 2011; Wiedinmyer et al. 2011; Larkin et al. 2014). The transport and deposition of black carbon to the Arctic (Larkin et al. 2012; Hao et al. 2016) can reduce surface albedo which enhances snow and ice melt (Evangeliou et al. 2016). Some wildland fire smoke plumes can penetrate the tropopause and mix into the stratosphere to affect ozone, carbon monoxide, carbon dioxide, reactive nitrogen, and water vapor concentrations at high altitudes (Jost et al. 2004; Chap. 6).

Accurate characterization of smoke plume dynamics can improve our understanding of smoke impacts on atmospheric radiation, weather, climate, and photochemical processes. Smoke particles modify atmospheric radiative transfer directly through light scattering and absorption of solar radiation, and indirectly by influencing cloud formation (Bauer and Menon 2012). This potentially changes atmospheric temperature, stability, and convection. Depending on the photochemical environment, ozone enhancement from biomass burning has been observed in many instances (Brey and Fischer 2016). Smoke plume temperature usually decreases with height, which affects photochemical reaction rates for ozone production (Lim et al. 2019). This further affects secondary organic carbon (SOC) production.

Smoke plume dynamics provide the scientific foundation for developing predictive models and tools appropriate for wildland fire applications. For example, plume rise is an important property determining the local and regional impacts of smoke. Emissions that remain within the mixing layer can be influenced by topography (Chap. 5) and the diurnal change in the mixing layer, affecting local areas near the fire site. In addition, a smoke plume can penetrate into the free atmosphere and high lofted smoke particles can be transported tens of kilometers, affecting regional air quality downwind. For example, locations in Idaho often receive smoke transported from California wildfires and ozone concentrations in eastern US cities can be affected by long-range transport of smoke from Canada and the western USA (Wilkins et al. 2020). Many smoke and air quality models require estimates of plume rise, whereby particulates and gases emitted from wildland fires are distributed into model-domain atmospheric layers above fires. Improvements in smoke modeling can be made by basing these estimates on a sound scientific understanding of how fire-induced convective plumes interact with the atmosphere, from the flaming regions immediately above the fire to much higher in the atmosphere.

4.1.2 Management Significance

The area burned by wildfire, fire severity, and the size of the largest fires have increased in the USA over the past 20 years (Dennison et al. 2014). Smoke from wildland fires has been associated with increased physician and emergency room visits, hospital admissions, and mortality (Chap. 7). Illnesses attributed to smoke exposure can also result in absences from work and school, affecting economic productivity and educational achievement, respectively. Fires and smoke also are increasingly affecting urban areas, leading to air pollution and visibility problems (Mass and Ovens 2018). Dispersal of smoke across roadways and airports can be a major concern as very high concentrations of fine particulates can significantly reduce motorist visibility during both day and night.

Modeling wildland fire smoke plume dynamics helps provide important information for decision makers and society (Chap. 8). An understanding of plume dynamics contributes to an accurate air quality forecast and helps air quality and land managers answer the most commonly asked public questions during smoke events: (1) Where is the smoke coming from?, (2) How long will it last?, (3) What are the pollutant concentrations?, and (4) Should I be concerned? Knowledge of the spatial and temporal smoke patterns helps air quality forecasters warn a potentially affected community of the likely location, magnitude, extent, and duration of smoke-related air quality impacts, enabling people to modify their behavior to reduce exposure to pollutants. This chapter therefore synthesizes existing research and knowledge on smoke plume dynamics. The objectives are to assess the current state of science, mainly based on studies conducted in the past decade, identify research gaps, provide a vision for improving smoke dynamics science, and describe emerging issues and challenges.

4.2 Current State of Science

4.2.1 Theoretical Framework

4.2.1.1 Conceptual Models

The simplest conceptual model of a plume begins with a point, or at least an axially symmetric, uniform source (Fig. 4.1a). This is similar to the smokestack and mathematical plume models developed for smokestack plumes (Briggs 1982), which are still the most common models used for plumes from wildfires. The plume simulated using these models depends on buoyancy flux, stability, vertical profile of wind, and distance in the downwind direction from a fire, expanding as it mixes with the environment.

4 Smoke Plume Dynamics

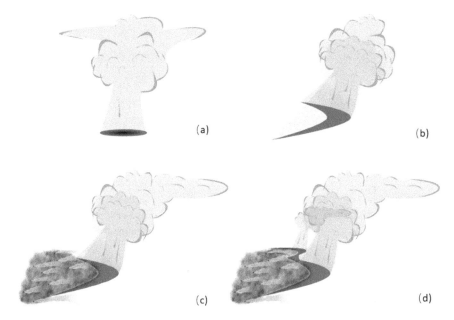

Fig. 4.1 Conceptual smoke plume models. **a** Axisymmetric, simple updraft plume model. **b** Simple wildland fire plume model; a fire with a flaming front and weaker flanks produces a single plume as wind moves the fire forward. **c** Wildfire plume model including residual smoldering behind the flaming front. **d** "Complete" wildfire plume model; multiple flaming head regions produce multiple updraft cores; smoldering combustion behind the front produces smoke, part of which is entrained into the updraft cores, part of which remains separate and near the ground

However, wildland fire practitioners and scientists have long recognized the need for a conceptual and mathematical model that better reflects the geometry and environment typical of a wildland fire in a complex atmosphere. The most basic wildfire-specific conceptual model with one primary updraft is shown in Fig. 4.1b, with a curved flaming front that is more vigorous at the downwind head than on the flanks. At present, there are no mathematical models comparable to Briggs (1982) that incorporate the specific features—variable heating and noncircular shape—that differentiate Fig. 4.1b from Fig. 4.1a.

This fire-specific model is still too simplistic, and the scientific and operational questions regarding plume dynamics require a more complex and physically complete model. Specifically, in a real fire, flaming and smoldering combustion often occur simultaneously and near one another. The area behind the flaming front often continues to smolder after the front passes (Fig. 4.1c), producing smoke that may be entrained into the rising plume, or remain near ground level and disperse more or less independently.

In most wildfires, and in some large prescribed fires, the geometry of the flaming front may result in multiple flaming fronts, each producing a distinct plume updraft, or core (Fig. 4.1d). Different cores may rise to different heights and distribute smoke into different, possibly overlapping, layers of the atmosphere. This requires a more complex model, accounting for varying scales of eddies and turbulence, effects of vegetation and fuel structure on fire and smoke, sloping terrain, and varying atmospheric stability.

In addition to being transported, the constituent gases and particulates in a fire plume undergo a number of chemical reactions, but these are not generally considered part of plume dynamics. Rather, they are commonly described as plume chemistry. One exception to this is the role of moisture, which is usually present in sufficient quantity, and with sufficient latent heat, to affect smoke rise and generate clouds. Although it constitutes a small part of the total mass, the presence of this moisture can modify the distribution of smoke and influence smoke chemistry (Chap. 6). Plume moisture depends on environmental conditions—both atmospheric and vegetative—which vary as a function of location, season, and time of day. A representative range can be 1 g kg^{-1} to 15 g kg^{-1} of dry air. This chapter does not consider chemistry other than moist thermodynamics, which relies on the presence of moisture in the plume (see Chap. 6).

4.2.1.2 Physical Processes

The major physical processes of smoke plume dynamics are shown in Fig. 4.2. Major properties include fire size, fire emissions, ambient atmospheric stratification, turbulent mixing, wind shear, and latent heat released from the condensation of water (Paugam et al. 2016). After being emitted, smoke particles and gases move through the vegetation canopy, the planetary boundary layer, and sometimes into the free atmosphere. The concentrations and three-dimensional dynamical spatial structure and temporal evolution of smoke plumes are determined by the processes of: eddies, turbulence, smoke–canopy interactions, plume rise, dispersion, transport, multiple updrafts, pyro-convection, entrainment of the ambient air, and smoke-induced radiative and cloud disturbances.

4.2.2 Smoke Measurements

4.2.2.1 Smoke Structure and Atmospheric Disturbances

Limited observations of the fire environment during various experimental fires, prescribed fires, and wildfires (Heilman et al. 2014; Clements and Seto 2015; Clements et al. 2019) have focused mostly on fire spread, fire behavior, and other aspects of the fire environment. Recent field studies have focused on fire–atmosphere interactions in grasslands (Clements and Seto 2015; Clements et al. 2019), and forested environments (Heilman et al. 2014; Seto et al. 2014; Clements and Seto 2015), with a small emphasis on ignition and combustion and a large emphasis on

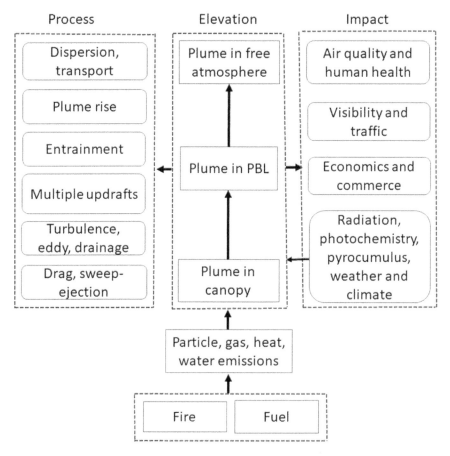

Fig. 4.2 Schematic illustration of smoke processes and relations, including components presented in other chapters of this book

near-surface atmospheric feedbacks. Clements et al. (2019) have summarized recent field campaigns to study fire spread and fire–atmosphere interactions.

The Rapid Deployments to Wildfires Experiment (RaDFIRE) (Clements et al. 2018) sampled 22 wildfire plumes using a mobile atmospheric profiling system (Clements and Oliphant 2014) equipped with a scanning Doppler Lidar. This campaign included observations of plume and wildfire phenomena including rotating convective plumes, plume interactions with stable layers, multilayered smoke detrainment, plume entrainment processes, pyro-cumulus and pyrocumulonimbus, and smoke-induced density currents formed by smoke-shading (Lareau and Clements 2015, 2016, 2017). The mobile Doppler Lidar systems were especially valuable because of their high resolution and ease of mobility. Figure 4.3 shows a series of Lidar images with detailed, coherent plume structures from the Lidar observations (Lareau and Clements 2017) taken during the El Portal fire near Yosemite National

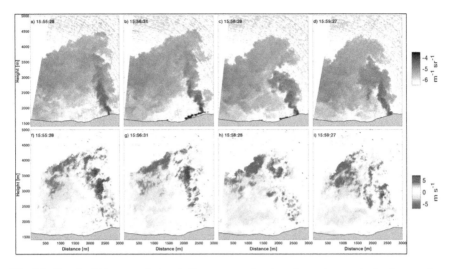

Fig. 4.3 Doppler Lidar range-height indicator scans showing plume rise, smoke backscatter (**a–d**), and radial velocity (**f–i**). Reddish shading indicates flow away from the Lidar, bluish shading indicates flow toward the Lidar; time is Pacific Daylight Time. Ambient wind is from right to left in the panels. From Lareau and Clements (2017) © American Meteorological Society, used with permission

Park (California) on July 28, 2014. Strong updraft cores and subsequent detrainment downwind of the upper plume are evident in the radial velocities (Fig. 4.3f–i). Near-surface indrafts of ~2–3 ms^{-1} extend approximately 1 km away from the column base.

Disadvantages of Doppler Lidars are (1) attenuation in optically thick plumes (Lareau and Clements 2016), (2) fairly slow scan speed (1–2°s^{-1}) to resolve plume structures when compared to radars (>10° s^{-1}), and (3) lack of polarization in the Lidar beam. Dual-polarized (or Dual pol) Doppler radars transmit and receive both horizontally and vertically orientated beams, providing measurements of the size and shape of particles (pyrometeors [large debris lofted above wildfires that are composed of the by-products of combustion of the fuels], ash, and debris) within the plume. McCarthy et al. (2018) demonstrated the use of mobile X-band dual-polarized Doppler radar for studying wildfire plumes. They sampled bushfire plumes in Australia, with correlation coefficients <0.5 between radar beams within smoke plume in horizontal and vertical directions, which is similar to what other studies have measured (Melnikov et al. 2009; Lang et al. 2014; Lareau and Clements 2016). This range makes it easy to distinguish between the smoke plume and hydrometeors, because correlation coefficients of rain are mostly greater than 0.6.

Extreme wildfire updrafts were measured during the 2016 Pioneer Fire in Idaho using an instrumented aircraft and airborne Doppler cloud radar (Rodriguez et al. 2020). Updraft velocities of 58 m s^{-1} were observed deep within the plume at altitudes of 2 km and 3 km above ground level. This updraft core was flanked on its edges with downdrafts of 30 m s^{-1}. The observed updrafts were below the cloud base and a developing pyrocumulonimbus, indicating that the primary mechanism for the strong updrafts was sensible heat flux associated with the fire front.

4.2.2.2 Plume Rise

A number of techniques have been used to determine the height of smoke plumes, including Lidar/radar detection, imaging, airborne observations, and plume sampling (Urbanski et al. 2010; Heilman et al. 2014). The smoke plume heights for 20 prescribed fires were measured using a ground-based ceilometer in the southeastern USA (Liu et al. 2012); the results indicated that the average smoke plume height was approximately 1 km, with plume heights trending upward from winter to summer. Lidar instruments aboard satellites have been used to detect smoke plume height (e.g., Kahn et al. 2008; Raffuse et al. 2012), including the Cloud-Aerosol Lidar with Orthogonal Polarization (CALIOP) aboard the Cloud-Aerosol Lidar and Infrared Pathfinder Satellite Observations (CALIPSO) satellite (Winker et al. 2007), and the Multi-angle Imaging SpectroRadiometer (MISR) aboard the National Aeronautics and Space Administration (NASA) TERRA satellite (Kahn et al. 2008). These satellite detections found different plume heights (from tropical to boreal fires) and different percentage of plumes that penetrate into the free atmosphere, where smoke particles can be transported long distances downwind. The mean heights of plumes from wildfires in the western and central USA were approximately 2 km and 3 km, respectively (Raffuse et al. 2012).

Several long-term plume-rise datasets have been compiled based on satellite detections for evaluating plume-rise modeling and fire–climate interactions. A five-year plume height climatology derived from MISR observations (Val Martin et al. 2010) was used to evaluate the performance of the dynamical smoke plume-rise model (Freitas et al. 2007). The model simulations generally underestimated the plume-height dynamic range observed by MISR and did not reliably identify plumes injected into the free troposphere. CALIPSO data combined with a trajectory model and the National Oceanic and Atmospheric Administration (NOAA) hazard mapping system generated daily plume-injection height (Soja et al. 2012). Hourly injection profiles of plumes were developed from all fires recorded globally for more than a decade, using a methodology that considers wildfire plumes similar to Convective Available Potential Energy computations (Sofiev et al. 2012, 2013). An observation-based global hourly fire plume-rise dataset for 2002–2012 used a modified 1D plume-rise model based on observed fire size and fire radiative power data as a function of plant functional type for different regions of the world (Wang et al. 2020). The dataset is especially valuable for fire–climate interaction modeling.

4.2.2.3 Plumes from Prescribed Fires

Prescribed fires are used for ecological restoration, fuel treatment, and habitat management in a variety of ecosystems (Chaps. 2, 8). Because prescribed fires and wildfires typically have different intensities, plume dynamics and environmental impacts of the emissions differ between the two fire types (Williamson et al. 2016). Although it is expected that health impacts from prescribed fire emissions will be less, smoke from prescribed burning can linger for relatively long periods of time, degrade local air quality, and pose a human health risk, due in part to the nature of plume dynamics.

Prescribed fires provide opportunities for safely measuring fire–fuel–atmosphere interactions within and near the fire environment that can help improve our understanding of how fuels, topography, forest canopies, and ambient and fire-induced atmospheric conditions affect fire and local smoke plume behavior. Recent prescribed fire experiments that assessed plume dynamics include:

- The Prescribed Fire Combustion and Atmospheric Dynamics Research Experiment (RxCADRE) conducted in Florida (Clements et al. 2016; Ottmar et al. 2016; Peterson and Hardy 2016).
- Sub-canopy fire experiments in New Jersey and North Carolina (Heilman et al. 2013, 2015; Strand et al. 2013; Seto et al. 2014).
- The FireFlux grassfire experiments conducted over flat terrain in Texas (Clements 2010; Clements et al. 2019).
- Grassfire experiments in complex California terrain (Seto and Clements 2011; Charland and Clements 2013; Clements and Seto 2015).
- Numerous low-intensity fire experiments focused on plume heights, superfog formation, and validation of smoke dispersion models in Florida, Georgia, and Alabama (Achtemeier 2005, 2009; Liu et al. 2009; Achtemeier et al. 2011).

These studies found that even low-intensity prescribed fires can modify ambient atmospheric flow near the fires due to buoyancy at the fire front. Atmospheric turbulence can be maximized just above the canopy and is typically anisotropic, with horizontal turbulent mixing of smoke plumes exceeding vertical mixing. Maximum plume updraft speeds during low-intensity sub-canopy fires typically occur near the canopy top, although downdrafts can also occur in a sub-canopy fire environment. Entrainment of cooler ambient air often occurs on the back side of advancing fire fronts and their associated plumes.

Fire-induced surface-pressure perturbations at fire fronts can increase wind velocity at the base of plumes (Clements and Seto 2015; Clements et al. 2019). Smoke transport and dispersion during low-intensity fires in complex terrain at night are governed by large-scale atmospheric synoptic forcing and buoyancy-related drainage flows. The mixing of warm and moist smoke plumes generated by low-intensity fires with cool and moist ambient nighttime air can generate superfog plumes (a combination of smoke and water vapor that produces zero visibility over roadways; see below and Chap. 3) that are carried downwind of the fire.

4.2.2.4 Field Campaigns

In recent years, several comprehensive field campaigns that combined Lidar, radar, weather towers, aircraft, drones, and satellites to measure smoke plumes over burn sites and downwind from wildfires and prescribed burns have been implemented. These projects include (1) the Fire and Smoke Model Evaluation Experiment (FASMEE) (Prichard et al. 2019), (2) Fire Influence on Regional to Global Environments and Air Quality (FIREX-AQ), and (3) Western Wildfire Experiment for Cloud Chemistry, Aerosol Absorption and Nitrogen (WE-CAN). These projects have investigated physical, chemical, and optical properties of smoke plumes and the associated influences of fuels, fire behavior, fire energy, and meteorology on dynamics of near-source plumes and long-range smoke transport.

4.2.3 Smoke Plume Modeling

4.2.3.1 Plume Rise

It is important to obtain dynamic properties at the fire–atmosphere interface (e.g., heat fluxes, exit temperature, velocity) and evaluate their effects on plume rise in order to understand and develop schemes to predict the effects of complex plume structures (e.g., multiple updrafts) on plume rise. Both empirical and physics-based models are available for calculating plume rise (Liu et al., 2010; Paugam et al. 2016).

Empirical models are based on field and laboratory measurements using statistical methods or similarity theory, with algebraic expressions that require no time or space integration. The Briggs scheme originally developed for stack plumes (Briggs 1982) was adapted for fire plumes by including fire heat release (Pouliot et al. 2005) but has been shown to have systemic biases compared with satellite data (Raffuse et al. 2009). It has been used in the Community Multiscale Air Quality (CMAQ) (Byun and Schere 2006) regional air quality model (Baker et al. 2018). A fire plume-rise height parameterization was developed using observation-based plume-rise data for a 10-year period for 15 global wildfire regions (Wang et al. 2020), including nearly 30 parameters, mostly from climate models. Regression models have been used to calculate plume rise of prescribed fires based on measured plume-rise data in the southeastern USA (Liu 2014).

Physics-based models consist of differential equations governing fluxes of mass, momentum, and energy, with solutions found through time and/or space integration. Among the first dynamical models are (1) a one-dimensional model based on the dynamic entrainment plume model that simulates the time evolution of plume rise and determines the final injection layer (Freitas et al. 2007), (2) a parameterization based on an eddy diffusivity/mass flux scheme for modeling shallow convection and dry convection (Rio et al. 2010), and (3) a vertical static model using a concept similar to Convective Available Potential Energy computations (Sofiev et al. 2012, 2013). An atmospheric modeling framework with different plume-rise parameterizations for a well-constrained prescribed burn found that model results were significantly improved when fire emissions were distributed below the plume top following a Gaussian distribution (Mallia et al. 2018). Daysmoke, a hybrid of the empirical and dynamic smoke models, simulates smoke particle movements using statistical and stochastic relations.

4.2.3.2 Dispersion and Transport

Modeling tools of varying complexity are available for simulating and predicting transport and dispersion of smoke from wildland fires (Goodrick et al. 2013), including:

- Box models such as the Ventilation Index and the Atmospheric Dispersion Index.
- Gaussian plume models such as VSMOKE and the Simple Approach Smoke Estimation Model (SASEM).
- Lagrangian puff or particle models such as CALPUFF (Scire et al. 2000), the Hybrid Single-Particle Lagrangian Integrated Trajectory (HYSPLIT) model (Stein et al. 2015), FLEXPART (Brioude et al. 2013), Daysmoke (Achtemeier et al. 2011), and Planned Burn-Piedmont (PB-P) (Achtemeier 2005).
- Eulerian grid models such as CMAQ, AERO-RAMS (Wang et al. 2006), and the chemistry version of the Weather Research and Forecasting model (WRF-Chem) (Grell et al. 2005).

Many of the Lagrangian puff/particle and Eulerian grid models rely on observed or model-predicted atmospheric variables (e.g., wind fields, temperatures, moisture, turbulence) at different spatial and temporal resolutions to generate predictions of smoke transport and dispersion. For example, CMAQ, the Advanced Regional Prediction System (ARPS) (Xue et al. 2000, 2001), and its canopy sub-model variant (ARPS-Canopy) (Kiefer et al. 2013) have been coupled to the FLEXPART particle dispersion model to allow for mesoscale, boundary-layer, and canopy-scale simulations/predictions of smoke transport and dispersion (Charney et al. 2019).

The WRF model and its atmospheric chemistry (WRF-Chem), large-eddy-simulation (WRF-LES) (Mirocha et al. 2010) and fire (WRF-Fire) variants (see also Chap. 3) have been used to investigate:

- Plume transport and dispersion-related issues, such as the effects of plume dynamics on fire weather (Grell et al. 2011).
- Vertical and horizontal plume transport during boreal wildfires (Thomas et al. 2017).
- The impact of smoke plumes transported from California, Oregon, and Washington on elevated ozone and $PM_{2.5}$ in other locations in the western USA (Miller et al. 2019).
- The effects of vortices on plume dispersion and plume rise (Cunningham and Goodrick 2012).

The WRF model has also been coupled with CMAQ to provide smoke transport and dispersion assessments of fire events (e.g., Zou et al. 2019a). Before being superseded by the WRF model in 2010, the 5th Generation Penn State/National Center for Atmospheric Research Mesoscale Model (MM5) was used extensively with CMAQ (e.g., Liu et al. 2009), CALPUFF (e.g., Jain et al. 2007; Larkin et al. 2009), and Daysmoke (e.g., Liu et al. 2010) to conduct smoke transport and dispersion assessments in the USA (Fig. 4.4).

4.2.3.3 Nighttime Smoke

Plume dynamics can differ greatly between nighttime and daytime. At night, fire combustion shifts toward smoldering and low-intensity flaming. Decreasing surface temperatures favor the development of a nocturnal surface inversion, a stable layer in the atmosphere that suppresses convective motion. The combination of less buoyant smoke and suppressed convection leads to smoke being trapped near the surface, resulting in degraded air quality and visibility. Simulations with PB-P (Achtemeier 2005) indicate that dispersion of the nonbuoyant nocturnal smoke is largely driven by drainage flows. As the land surface cools and forms a shallow layer of adjacent cool air, small pressure gradients develop, forcing the cooled air downslope. The cool air accumulates in valleys and either pools there or continues flowing down-valley, depending on the slope of the valley floor.

The risk associated with accidents due to impaired visibility at night is amplified under certain conditions. Under a nocturnal inversion, smoldering combustion releases water vapor and particulates that act as cloud condensation nuclei (Engelhart et al. 2012) into the atmosphere. This warm, moist, smoky air mixes with cooler ambient air to form a saturated or potentially supersaturated air mass often referred to as superfog (Achtemeier 2006). The combination of saturated to supersaturated conditions, with an abundance of cloud condensation nuclei particles, creates an air mass dominated by very small droplets that scatter light and can greatly reduce visibility.

A simple two-part model based on radiational cooling of the smoke air mass to its dewpoint temperature, and nongradient mixing of the smoke and ambient air masses, was developed to simulate superfog formation (Achtemeier 2008). The simulation showed that the liquid water content of the smoke–fog mixtures was much higher than for natural fog and smoke, and for fog flowing northward along drainages from a prescribed burn 3.2 km to the south of a highway. Bartolome et al. (2019) coupled a thermodynamic model similar to that of Achtemeier (2008) with a two-dimensional boundary-layer model, which describes transport and turbulent mixing processes that control the persistence of superfog as it disperses from a burned area. Boundary-layer growth predictions from the model were verified with laboratory experiments, describing superfog development for two events.

Drainage flows are driven by cooling of sloping terrain, and surface type and moisture content affect cooling rates, influencing pollutant dispersion. Vegetation and terrain heterogeneities alter atmospheric dispersion patterns that switch from

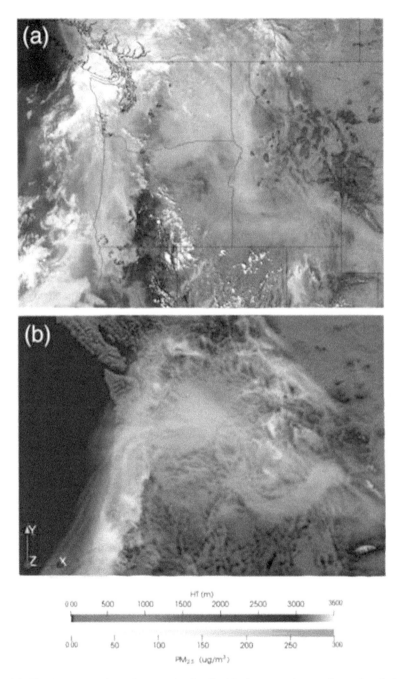

Fig. 4.4 Fire hotspots and smoke over the Pacific Northwest region on September 5, 2017: **a** satellite imagery and **b** WRF-CMAQ PM$_{2.5}$ simulation (Zou et al. 2019a)

non-Gaussian to near-Gaussian behavior as soil moisture increases (Wu et al. 2009). Land-cover heterogeneity (e.g., a mosaic of dry areas and wet areas) can produce complex patterns of differential heating that induce complex atmospheric circulations at meso- and microscales. These circulation patterns are primarily responsible for the development of non-Gaussian dispersion patterns. Soil and vegetation have the biggest effect on atmospheric dispersion in the nocturnal boundary layer.

4.2.4 Interactive Processes

4.2.4.1 Terrain Interactions

The effects of terrain heterogeneity on atmospheric and smoke plume dynamics have been examined extensively in both observational and numerical modeling studies. These studies have shown that atmospheric boundary-layer, tropospheric wind, temperature, and moisture fields are strongly influenced by surface-elevation variations, land–water boundaries, and land-use/land-cover patterns. Terrain-induced phenomena such as katabatic winds (i.e., drainage flows), anabatic winds, föhn and chinook winds, terrain channeling of flows, land/sea breezes, and urban heat islands affect how smoke from wildland fires is transported and dispersed (Liu et al. 2009; Sharples 2009; Lu et al. 2012; Kiefer et al. 2019; Miller et al. 2019). Many of these same phenomena can directly affect the spread of wildland fires across the landscape, which in turn affects when and where fire emissions for subsequent transport and dispersion occur (Clements 2011).

4.2.4.2 Canopy Interactions

The behavior of wildland fires in forested environments and the associated dispersion of smoke through forest vegetation depends on local ambient, fire-induced, and canopy-induced meteorological conditions (Heilman et al. 2013; Strand et al. 2013; Mueller et al. 2014). Forest vegetation acts as a drag on ambient and fire-induced winds (Massman et al. 2017; Charney et al. 2019; Moon et al. 2019), and horizontal and vertical wind shear patterns are a source of turbulence generation (Kiefer et al. 2015; Heilman et al. 2017). Vegetation can also influence the lower atmospheric boundary-layer thermal environment, which determines the amount of turbulence generated or dissipated through buoyancy (Kiefer et al. 2015; Charney et al. 2019). Finally, vegetation can generate wake-generated turbulence and enhanced dissipation of turbulence as canopy elements break down flow eddies to smaller sizes.

Observational and modeling studies that assess the potential effects of fire–canopy–atmosphere interactions on smoke plume behavior conclude that:

- Maximum increases in the energy of turbulent circulations as a result of fire spread can occur near the top of forest canopies, thereby enhancing the dispersion of smoke plumes (Heilman et al. 2015, 2019).
- Horizontal mixing of smoke plumes within forest vegetation layers due to turbulent circulations is often stronger than vertical mixing, particularly near the surface, both before and after the passage of fire fronts (Seto et al. 2013; Heilman et al. 2015, 2019).
- Distribution of turbulent velocity near wildland fires is highly skewed, making turbulence regimes in such environments non-Gaussian and calling into question the application of smoke modeling tools that assume Gaussian turbulence for diffusing smoke (Heilman et al. 2017).
- Dense forest canopies (high plant-area densities) can lead to more upright smoke plumes above and in the vicinity of combustion zones, corresponding with larger plume heights (Charney et al. 2019).

4.2.4.3 Integrated and Interactive Systems

Several integrated and interactive systems of fire behavior, smoke, and meteorology are powerful tools for understanding the coupled processes linking smoke, fire, and the atmosphere, as well as informing different aspects of smoke management decision making. Although they operate on different scales and focus on different processes, FIRETEC (Linn and Cunningham 2005), WFDS (Mell et al. 2007), CAWFE (Coen 2013), Meso-NH ForeFire (Filippi et al. 2009), WRF-SFIRE (Mandel et al. 2011), and WRF-SFIRE-CHEM (e.g., Kochanski et al. 2015) are coupled models with the potential to resolve plume dynamics associated with fire behavior.

Physics-based computational fluid dynamics (CFD) fire models such as FIRETEC and WFDS have high spatial resolution (a few meters), requiring a large amount of computational resources. As a consequence, their use in wildfire-scale problems exceeding approximately 40 ha is not feasible. In an effort to improve fire and smoke modeling capabilities while maintaining modest computational cost and input requirements, so-called hybrid models have been developed. Models such as CAWFE, Meso-NH ForeFire, and WRF-SFIRE couple a CFD-type weather model with a simplified fire model to account for first-order fire–atmosphere interactions. Such models calculate plume rise and dispersion but rely on parameterized fire physics. As they treat smoke as a passive tracer, they can describe basic plume dynamics, but cannot account for interactions between smoke, atmospheric radiation, or chemistry. Efforts have been made toward developing integrated modeling systems that can take into account fire progression, emissions, plume rise, dispersion, and radiative and chemical impacts of smoke. For example, WRF-SFIRE-CHEM couples the chemical transport model WRF-CHEM (Grell et al. 2011) with the fire module SFIRE (Mandel et al. 2011).

In WRF-SFIRE-CHEM, a hybrid model that couples a CFD-type weather model with an empirical rather than a dynamical fire model, fire progression and emissions are driven by local meteorological and fuel conditions affected by fire itself, so the

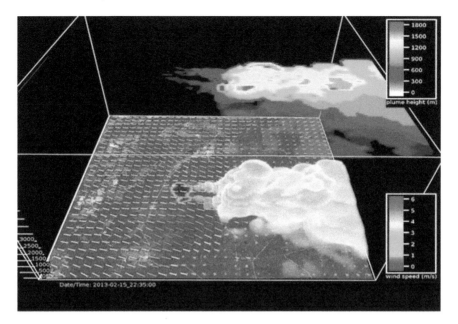

Fig. 4.5 WRF-SFIRE simulation of a prescribed burn at Fort Stewart, Georgia, on February 15, 2013. The color arrows represent wind speed (see right-hand color bar) and direction. The upper-level plane shows local plume heights (see right-hand color bar). From Liu et al. (2019)

fire progression is simulated in-line with fire emissions and chemistry. WRF-SFIRE-CHEM computes fire emissions and plume rise based on fire behavior, fuel moisture, and atmospheric conditions computed at each WRF time step (Fig. 4.5). Combustion rates are based on the mass of fuel consumed within each fire-grid point. Emission fluxes are the products of combustion rates and fuel-specific emission factors. Smoke emissions are represented as a sum of fluxes of chemical species and incorporated into the lowest WRF model layer. Smoke emissions are transported and undergo chemical transformations in the atmosphere according to modeled chemical mechanisms. Aerosol emissions are linked to the aerosol model (GOCART) and interact with atmospheric radiation and microphysics. Initial wildfire simulations suggest that WRF-SFIRE-CHEM can simulate elevated concentrations of NO_x and $PM_{2.5}$ of fire emissions associated with wildland fires (Kochanski et al. 2015) (Fig. 4.5).

4.2.4.4 Smoke–Radiation Interactions

The opacity of smoke from large fires affects solar irradiance as well as within-plume radiative transfer characterizations. Both gas and aerosol species contribute to this opacity and vary dynamically within the fire plume due to ongoing chemistry and aerosol physics after emission. Activation (or suppression) of clouds due to smoke can further complicate the understanding of smoke transport (Feingold et al. 2005).

Several sources of information are used to evaluate links between smoke properties and plume dynamics, including: (1) episodic collection of in situ gas and aerosol concentrations and the aerosol optical properties of smoke in different atmospheric regimes (e.g., Forrister et al. 2015; Liu et al. 2016; Selimovic et al. 2017; Ditas et al. 2018), (2) remote sensing of properties of large-scale burning pollution from satellite instrumentation, and (3) ground-based sun photometer and Lidar networks (e.g., Nikonovas et al. 2017). This information is incorporated into models via parametric simplifications and/or prescribed aerosol optical properties.

Numerical experiments have examined the local effects of smoke in complex terrain using an integrated framework coupled with the fire, atmosphere, and chemical transport model WRF-SFIRE-CHEM (Kochanski et al. 2015). Mallia et al. (2020) investigated smoke transport driven by small-scale topographical flows in a local wildfire event in Utah and found good agreement between simulated and observed spatial $PM_{2.5}$ patterns, including realistic representation of the drainage flow advecting smoke into a valley.

Simulations of the 2015 California fires that included the radiative impact of smoke successfully resolved local reductions in incoming solar radiation and surface temperatures associated with smoke shading. Additional sensitivity experiments demonstrated a positive feedback associated with radiative smoke effects: smoke cools the surface, stabilizes the atmosphere, and enhances local inversions, as well as reducing the planetary boundary-layer height and near-surface winds, leading to reduced ventilation and smoke accumulation. This radiatively driven mechanism results in positive feedback, manifesting a nonlinear increase in surface $PM_{2.5}$ concentrations as a function of increasing emissions (Kochanski et al. 2019).

4.2.4.5 Smoke–Fire Interactions

Plume dynamics are directly linked to fire behavior because fire-emitted heat and moisture fluxes control the development of the buoyant smoke column. Conversely, the lack of fire-emitted heat results in surface-smoke accumulation and limited dilution, as often observed during fires in the Southeast. As a consequence, changes in fireline intensity affect plume buoyancy and evolution of the smoke column. However, fire behavior is linked to atmospheric conditions via coupling at various timescales. At scales of seconds to minutes, the most important atmospheric driver is wind, which drives fire propagation and controls the tilt and dispersion of the smoke column.

Fire also modifies local weather conditions. Pyroconvective plumes generate indrafts into the base of rising smoke columns, accelerating winds in the vicinity of the fire front. Observational data (Heilman et al. 2015; Clements et al. 2019) and numerical experiments (Kochanski et al. 2013a; Kiefer et al. 2014) have shown that fire-affected winds may be over two times stronger than ambient winds. These perturbations in near-surface winds that control heat release also depend on atmospheric stability and vertical wind shear (Kochanski et al. 2013b). Just as fire behavior is controlled by two-way coupling between fire and the atmosphere, so is plume dynamics.

4.2.5 Smoke Decision Support Systems

Decision support systems for smoke management consist of one or more computer-based applications that assist managers in planning for and implementing prescribed fires and can predict potential downwind impacts from wildfires. For example, a user interface developed for personal computers (https://webcam.srs.fs.fed.us/tools/vsmoke) allows the VSMOKE atmospheric dispersion model to estimate the maximum hourly $PM_{2.5}$ and carbon monoxide, along with corresponding heat release rate (Anderson et al. 2004). A web-based version called VSMOKE-GIS displays a Google Map and results using the air quality index (see http://weather.gfc.state.ga.us/GoogleVsmoke/vsmoke-Good2.html).

The HYSPLIT trajectory model estimates downwind air quality impacts, facilitating an assessment of whether to implement a prescribed fire within ~2 days. Smoke management personnel typically use one of two decision support systems that utilize HYSPLIT. The first version uses the same user interface as VSMOKE, but the user interface also formats the Fire Emission Production Simulator (FEPS) (Anderson et al. 2004) hourly $PM_{2.5}$ emission and plume rise estimates to produce the input files needed by PC HYSPLIT. The second version using HYSPLIT is the BlueSky Playground modeling framework (Larkin et al. 2009), which is used in the Montana–Idaho Airshed Coordinating Group's decision support system. On the Internet, a user completes the inputs and runs the Fuels and Fire Tools (which includes FEPS) to estimate $PM_{2.5}$ emission and plume rise, prior to running HYSPLIT. Both the BlueSky framework and the PC HYSPLIT version produce outputs that are viewed in Google Earth. BlueSky displays estimated $PM_{2.5}$ concentrations, whereas PC HYSPLIT displays hourly results using an air quality index.

Resource managers use VSMOKE and HYSPLIT to assess potential smoke impacts during the daytime. PB-P, a web-based application (https://piedmont.dri.edu), is used to evaluate the flow of nighttime smoke and whether fog may form but requires field evaluations to earn confidence in its predictions. Users are encouraged to support decisions by obtaining spot weather forecasts within 5 km of the burn for certain weather and dispersion conditions (Long et al. 2014). If PB-P results and/or most of the conditions indicate potential for fog formation on roadways, then mitigation measures can be implemented.

Aside from the models outlined above, significant effort has been made toward operational implementation of hybrid fire–atmosphere models for integrated fire spread and smoke forecasting. Recent implementations of such models (Jimenez et al. 2018; Giannaros et al. 2020) have a potential to improve operational smoke forecasting by linking fire and smoke modeling. This type of system can simplify using coupled models for smoke forecasting by utilizing simple web portals for easy model initialization and online presentation of model results (Mandel et al. 2019).

4.3 Gaps in Understanding Plume Dynamics

4.3.1 Measurements

Although there have been various observations of plume structures during prescribed fires (Liu et al. 2012; Clements and Seto 2015; Clements et al. 2019), few observations exist for deep plumes from large wildfires. An exception is the Rapid Deployments to Wildfires Experiment (4.2.2.1; Clements et al. 2018), which sampled 22 wildfire plumes using a mobile Doppler Lidar (Clements and Oliphant 2014). Although this study sampled different wildfires, it was limited by a lack of real-time fire behavior observations.

There is an operational and research need for coincident measurement of fire, smoke, and atmospheric structures to better understand fire–atmosphere interactions and plume dynamics. To date, no datasets link airborne infrared imagery of fire-front properties (e.g., flame intensity and length, front spread rate, heat release) to vertical velocities, so our understanding of plume structures and what happens on the ground is limited. Direct measurements of vertical velocities in deep wildfire plumes, which are needed to better constrain modeled smoke injection heights and dispersion, are limited to one study (Clements et al. 2018).

Our understanding of deep wildfire plumes is also affected by having few observations of the microphysical properties of plume particles. These observations require in situ airborne sampling and/or remote sensing measurements, using dual-polarized Doppler radars. McCarthy et al. (2018) documented the dual-polarized features associated with bushfires in Australia, showing that the correlation coefficient is a potential indicator for ash and debris detection. Observational studies are needed using multi-wavelength radars to better understand the size and distribution of pyrometeors (large debris lofted above wildfires that are composed of the by-products of combustion of the fuels) in wildfire plumes. Furthermore, to better understand plume dynamics and their effects on fire behavior, a coordinated meteorological field program utilizing ground-based and airborne remote sensing and in situ sampling technologies targeting large, active wildfires is needed.

4.3.2 Plume Rise

Modeling of smoke plume rise has been evaluated primarily with multiple-angle satellite products. Although many tools have been developed for plume-rise modeling, less attention has been paid to modeling of vertical concentration profiles. Smoke profiles are generally specified, rather than resolved, based on fire dynamics and local weather conditions.

Smoke measurements have indicated the existence of multiple, simultaneous updrafts within a smoke plume. Multiple-core updrafts have smaller updraft velocities and diameters and are more affected by entrainment than single-core updrafts, so they are less efficient in vertical smoke movement (Achtemeier et al. 2012). The number of updraft cores is a critical factor for describing plume rise (Liu et al. 2010). Some models have developed parameterization to include multiple core numbers in heat flux calculations or explicitly simulate multiple plumes. However, additional progress is needed in: (1) quantifying updraft; (2) understanding contributing factors for ignition patterns, vegetation structure, fire spread, and atmospheric processes; and (3) understanding the evolution of updrafts in the atmosphere (including mergence).

4.3.3 Dispersion and Transport Modeling

Although the fundamental science governing atmospheric transport and dispersion is fairly well-established in most smoke models, the evolution of strongly buoyant smoke plumes is poorly described (Goodrick et al. 2013). Therefore, simpler approximation schemes on coarser scales (e.g., WFDS-Level Set and WRF-SFIRE) are used (Ottmar et al. 2017). This is due in part to a lack of computational capacity, especially for operational purposes, but measurements are also lacking for key inputs of fuels, fire, and meteorology to support plume model development.

The successful evaluation and validation of modeling tools depend on availability of observational data across a wide range of spatial and temporal scales. Closing the gaps in our understanding of plume dynamics, transport, and dispersion is contingent on establishing new observational datasets upon which models can be evaluated and model output can be verified. Without ample observational data collected during actual wildland fire events or in controlled laboratory environments, the uncertainty and errors in model simulations of plume dynamics, transport, and dispersion are difficult to quantify.

4.3.4 Nighttime Smoke

Vertical and horizontal resolution is the primary challenge for modeling nighttime smoke drainage and potential superfog conditions. As large-scale forcing from synoptic weather systems weakens, details of the local environment are increasingly important. Tools such as PB-P account for the influence of local topography by using digital elevation models to resolve topographic variations at a horizontal grid size of 30 m. Although greater topographic resolution is needed to simulate drainage flows, a less obvious need includes land-cover types and surface-moisture conditions (micrometeorology).

4.3.5 Physics-Based Fire Models

As stated in Sect. 4.2.4.3, physics-based CFD fire models such as FIRETEC and WFDS can resolve the complexities of coupling fire dynamics with atmospheric dynamics; however, their computational costs and input data requirements make operational applications infeasible. Because these models focus on small-scale processes important from the combustion standpoint, they lack capabilities in terms of aerosol physics, microphysics, and chemistry which become important at the larger scales typical for wildland fires.

4.3.6 Smoke Management for Prescribed Fires

Managing smoke from prescribed fires requires technical specialists to work with fire managers to predict and effectively communicate likely smoke effects (Chaps. 7, 8). Occasionally, when a prescribed fire is conducted using mass ignition and no local smoke impacts are predicted or reported, the burn manager will be surprised to receive complaints of smoke from a location far downwind of the burn unit. Without implementing smoke prediction, it is hard to know how high the plume will rise and if fine particulates will travel a long distance from the fire. To effectively implement smoke prediction, we need to understand the strengths and weaknesses of smoke models, which are listed in NWCG (2020). In addition, managers need to be able to estimate multiple emissions and plume rise from co-occurring fires which will require input from both empirical data and model output.

4.4 Vision for Improving Plume Dynamics Science

Both conceptual understanding and practical ability to accurately model wildland fire plumes are poised to make significant advancements. Improvements will be driven by a combination of increasing computing power, new observational techniques, new integrated observational campaigns, and greater recognition of the need for such improvements. We discuss these factors below and provide a vision for improving smoke plume research as a component of a broader perspective for fire and smoke science (Chaps. 2, 3, 5, 6, 7).

4.4.1 New Research on Observational and Computational Capabilities

Routine observations of plumes are currently limited to satellite observations of plume tops, with the longest time series coming from the polar-orbiting MISR (Diner et al. 1998) stereo imaging instrument. MISR is capable of imaging significant portions of the globe once per day, but the overflight time may not be optimal in terms of obtaining maximum plume height. The polar-orbiting CALIOP (Hunt et al. 2009) satellite Lidar system provides a vertically distributed glimpse into smoke plumes that fall directly under its orbital path. However, such observations are limited and often do not intersect the plume directly over the fire, further limiting their usefulness. The MISR twin satellite system and CALIOP Lidar capabilities are highly specialized and serve as research technologies, with operational capability still unavailable.

A new methodology (Lyapustin et al. 2019) for determining plume heights directly from aerosol and gas products of the Moderate Resolution Imaging Spectroradiometer (MODIS) (Justice et al. 2002, 2010) has the potential to make plume-top observations more routine. MODIS is based on two polar-orbiting satellites, and if it is applicable to other platforms, it could lead to operational implementation. A major advance in observing the development and evolution of plumes is possible if this technology can be applied to the new geostationary GOES-16 and GOES-17 (Schmidt 2020) Advanced Baseline Imager (ABI) (Schmit et al. 2005, 2008, 2017) imagery. Specifically, GOES-16 and GOES-17 provide imagery rapidly, and application of such a system to these platforms may allow for observation of plume-top development every 5 min throughout the day (perhaps every minute in some cases). This type of near real-time series observation on a routine basis would provide more than an order of magnitude additional observations of plume tops than are currently available, providing insight on how the plume is changing over short time intervals.

NASA is launching new missions to increase the capacity to detect air pollutants from different sources, including wildfires. The Multi-Angle Imager for Aerosols (MAIA) mission (https://maia.jpl.nasa.gov) is focused on understanding how different types of pollutants affect human health. The MAIA mission will study 12 specific locations in the world with dense population, available health records, and available ground-based air monitor data. Two of the locations, Los Angeles and Atlanta, are often affected by smoke, from wildfires and prescribed burns, respectively. The multi-angle data are useful for determining smoke plume heights. MAIA will pass over a specific location once a day in late morning. Another mission, Tropospheric Emissions: Monitoring of Pollution (TEMPO) (http://tempo.si.edu), will measure particles and gases in the troposphere (lowest layer of the atmosphere), but at an hourly frequency.

Planning for FASMEE (Prichard et al. 2019) has identified a comprehensive set of observations that could be obtained through large-scale planned burns (Liu et al. 2019). It may be possible to use multiple synchronized ground Lidar units with their directional measurements intersecting at a fire; the intersection would provide a virtual vertical tower providing details on the movement of air and aerosols at the center of the plume. This capability is planned for future FASMEE burns.

Some citizen-science efforts have contributed to knowledge about plume dynamics, including a project in Canada that used trained volunteers with equipment to measure plume heights. However, newer technology may make such data easier to obtain and available in greater quantities. Cellphones can take photographs or video with high resolution and record the geolocation of the phone, so it is conceivable that an app would allow citizen scientists or agency personnel to quickly and accurately collect numerous photos. This type of image database would provide the potential to develop novel analytical techniques, using automated algorithms or distributed human-powered image interpretation.

As advanced models of plumes require substantial computing power, current coupled CFD-based models run too slowly for operational and decision support. Cloud computing and improved computer processing may allow advanced models to become practical for applications, either directly or in optimized variants. It is also possible that the vertical distribution of emissions can be reduced to a number of typical structures that could be derived from CFD modeling efforts, then related back to simpler quantities for faster application. Combinations of atmospheric profile, fuelbed type and conditions (e.g., moisture), fire size, fire shape, and regional topography may control the type of vertical allotment sufficiently that the cached results may be used within a smoke forecasting system, without the need to perform a new run of the full CFD model.

4.4.2 New Approaches and Tools

Any effort to substantially improve our understanding of fire plumes needs to be multidisciplinary and integrated across modeling and field research. Incorporation of modelers and preliminary model results into the planning of observational campaigns that can help pinpoint areas where observations are most critical and ensure that time and space scales and resolutions of the observations are in sync with model analysis and development. The Department of Defense Strategic Environmental Research and Development Program field campaigns exemplify how to apply this approach for program-level direction and support. This includes forming an Integrated Research Management Team to coordinate/facilitate research integration and to act as an interface/liaison between the host unit and researchers.

A set of intensively observed fires and a more limited set of broadly obtained observations are needed to inform our understanding of plume dynamics. This is similar to how fuelbed maps have been developed, with intensively measured specific plots combined with satellite observations to apply the plot observations across the map (Chap. 2). However, applications are needed on a scale of at least an order of magnitude more complex than what is used in creating fuelbed maps. Because spatial variability of fuels affects smoke plumes, fuelbeds will need to include appropriate spatial statistics (e.g., spacing between areas of higher or lower fuel density).

4.4.3 New Projects

Flexibility by agencies, institutions, researchers, and resource managers can help to facilitate timely advances. A good example is the initial planning stages of the FASMEE project (Ottmar et al. 2017; Liu et al 2019; Prichard et al 2019), which has incorporated modelers and observational lists from a wide range of research groups. Coordination among the FASMEE effort (Joint Fire Science Program), the FIREX-AQ aircraft campaign (NOAA, NASA), and the WE-CAN project (National Science Foundation) has demonstrated interagency and funding collaboration. FASMEE has provided ground-based observations in support of the FIREX-AQ Western wildfire aircraft campaign, and FIREX-AQ conducted detailed airborne observations of a prescribed burn in Florida.

The FIREX-AQ and WE-CAN experiments have produced observations that will allow researchers to analyze and improve the representation of smoke chemistry for the next several years (Chap. 6). FASMEE has produced an initial set of observations from two prescribed burns, with additional burns planned in coming years. The FASMEE observations are just beginning, and there is ongoing discussion for continued interagency collaboration on future burns.

New and developing efforts on the operational side have the potential to produce more data for model development. Specifically, deployment of Air Resource Advisors (ARAs) to wildfires under the Interagency Wildland Fire Air Quality Response Program (IWFAQRP) (Chap. 8; Lahm and Larkin 2020) has the potential to collect and aggregate operational fire information available to the assigned Incident Command Teams that can be used for retrospective studies. For instance, ARAs may be able to collect photographs of plumes and record requisite metadata. Datasets and tools developed for real-time distribution through the IWFAQRP are providing particulate monitoring data from permanent in situ networks and temporary monitors, including monitors deployed by ARAs that could be guided, in part, by the needs of smoke and plume model evaluation.

4.4.4 Recent Policies and Integration with Smoke Impacts Research

The National Cohesive Wildland Fire Management Strategy initiated in 2009 facilitates collaboration among stakeholders and across ecosystems in the USA to utilize the best science to address socially relevant wildland fire issues. The strategy focuses on resilient landscapes, fire-adapted communities, and safe and effective wildfire response. Plume dynamics and the interface of the fire environment are critical to all three goals. The USDA Shared Stewardship Strategy also provides impetus for improved knowledge about plume dynamics, recognizing that partnerships with states and private landowners are needed to address the problem of elevated fuels in fire-prone landscapes.

The plume dynamics portion of smoke science will contribute to the effectiveness of these strategies as smoke increasingly affects a range of critical social values (Chap. 7). Understanding the intersection of the physics of plume development with subsequent dispersion is needed to accurately predict air pollution and human health effects on the public and firefighters. Plume impacts and chemical constituents are critical for understanding air pollution and human health impacts, but the influence of complex chemistry and atmospheric conditions is also tied to visibility concerns. Smoke can be a contributing factor in roadway accidents, and accurate predictions of the location of smoke impacts are critical. With the recent increase in the use of prescribed fire and area burned by wildfire, more smoke-related incidents are expected, meaning that better information is needed on low-level plume movement, fumigation, and subsidence.

Recent research is helping to improve understanding of plume dynamics and must be integrated into operational tools that can be accessed and supported by ARAs, meteorologists, air quality specialists, prescribed fire practitioners, public health specialists, and policy makers. In addition to research needs, investment is also needed for validation and operational testing that can lead to applications.

Authorization of the IWFAQRP and use of ARAs on wildfire incidents to predict public health impacts of smoke are an example of how scientific information can address policy issues. Since the 1980s, some states have recognized the need to solve smoke issues and have implemented States Implementation Plans that address smoke management and emission reduction. The Northwest Fire Summit of 2019 noted that potential deaths from wildfire smoke likely far exceed those directly caused by the wildfire itself, but opportunities to study wildland fire smoke and operational response have been rare. A better scientific understanding of the health impacts of smoke is needed by practitioners engaged in managing smoke and air quality.

4.5 Emerging Issues and Challenges

4.5.1 *Coupled Modeling Systems*

There is an ongoing effort to develop high-resolution dynamical systems that can account for interactions among atmospheric processes, fire behavior, fire emissions, and smoke dynamics (Liu et al. 2019). Current fire–smoke–atmospheric models such as WRF-SFIRE-CHEM use the Rothermel fire spread model. The next-generation coupled model will use high-resolution dynamical fire models such as FIRETEC (Chap. 3). Development of next-generation smoke research and forecasting systems requires coordinated measurements across fuels, fire behavior and energy, smoke and meteorology, emissions, and chemistry. More powerful computation capacity will be needed to make the coupled systems practical for real-time, operational applications.

4.5.2 Improving Modeling Tools with Field Campaign Data

Comprehensive field research campaigns including FASMEE, FIREX-AQ, and WE-CAN for evaluation of smoke modeling tools will advance our understanding of the complex fire–atmosphere system. They will also help evaluate how well specific models perform under real-world applications, the level of model uncertainties, and which sources of uncertainty can be improved. The outcomes are expected to (1) improve scientific knowledge of the physically coupled fuels–fire–smoke–chemistry system; (2) develop exportable methodologies for measuring fuels for fire spread, fuel consumption, and fire emissions models; (3) develop insights on processes that drive the spatial organization of fire energy and emissions; (4) improve existing operational fire and smoke models; and (5) develop advanced models based on data collected on fuels, fire, meteorology, smoke plumes, and smoke chemistry.

4.5.3 Real-Time Smoke Transport Modeling and Prediction

Over the past 20 years, smoke from large wildfires has affected metropolitan areas of the western USA for extended periods and, in some cases, has been transported thousands of kilometers across North America (Navarro et al. 2016). Accurate predictions of smoke transport are needed to inform effective mitigation (e.g., reduce outdoor activities, close highways, acquire respirators) (Chap. 7). Real-time prediction of smoke transport is critical (O'Neill et al. 2019) and can be assisted by dynamical coupled smoke modeling systems.

NOAA continues to improve its hazard mapping system with the latest fire and smoke monitoring methods and satellite data (www.usfa.fema.gov/operations/infogr ams/011421.html). The product provides near real-time maps, fire data statistics, and datasets for monitoring wildfire and smoke positions. The NOAA High-Resolution Rapid Refresh-Smoke (HRRR-Smoke) produces a new weather and smoke forecast every hour.

4.5.4 Smoke from Duff Burning Under Drought Conditions

It is typically difficult to burn duff (the layer of decomposing organic materials lying below the litter layer of freshly fallen twigs, needles, and leaves and immediately above the mineral soil) because of high fuel moisture (Varner et al. 2009; Ottmar 2014). However, under persistent drought conditions, duff will burn readily. Most of the deep duff layer was burned in the 2016 Rough Ridge fire in northern Georgia, which contributed to unexpectedly high fire emissions that dispersed into metro Atlanta (Zhao et al. 2019). Current tools likely underestimate duff in some regions. Better quantification of the duff layer is needed for accurate prediction of emissions as well as for peat soils (Chap. 2), which have contributed to globally significant smoke events in terms of public health, climate, and regional air quality (Watts 2013).

4.5.5 Smoke Plume Dynamics and Climate Change

Increasing drought associated with climate change will increase wildfire occurrence and emissions, affecting smoke dynamics. Ford et al. (2018) simulated the impacts of climate change on air quality, visibility, and premature deaths, concluding that fire-related $PM_{2.5}$ would increase in the middle and late twenty-first century. Altered atmospheric thermal structure, winds, and precipitation as a result of climate change could also affect smoke dynamics. For example, fuel moisture is projected to decrease in most US regions (Liu 2017), leading to more heat release from fires. Warming due to the greenhouse effect is larger on the ground than in the atmosphere, which will reduce atmospheric stability (Tang et al. 2015). Changes in both heat release and stability will allow smoke plumes to rise to higher elevations.

4.5.6 Smoke Dynamics in the Earth System

Fire, smoke, ecosystems, and climate are interactive components of the Earth system (Bowman et al. 2009; Andela et al. 2017; Liu 2018). Smoke–climate interactions have long been part of climate modeling, which has shown that the radiative effects of some particles can affect: (1) Hadley circulation and precipitation in the tropics (Allen et al. 2012; Tosca et al. 2015), (2) regional climate and weather patterns in the middle latitudes (e.g., Grell et al. 2011), and (3) radiation–ice–temperature feedbacks in the polar region (Keegan et al. 2014; Winiger et al. 2016). Earth system models (Hurrell et al. 2013; Malavelle et al. 2019) include atmospheric models and dynamic global vegetation models to simulate environmental conditions for wildfires and atmospheric radiation and climatic effects of fire carbon and particle emissions and calculate fire-induced disturbances in land-cover and land–air fluxes. Earth system models have greater capacity for modeling interactions of wildland fires and smoke particles (Li et al. 2013, 2014; Unger and Yue 2014; Zou et al. 2019b). Improvements are needed to incorporate global fuel systems and provide dynamical fire emissions for smoke modeling and interactions with atmospheric processes.

4.6 Conclusions

Large wildfires have increased in the USA, and smoke has degraded air quality and visibility in large areas. Recent advances in smoke measurements, model development, and operational decision support tools have increased our understanding of smoke dynamics and ability to provide information to resource managers. Field campaigns focused on smoke–atmosphere interactions have revealed complex smoke structures and processes, and smoke-induced atmospheric disturbances and satellite imagery have been used to develop long-term global smoke plume height datasets. Integrated and interactive coupled models of smoke, fire, atmosphere, and canopies have been developed and applied to simulate smoke processes and mechanisms for air quality assessment, fire management, and climate change studies.

The largest gaps in smoke dynamics science are (1) the lack of high-resolution, dynamical fire, smoke, and atmospheric coupled systems; and (2) simultaneous measurements of these components, especially for wildfires. The following improvements are also needed:

- Smoke modeling needs to simulate buoyant dominant smoke processes.
- Integrated, multidisciplinary observational data across multiple temporal and spatial scales are needed to evaluate simulations of dynamical smoke processes and validate predictions.
- Improved methods are needed for modeling vertical plume distributions and multiple updrafts.
- The impacts of topography and the canopy on nighttime smoke need to be better described.
- Better predictions are needed for local smoke effects from prescribed fire.

Plume dynamics science is likely to improve through the development of new directions and strategies. New research directions include (1) increasing observational and computational abilities, using integrative tools with varied observation levels; (2) implementing field campaigns and operational management projects; and (3) implementing fire and stewardship strategies that help transition smoke dynamics science into operational tools for air quality and public health management.

4.7 Key Findings

- The focus of field experiments has changed recently from fire behavior to smoke–atmosphere interactions. Measurement techniques such as mobile atmospheric profiling systems equipped with scanning Doppler Lidar have revealed complex smoke structure and processes and smoke-induced atmospheric disturbances.
- Multiple-year smoke plume height datasets with regional and global coverages have been developed based on satellite multiple-angle detection and other techniques. The datasets are valuable for modeling some impacts and fire–climate interactions.
- A major advance in smoke modeling has been the development and application of integrated and interactive coupled models of smoke, fire, atmosphere, and canopy; smoke operational and decision support systems; and plume-rise models for wildland fires.

- Applications of these models have improved our understanding of smoke processes and mechanisms, such as factors determining plume evolution, feedbacks to ambient conditions, impacts of multiple updrafts on fire energy and plume rise, and formation of superfog. Modeling tools also provide concentrations and spatial and temporal distributions of emissions from wildland fire for air quality assessment and fire management.
- Recent improvements in our understanding of smoke dynamics include development of high-resolution dynamical coupled smoke research and forecast systems, smoke model evaluation and improvement using data from field campaigns, real-time smoke prediction, consideration of smoke from duff burning, and assessments of future smoke dynamics with respect to changing climate and the Earth system.
- Gaps in research on smoke plume dynamics can be specifically addressed through enhanced measurements of wildfire smoke, model improvement of buoyant dominant lines, model evolution using observational data across large spatial and temporal scales, development of vertical plume distributions, description of multiple updrafts, better description of how topography and canopy affect night smoke, and improvements in smoke predictions for prescribed fire.
- Plume dynamics science will generally improve through new research on observational capabilities and computational ability, new approaches and tools of integration, varied levels of observations, new partnerships, and new field campaigns. These will help transition smoke dynamics science into operational tools for air quality and public health management.

Acknowledgements The authors thank the reviewers for their valuable and constructive comments and suggestions; Tom Pierce, Talat Odman, and Bret Anderson for participating in discussions; and Yvonne Shih for various forms of assistance.

References

Achtemeier GL (2005) Planned burn-piedmont. A local operational numerical meteorological model for tracking smoke on the ground at night: model development and sensitivity tests. Int J Wildland Fire 14:85–98

Achtemeier GL (2006) Measurements of moisture in smoldering smoke and implications for fog. Int J Wildland Fire 15:517–525

Achtemeier GL (2008) Effects of moisture released during forest burning on fog formation and implications for visibility. J Appl Meteorol Climatol 47:1287–1296

Achtemeier GL (2009) On the formation and persistence of superfog in woodland smoke. Meteorol Appl 16:215–225

Achtemeier GL, Goodrick SA, Liu YQ et al (2011) Modeling smoke plume-rise and dispersion from southern United States prescribed burns with Daysmoke. Atmosphere 2:358–388

Achtemeier GL, Goodrick SA, Liu YQ (2012) Modeling multiple-core updraft plume rise for an aerial ignition prescribed burn by coupling Daysmoke with a cellular automata fire model. Atmosphere 3:352–376

Allen RJ, Sherwood SC, Norris JR, Zender CS (2012) Recent northern hemisphere tropical expansion primarily driven by black carbon and tropospheric ozone. Nature 485:350–354

Andela N, Morton DC, Giglio L et al (2017) A human-driven decline in global burned area. Science 30:1356–1362

Anderson GK, Sandberg DV, Norheim RA (2004) Fire emission production simulator (FEPS), user's guide (version 1.0). Seattle: U.S. forest service, pacific northwest research station. http://www.fs.fed.us/pnw/fera/feps. 20 March 2020

Baars H, Ansmann A, Althausen D et al (2011) Further evidence for significant smoke transport from Africa to Amazonia. Geophys Res Lett 38:L20802

Baker K, Woody M, Valin L et al (2018) Photochemical model evaluation of 2013 California wildfire air quality impacts using surface, aircraft, and satellite data. Sci Total Environ 637:1137–1149

Bartolome C, Princevac M, Weise DR et al (2019) Laboratory and numerical modeling of the formation of superfog from wildland fires. Fire Saf J 106:94–104

Bauer SE, Menon S (2012) Aerosol direct, indirect, semidirect, and surface albedo effects from sector contributions based on the IPCC AR5 emissions for preindustrial and present-day conditions. J Geophys Res: Atmos 117:1–15

Bowman D, Balch J, Artaxo P et al (2009) Fire in the earth system. Science 324:481–484

Brey SJ, Fischer EV (2016) Smoke in the city: how often and where does smoke impact summertime ozone in the United States? Environ Sci Technol 50:1288–1294

Briggs GA (1982) Plume rise predictions. In: Haugen D (ed) Lectures on air pollution and environmental impact analysis. American Meteorological Society, Boston, pp 59–111

Brioude J, Arnold D, Stohl A et al (2013) The Lagrangian particle dispersion model FLEXPART-WRF version 3.1. Geosci Model Dev 6:1889–1904

Byun D, Schere KL (2006) Review of the governing equations, computational algorithms, and other components of the models-3 community multiscale air quality (CMAQ) modeling system. Appl Mech Rev 59:51–77

Charland AM, Clements CB (2013) Kinematic structure of a wildland fire plume observed by Doppler lidar. J Geophys Res: Atmos 118:1–13

Charney JJ, Kiefer MT, Zhong S et al (2019) Assessing forest canopy impacts on smoke concentrations using a coupled numerical model. Atmosphere 10:273

Clements CB (2010) Thermodynamic structure of a grass fire plume. Int J Wildland Fire 19:895–902

Clements CB (2011) Effects of complex terrain on extreme fire behavior. In: Synthesis of knowledge of extreme fire behavior: vol I for fire managers. (General technical report PNW-GTR-854. pp. 5–24). Portland: U.S. Forest Service, Pacific Northwest Research Station

Clements CB, Kochanski AK, Seto D et al (2019) The FireFlux II experiment: a model-guided field experiment to improve understanding of fire-atmosphere interactions and fire spread. Int J Wildland Fire 28:308–326

Clements CB, Lareau NP, Kingsmill DE et al (2018) The rapid deployments to wildfires experiment (RaDFIRE): Observations from the fire zone. Bull Am Meteor Soc 99:2539–2559

Clements CB, Lareau NP, Seto D et al (2016) Fire weather conditions and fire–atmosphere interactions observed during low-intensity prescribed fires -RxCADRE 2012. Int J Wildland Fire 25:90–101

Clements CB, Seto D (2015) Observations of fire–atmosphere interactions and near-surface heat transport on a slope. Bound-Layer Meteorol 154:409–426

Clements CB, Oliphant AJ (2014) The California state university mobile atmospheric profiling system: a facility for research and education in boundary layer meteorology. Bull Am Meteor Soc 95:1713–1724

Coen JL (2013) Modeling wildland fires: a description of the coupled atmosphere-wildland fire environment model (CAWFE) (Technical Note NCAR/TN-500+STR). Boulder: University corporation for atmospheric research. https://opensky.ucar.edu/islandora/object/technotes%3A511. 20 March 2020

Cunningham P, Goodrick SL (2012) High-resolution numerical models for smoke transport in plumes from wildland fires. In: Qu JJ, Sommers W, Yang R, Riebau A, Kafatos M (eds) Remote sensing and modeling applications to wildland fires. Tsinghua University Press, Beijing, pp 74–88

Dennison PE, Brewer SC, Arnold JD, Moritz MA (2014) Large wildfire trends in the western United States, 1984–2011. Geophys Res Lett 41:2928–2933

Diner DJ, Beckert JC, Reilly TH et al (1998) Multi-angle imaging spectro radiometer (MISR) instrument description and experiment overview. IEEE Trans Geosci Remote Sens 36:1072–1087

Ditas J, Ma N, Zhang YX et al (2018) Strong impact of wildfires on the abundance and aging of black carbon in the lowermost stratosphere. ProC National Acad Sci USA 115:E11596–E11603

Engelhart GJ, Hennigan CJ, Miracolo MA et al (2012) Cloud condensation nuclei activity of fresh primary and aged biomass burning aerosol. Atmos Chem Phys 12:7285–7293

Evangeliou N, Balkanski Y, Hao WM et al (2016) Wildfires in northern Eurasia affect the budget of black carbon in the Arctic—a 12-year retrospective synopsis (2002–2013). Atmos Chem Phys 16:7587–7604

Feingold G, Jiang H, Harrington JY (2005) On smoke suppression of clouds in Amazonia. Geophys Res Lett 32:1–4

Filippi BJ, Bosseur F, Mari C et al (2009) Coupled atmosphere-wildland fire modelling. J Adv Model Earth Syst 1:11

Ford B, Val Martin M, Zelasky SE et al (2018) Future fire impacts on smoke concentrations, visibility, and health in the contiguous United States. GeoHealth 2:229–247

Forrister H, Liu JM, Scheuer E et al (2015) Evolution of brown carbon in wildfire plumes. Geophys Res Lett 42:4623–4630

Freitas SR, Longo KM, Chatfield R et al (2007) Including the sub-grid scale plume rise of vegetation fires in low resolution atmospheric transport models. Atmos Chem Phys 7:3385–3398

Giannaros TM, Lagouvardos K, Kotroni V (2020) Performance evaluation of an operational rapid response fire spread forecasting system in the southeast Mediterranean (Greece). Atmosphere 11:1264

Goodrick SL, Achtemeier GL, Larkin NK et al (2013) Modelling smoke transport from wildland fire: a review. Int J Wildland Fire 22:83–94

Grell GA, Peckham SE, Schmitz R et al (2005) Fully coupled "online" chemistry within the WRF model. Atmos Environ 39:6957–6975

Grell G, Freitas SR, Stuefer M, Fast J (2011) Inclusion of biomass burning in WRF-Chem: Impact of wildfires on weather forecasts. Atmos Chem Phys 11:5289–5303

Hao WM, Petkov A, Nordgren BL et al (2016) Daily black carbon emissions from fires in northern Eurasia for 2002–2015. Geosci Model Dev 9:4461–4474

Heilman WE, Bian X, Clark KL et al (2017) Atmospheric turbulence observations in the vicinity of surface fires in forested environments. J Appl Meteorol Climatol 56:3133–3150

Heilman WE, Clements CB, Seto D et al (2015) Observations of fire-induced turbulence regimes during low-intensity wildland fires in forested environments: Implications for smoke dispersion. Atmospheric Science Letters 16:453–460

Heilman WE, Clements CB, Zhong S et al (2019) Atmospheric turbulence. In: Manzello SL (ed) Encyclopedia of wildfires and wildland-urban interface (WUI) fires. Springer, Cham, Switzerland, pp 1–17

Heilman WE, Liu Y, Urbanski S et al (2014) Wildland fire emissions, carbon, and climate: Plume rise, atmospheric transport and chemistry processes. For Ecol Manage 317:70–79

Heilman WE, Zhong S, Hom JL et al (2013) Development of modeling tools for predicting smoke dispersion from low-intensity fires (Final report, Project 09-1-04-1a). Boise: Joint Fire Science Program. http://www.firescience.gov/projects/09-1-04-1/project/09-1-04-1_final_report.pdf. 20 March 2020

Hunt WH, Winker DM, Vaughan MA et al (2009) CALIPSO lidar description and performance assessment. J Atmos Oceanic Tech 26:1214–1228

Hurrell JW, Holland MM, Gent PR (2013) The community earth system model: a framework for collaborative research. Bull Am Meteor Soc 94:1339–1360

Jain R, Vaughan J, Heitkamp K et al (2007) Development of the clearsky smoke dispersion forecast system for agricultural field burning in the Pacific Northwest. Atmos Environ 41:6745–6761

Jiménez PA, Muñoz-Esparza D, Kosović BA (2018) High resolution coupled fire–atmosphere forecasting system to minimize the impacts of wildland fires: applications to the Chimney tops II wildland event. Atmosphere 9:197

Jost HJ, Drdla K, Stohl A et al (2004) In-situ observations of mid-latitude forest fire plumes deep in the stratosphere. Geophys Res Lett 31:L11101

Justice CO, Giglio L, Korontzi S et al (2002) The MODIS fire products. Remote Sens Environ 83:244–262

Justice CO, Giglio L, Roy D et al (2010) MODIS-derived global fire products. In: Ramachandran B, Justice CC, Abrams M (eds) Land remote sensing and \global environmental change. Remote sensing and digital image processing, vol 11. New York: Springer, pp 661–679.

Kahn RA, Chen Y, Nelson DL et al (2008) Wildfire smoke injection heights: two perspectives from space. Geophys Res Lett 35:L04809

Keegan KM, Albert MR, McConnell JR, Baker I (2014) Climate change and forest fires synergistically drive widespread melt events of the greenland ice sheet. Proc National Acad Sci, USA 111:7964–7967

Kiefer MT, Zhong S, Heilman WE et al (2013) Evaluation of an ARPS-based canopy flow modeling system for use in future operational smoke prediction efforts. J Geophys Res: Atmos 118:6175–6188

Kiefer MT, Heilman WE, Zhong S et al (2014) Multiscale simulation of a prescribed fire event in the New Jersey Pine Barrens using ARPS-CANOPY. J Appl Meteorol Climatol 53:793–812

Kiefer MT, Heilman WE, Zhong S et al (2015) Mean and turbulent flow downstream of a low-intensity fire: influence of canopy and background atmospheric conditions. J Appl Meteorol Climatol 54:42–57

Kiefer MT, Charney JJ, Zhong S et al (2019) Evaluation of the ventilation index in complex terrain: a dispersion modeling study. J Appl Meteorol Climatol 58:551–568

Kochanski AK, Jenkins MA, Mandel J et al (2013a) Evaluation of WRF-Sfire performance with field observations from the fireflux experiment. Geosci Model Dev 6:1109–1126

Kochanski AK, Jenkins MA, Sun R et al (2013b) The importance of low-level environmental vertical wind shear to wildfire propagation: proof of concept. J Geophys Res: Atmos 118:8238–8252

Kochanski AK, Jenkins MA, Yedinak K et al (2015) Toward an integrated system for fire, smoke and air quality simulations. Int J Wildland Fire 25:534–546

Kochanski AK, Mallia DV, Fearon MG et al (2019) Modeling wildfire smoke feedback mechanisms using a coupled fire–atmosphere model with a radiatively active aerosol scheme. J Geophys Res: Atmos 124:9099–9116

Lahm P, Larkin N (2020) The interagency wildland fire air quality response program. EM Magazine (June). Pittsburgh: Air & Waste Management Association

Lareau NP, Clements CB (2015) Cold smoke: smoke-induced density currents cause unexpected smoke transport near large wildfires. Atmos Chem Phys Discuss 15:17945–17966

Lareau NP, Clements CB (2016) Environmental controls on pyrocumulus and pyrocumulonimbus initiation and development. Atmos Chem Phys 16:4005–4022

Lareau NP, Clements CB (2017) The mean and turbulent properties of a wildfire convective plume. J Appl Meteorol Climatol 56:2289–2299

Larkin NK, DeWinter JL, Strand TM et al (2012) Identification of necessary conditions for Arctic transport of smoke from U.S. fires (Final report, Project 10-S-02-1). Boise: U.S. Joint fire science program. http://www.firescience.gov/projects/09-1-04-1/project/09-1. 20 March 2020

Larkin NK, O'Neill SM, Solomon R et al (2009) The BlueSky smoke modeling framework. Int J Wildland Fire 18:906–920

Larkin NK, Raffuse SM, Strand TM (2014) Wildland fire emissions, carbon, and climate: US emissions inventories. For Ecol Manage 317:61–69

Li F, Levis S, Ward DS (2013) Quantifying the role of fire in the earth system—part 1: improved global fire modeling in the community earth system model (CESM1). Biogeosciences 10:2293–2314

Li F, Bond-Lamberty B, Levis S (2014) Quantifying the role of fire in the earth system-part 2: impact on the net carbon balance of global terrestrial ecosystems for the 20th century. Biogeosciences 11:1345–1360

Lim CY, Hagan DH, Coggon MM et al (2019) Secondary organic aerosol formation from the laboratory oxidation of biomass burning emissions. Atmos Chem Phys 19:12797–12809

Linn RR, Cunningham P (2005) Numerical simulations of grass fires using a coupled atmosphere-fire model: basic fire behavior and dependence on wind speed. J Geophys Res: Atmos 110:D13107

Liu YQ (2014) A regression model for smoke plume rise of prescribed fires using meteorological conditions. J Appl Meteorol Climatol 53:1961–1975

Liu YQ, Achtemeier GL, Goodrick SL, Jackson WA (2010) Important parameters for smoke plume rise simulation with daysmoke. Atmos Pollut Res 1:250–259

Liu YQ, Goodrick S, Achtemeier G et al (2009) Smoke incursions into urban areas: simulation of a georgia prescribed burn. Int J Wildland Fire 18:336–348

Liu YQ, Goodrick S, Achtemeier G et al (2012) Smoke plume height measurement of prescribed burns in the southeastern United States. Int J Wildland Fire 22:130–147

Liu YQ, Goodrick S, Heilman W (2014) Wildland fire emissions, carbon, and climate: wildfire-climate interactions. For Ecol Manage 317:80–96

Liu JC, Wilson A, Mickley LJ et al (2017) Wildfire-specific fine particulate matter and risk of hospital admissions in urban and rural counties. Epidemiology 28:77–85

Liu X, Zhang Y, Huey LG et al (2016) Agricultural fires in the southeastern U.S. during SEAC 4 RS: emissions of trace gases and particles and evolution of ozone, reactive nitrogen, and organic aerosol. J Geophys Res: Atmos 121:7383–7414

Liu YQ (2017) Responses of dead forest fuel moisture to climate change. Ecohydrology 10(2):e1760

Liu YQ (2018) New development and application needs for earth system modeling of fire–climate–ecosystem interactions. Environ Res Lett 13:011001

Liu YQ, Kochanski A, Baker KR et al (2019) Fire behaviour and smoke modelling: model improvement and measurement needs for next-generation smoke research and forecasting systems. Int J Wildland Fire 28:570–588

Lyapustin A, Wang Y, Korkin S et al (2019) MAIAC thermal technique for smoke injection height from MODIS. IEEE Geosci Remote Sens Lett 17:730–734

Long A, Weiss J, Princevac M, Bartolome C (2014) Superfog: state of the science (Southern Fire Exchange Fact Sheet 2014–2). http://southernfireexchange.org/SFE_Publications/factsheets/2014-2.pdf. 20 March 2020

Lu W, Zhong S, Charney JJ et al (2012) WRF simulation over complex terrain during a southern California wildfire event. J Geophys Res: Atmos 117:D5

Malavelle FF, Haywood JM, Mercado LM et al (2019) Studying the impact of biomass burning aerosol radiative and climate effects on the Amazon rainforest productivity with an earth system model. Atmos Chem Phys 19:1301–1326

Mallia DV, Kochanski AK, Kelly KE et al (2020) Evaluating wildfire smoke transport within a coupled fire-atmosphere model using a high-density observation network for an episodic smoke event along Utah's Wasatch Front. J Geophys Res: Atmos 125(20):e2020JD032712

Mallia DV, Kochanski AK, Urbanski SP, Lin JC (2018) Optimizing smoke and plume rise modeling approaches at local scales. Atmosphere 9:166

Mandel J, Beezley JD, Kochanski AK (2011) Coupled atmosphere-wildland fire modeling with WRF 3.3 and SFIRE 2011. Geosci Model Dev 4:591–610

Mandel J, Vejmelka M, Kochanski AK et al (2019) An interactive data-driven HPC system for forecasting weather, wildland fire, and smoke. 2019 IEEE/ACM HPC for urgent decision making (UrgentHPC). Piscataway, Institute of Electrical and Electronics Engineers, pp 35–44

Mass CF, Ovens D (2018) The Northern California wildfires of October 8–9, 2017: the role of a major downslope wind event. Bull Am Meteor Soc 100:235–256

Massman WJ, Forthofer JM, Finney MA (2017) An improved canopy wind model for predicting wind adjustment factors and wildland fire behavior. Can J for Res 47:594–603

McCarthy N, McGowan H, Guyot A, Dowdy A (2018) Mobile X-pol radar: a new tool for investigating pyroconvection and associated wildfire meteorology. Bull Am Meteor Soc 99:1177–1195

Mell W, Jenkins M, Gould J, Cheney P (2007) A physics-based approach to modelling grassland fires. Int J Wildland Fire 16:1–22

Melnikov VM, Zrnic DS, Rabin RM (2009) Polarimetric radar properties of smoke plumes: a model. J Geophys Res 114:D21204

Miller C, O'Neill S, Rorig M, Alvarado E (2019) Air-quality challenges of prescribed fire in the complex terrain and wildland urban interface surrounding Bend, Oregon. Atmosphere 10:515

Mirocha JD, Lundquist J, Kosovic B (2010) Implementation of a nonlinear subfilter turbulence stress model for large-eddy simulation in the advanced research WRF model. Mon Weather Rev 138:4212–4228

Moon K, Duff TJ, Tolhurst KG (2019) Sub-canopy forest winds: understanding wind profiles for fire behavior simulation. Fire Saf J 105:320–329

Mueller E, Mell W, Simeoni A (2014) Large eddy simulation of forest canopy flow for wildland fire modeling. Can J for Res 44:1534–1544

National Wildfire Coordinating Group (NWCG) (2020) Smoke management guide for prescribed fire (PMS 420-3, NFES 001279). https://www.nwcg.gov/sites/default/files/publications/pms420-3.pdf. 25 Feb 2021

Navarro KM, Cisneros R, O'Neill SM et al (2016) Air-quality impacts and intake fraction of PM2.5 during the 2013 Rim megafire. Environ Sci Technol 50:11965–11973

Nikonovas T, North PRJ, Doerr SH (2017) Particulate emissions from large North American wildfires estimated using a new top-down method. Atmos Chem Phys 10:6423–6438

O'Neill SM, Diao MH, Raffuse SM et al (2019) 2017 Northern California wildfires–A NASA health and air quality applied sciences team (HAQAST) tiger team (AGU fall meeting presentation). American Geophysical Union, Washington, DC

Ottmar RD (2014) Wildland fire emissions, carbon, and climate: modeling fuel consumption. For Ecol Manage 317:41–50

Ottmar R, Brown TJ, French NHF, Larkin NK (2017) Fire and smoke model evaluation experiment (FASMEE). Study plan, Joint fire science program project #15-S-01–01. https://www.fasmee.net/study-plan. 20 March 2020

Ottmar RD, Hiers JK, Butler BW et al (2016) Measurements, datasets and preliminary results from the RxCADRE project—2008, 2011 and 2012. Int J Wildland Fire 25:1–9

Paugam R, Wooster W, Freitas S, Val Martin M (2016) A review of approaches to estimate wildfire plume injection height within large-scale atmospheric chemical transport models. Atmos Chem Phys 16:907–925

Peterson DL, Hardy CC (2016) The RxCADRE study: a new approach to interdisciplinary fire research. Int J Wildland Fire 25(1):i

Pouliot G, Pierce T, Benjey W et al (2005) Wildfire emission modeling: integrating BlueSky and SMOKE. Presentation. In: Proceedings of the 14th international emission inventory conference. Washington, DC: U.S. Environmental Protection Agency, pp 11–14

Prichard S, Larkin N, Ottmar R et al (2019) The fire and smoke model evaluation experiment—a plan for integrated, large fire–atmosphere field campaigns. Atmosphere 10:66

Pyroconvective updrafts during a megafire. Geophys Res Lett 47:e2020GL089001

Raffuse SM, Craig KJ, Larkin NK et al (2012) An evaluation of modeled plume injection height with satellite-derived observed plume height. Atmosphere 3:103–123

Rio C, Hourdin F, Chédin A (2010) Numerical simulation of tropospheric injection of biomass burning products by pyro-thermal plumes. Atmos Chem Phys 10:3463–3478

Rodriguez B, Lareau NP, Kingsmill DE, Clements CB (2020) Extreme

Schmidt C (2020) Monitoring fires with the GOES-R series. In: Goodman SJ, Schmit TJ, Daniels J, Redmon RJ (eds) The GOES-R series: a new generation of geostationary environmentalsSatellites. Elsevier, Amsterdam, pp 145–163

Schmit TJ, Gunshor MM, Menzel WP et al (2005) Introducing the next-generation advanced baseline imager on GOES-R. Bull Am Meteor Soc 86:1079–1096

Schmit TJ, Griffith P, Gunshor MM et al (2017) A closer look at the ABI on the GOES-R series. Bull Am Meteor Soc 98:681–698

Schmit TJ, Li J, Li J et al (2008) The GOES-R advanced baseline Imager and the continuation of current sounder products. J Appl Meteorol Climatol 47:2696–2711

Scire JS, Strimaitis DG, Yamartino RJ (2000) A user's guide for the CALPUFF dispersion model (version 5). Earth Tech Inc., Concord

Selimovic V, Yokelson RJ, Warneke C et al (2017) Aerosol optical properties and trace gas emissions by PAX and OPFTIR for laboratory-simulated western US wildfires during FIREX. Atmos Chem Phys 18:2929–2948

Seto D, Clements CB (2011) Fire whirl evolution observed during a valley wind-sea breeze reversal. J Combust 2011:569475

Seto D, Clements CB, Heilman WE (2013) Turbulence spectra measured during fire front passage. Agric for Meteorol 169:195–210

Seto D, Strand TM, Clements CB et al (2014) Wind and plume thermodynamic structures during low-intensity subcanopy fires. Agric Meteorol 198–199:53–61

Sharples JJ (2009) An overview of mountain meteorological effects relevant to fire behavior and bushfire risk. Int J Wildland Fire 18:737–754

Sofiev M, Ermakova T, Vankevich R (2012) Evaluation of the smoke-injection height from wildland fires using remote-sensing data. Atmos Chem Phys 12:1995–2006

Sofiev M, Vankevich R, Ermakova T, Hakkarainen J (2013) Global mapping of maximum emission heights and resulting vertical profiles of wildfire emissions. Atmos Chem Phys 13:7039–7052

Soja A, Fairlie T, Westberg D, Pouliot G (2012) Biomass burning plume injection height using CALIOP, MODIS and the NASA langley trajectory model. In: Proceedings of the 2012 international emission inventory conference. Washington, DC: U.S. Environmental Protection Agency

Stein AF, Draxler RR, Rolph GD et al (2015) NOAA'S HYSPLIT atmospheric transport and dispersion modeling system. Bull Am Meteor Soc 96:2059–2077

Strand T, Larkin N, Rorig M et al (2011) PM2.5 measurements in wildfire smoke plumes from fire seasons 2005–2008 in the Northwestern United States. J Aerosol Sci 42:143–155

Strand TM, Rorig M, Yedinak K et al (2013) Sub-canopy transport and dispersion of smoke: a unique observation dataset and model evaluation (Final report. Project 09-1-04-2). Boise: U.S. joint fire science program. https://www.firescience.gov/projects/09-1-04-2/project/09-1-04-2_final_report.pdf. 20 March 2020

Tang Y, Zhong SY, Luo LF et al (2015) The potential impact of regional climate change on fire weather in the United States. Ann Am Assoc Geogr 105(1):1–21

Thomas JL, Polashenski CM, Soja AJ et al (2017) Quantifying black carbon deposition over the greenland ice sheet from forest fires in Canada. Geophys Res Lett 44:7965–7974

Tosca MG, Diner DJ, Garay MJ, Kalashnikova OV (2015) Human-caused fires limit convection in tropical Africa: first temporal observations and attribution. Geophys Res Lett 42:6492–6501

Unger N, Yue X (2014) Strong chemistry-climate feedbacks in the Pliocene. Geophys Res Lett 41:527–533

Urbanski S, Kovalev VA, Hao WM et al (2010) Lidar and airborne investigation of smoke plume characteristics: Kootenai creek fire case study. In: Proceedings of 25th international laser radar conference. St. Petersburg: Publishing House of Russian Academy of Sciences, Siberian Branch, Institute of Atmospheric Optics, pp 1051–1054

Val Martin M, Logan JA, Kahn RA et al (2010) Smoke injection heights from fires in North America: analysis of 5 years of satellite observations. Atmos Chem Phys 10:1491–1510

Varner JM, Putz FE, O'Brien JJ et al (2009) Post-fire tree stress and growth following smoldering duff fires. For Ecol Manage 258:2467–2474

Wang J, Christopher SA, Nair US et al (2006) Mesoscale modeling of central American smoke transport to the United States: 1. 'Top-down' assessment of emission strength and diurnal variation impacts. J Geophys Res 111:D05S17

Wang YH, Ke ZM, Zou YF, Liu YQ (2020) Global wildfire plume-rise dataset and parameterizations for climate model applications. Earth Space Sci Open Archive. https://doi.org/10.1002/essoar.105 03128.1. 29 March 2021

Watts AC (2013) Organic soil combustion in cypress swamps: moisture effects and landscape implications for carbon release. For Ecol Manage 294C:178–187

Wiedinmyer C, Akagi SK, Yokelson RJ et al (2011) The fire inventory from NCAR (FINN): a high resolution global model to estimate the emissions from open burning. Geosci Model Dev 4:625–664

Wilkins JL, Foy B, Thompson AM et al (2020) Evaluation of stratospheric intrusions and biomass burning plumes on the vertical distribution of tropospheric ozone over the midwestern United States. J Geophys Res: Atmos 125(18):e2020JD32454

Williamson GJ, Bowman DMJS, Price OF et al (2016) A transdisciplinary approach to understanding the health effects of wildfire and prescribed fire smoke regimes. Environ Res Lett 11:125009

Winiger P, Andersson A, Eckhardt S et al (2016) The sources of atmospheric black carbon at a European gateway to the Arctic. Nat Commun 7:12776

Winker DM, Hunt WH, McGill MJ (2007) Initial performance assessment of CALIOP. Geophys Res Lett 34:L19803

Wu Y, Nair US, Pielke RA et al (2009) Impact of land surface heterogeneity on mesoscale atmospheric dispersion. Bound Layer Meteorol 133:367–389

Xue M, Droegemeier KK, Wong V (2000) The advanced regional prediction system (ARPS)—a multi-scale nonhydrostatic atmospheric simulation and prediction model. Part I: model dynamics and verification. Meteorol Atmos Phys 75:161–193

Xue M, Droegemeier KK, Wong V et al (2001) The advanced regional prediction system (ARPS)—a multi-scale nonhydrostatic atmospheric simulation and prediction tool. Part II: model physics and applications. Meteorol Atmos Phys 76:143–165

Zhao FJ, Liu YQ, Goodrick S et al (2019) The contribution of duff consumption to fire emissions and air pollution of the rough ridge fire. Int J Wildland Fire 28:993–1004

Zou Y, O'Neill SM, Larkin NK et al (2019a) Machine learning-based integration of high-resolution wildfire smoke simulations and observations for regional health impact assessment. J Environ Res Public Health 16:2137

Zou Y, Wang Y, Ke Z et al (2019b) Development of a Region-specific ecosystem feedback fire (RESFire) model in the community earth system model. J Advan Model Earth Syst 11:417–445

Open Access This chapter is licensed under the terms of the Creative Commons Attribution 4.0 International License (http://creativecommons.org/licenses/by/4.0/), which permits use, sharing, adaptation, distribution and reproduction in any medium or format, as long as you give appropriate credit to the original author(s) and the source, provide a link to the Creative Commons license and indicate if changes were made.

The images or other third party material in this chapter are included in the chapter's Creative Commons license, unless indicated otherwise in a credit line to the material. If material is not included in the chapter's Creative Commons license and your intended use is not permitted by statutory regulation or exceeds the permitted use, you will need to obtain permission directly from the copyright holder.

Chapter 5
Emissions

Shawn P. Urbanski, Susan M. O'Neill, Amara L. Holder, Sarah A. Green, and Rick L. Graw

Abstract This chapter assesses the current state of the science regarding the composition, intensity, and drivers of wildland fire emissions in the USA and Canada. Globally and in the USA wildland fires are a major source of gases and aerosols which have significant air quality impacts and climate interactions. Wildland fire smoke can trigger severe pollution episodes with substantial effects on public health. Fire emissions can degrade air quality at considerable distances downwind, hampering efforts by air regulators to meet air standards. Fires are a major global source of aerosols which affect the climate system by absorbing and scattering radiation and by altering optical properties, coverage, and lifetime of clouds. A thorough understanding of fire emissions is essential for effectively addressing societal and climate consequences of wildland fire smoke.

Keywords Emission factors · Emissions · Emission inventories · Smoke

S. P. Urbanski (✉)
U.S. Forest Service, Rocky Mountain Research Station, Missoula, MT, USA
e-mail: shawn.p.urbanski@usda.gov

S. M. O'Neill
U.S. Forest Service, Pacific Northwest Research Station, Seattle, WA, USA
e-mail: susan.oneill@usda.gov

A. L. Holder
Office of Research and Development, U.S. Environmental Protection Agency, Research Triangle Park, Durham, NC, USA
e-mail: holder.amara@epa.gov

S. A. Green
Department of Chemistry, Michigan Technological University, Houghton, MI, USA
e-mail: sgreen@mtu.edu

R. L. Graw
U.S. Forest Service, Pacific Northwest Regional Office, Portland, OR, USA
e-mail: rick.graw@usda.gov

This is a U.S. government work and not under copyright protection in the U.S.; foreign copyright protection may apply 2022
D. L. Peterson et al. (eds.), *Wildland Fire Smoke in the United States*,
https://doi.org/10.1007/978-3-030-87045-4_5

5.1 Introduction

Wildland fire smoke contains hundreds of gases (Urbanski 2014; Hatch et al. 2015) and aerosols diverse in size, composition, and morphology (Reid et al. 2005a, b) (Box 5.1).[1] Globally and in the USA wildland fires are a major source of gases and aerosols (Bond et al. 2013; Werf et al. 2017), and the production, dispersion, and transformation of fire emissions have significant air quality impacts and climate interactions. Wildfire smoke can trigger severe, multi-week pollution episodes over large areas with substantial impacts on public health (Chap. 7). Wildland fires are a major source of fine particulate matter $PM_{2.5}$ (particulates with an aerodynamic diameter <2.5 μm) (Lu et al. 2016; Brey et al. 2018) and can contribute to ozone (O_3) production (McClure and Jaffe 2018), both of which are criteria pollutants regulated under the U.S. Clean Air Act. Aerosols from fires affect the climate system by absorbing and scattering radiation (Bond et al. 2013); altering optical properties, coverage, and lifetime of clouds (Lohmann and Feichter 2005; Koch and Genio 2010); and lowering snow and ice albedo in the Arctic (Hansen and Nazarenko 2004).

Box 5.1 Biomass Burning Aerosol

The terms aerosol, particle, and particulate matter (PM) are used interchangeably in atmospheric sciences and in this chapter. Atmospheric aerosols are liquid and/or solid particles dispersed in air. Aerosols are often described according to aerodynamic size thresholds:

Aerodynamic diameter (D) (μm)	Nomenclature	Term
<0.1	$PM_{0.1}$	Ultrafine
<1	PM_1	Submicron
<2.5	$PM_{2.5}$	Fine
2.5–10	$PM_{2.5}-PM_{10}$	Coarse
<10	PM_{10}	

The particle count and mass in fresh smoke from wildland fires is predominantly PM_1 (Reid et al. 2005b, Sect. 5.2.2.1). For context, a typical cloud droplet has a diameter of ~20 μm, the width of human hair is ~50 μm (see Fig. 7.1), and the diameter of a typical raindrop is ~2000 μm. $PM_{2.5}$ and PM_{10} are among the six criteria pollutants for which the USEPA has set National Ambient Air Quality Standards under the federal Clean Air Act. The relationship between particle size and health impacts is discussed in Chap. 7.

---

[1] The terms aerosol, particle, and particulate matter (PM) are used interchangeably in atmospheric sciences and in this chapter.

In addition to size, aerosols are also classified according to composition: organic (OA), non-refractory (non-light absorbing and non-volatilizing), inorganic (sulfate, SO_4^{2-}; nitrate, NO_3^-; ammonium, NH_4^+; and chloride, Cl^-), black carbon (BC), and many other trace elements (e.g., K, Ca, Mg). The terms BC, rBC (refractory BC), elemental carbon, and soot are often used interchangeably to refer to light-absorbing carbonaceous particles with a graphitic-like structure (Buseck et al. 2014; Lack et al. 2014). OA is a mixture of thousands of chemical species (Gilardoni 2017), many of which absorb light preferentially in the UV wavelength range and are labeled as "brown carbon." The carbon fraction of OA is referred to as organic carbon (OC). OA dominates the composition of particles in fresh smoke, comprising >60% of PM_1 mass as seen below:

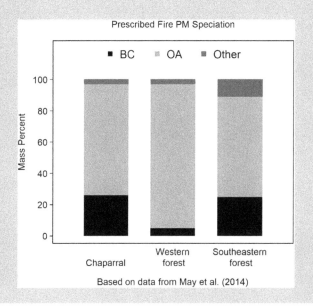

Understanding emissions—the composition and intensity of smoke—is vital for addressing the wide spectrum of decision support needs initiated by wildland fire smoke. Accurately characterizing the dependence of emissions on fuels, fire behavior, and environmental conditions is a key to improving basic smoke management practices and facilitating use of prescribed fire. Emissions are essential input to smoke forecasting systems relied upon by public health officials, air quality forecasters, and fire management teams to mitigate the impacts of wildland fire smoke on public health and safety. Air regulators need better fire emission estimates to quantify the contribution of wildland fires to air pollution and thereby inform decision making about control and regulation of anthropogenic air pollution sources. Robust emission estimates are also needed to quantify the contribution of fires to urban air pollution, assess human smoke exposure, and elucidate the role of smoke in climate forcing.

This chapter assesses the current state of the science on emissions from wildland fires in the USA and Canada. The chapter opens with a summary of current knowledge regarding the composition, intensity, and drivers of emissions. Next, we review emission datasets and tools available for smoke forecasting, regulatory activities, smoke management, and research. The chapter concludes with a discussion of critical gaps in our understanding of emissions.

5.2 Current State of the Science

5.2.1 Fuel Properties, Combustion Processes, and Emissions

The relative abundance of pollutants in fresh smoke (smoke which has not experienced significant photochemical processing, generally less than ~30 min old; see Akagi et al. 2011) is quantified with emission factors (EFs). EFs are determined by measuring the concentration of gases and aerosols in fresh smoke and in the ambient air outside the smoke plume. For a chemical species X, the concentration difference between the fresh smoke plume and background air defines the excess mixing ratio, $\Delta X = X_{\text{plume}} - X_{\text{background}}$. The EF for species X (EFX), the mass of X emitted per mass of dry biomass consumed, can be calculated from ΔX using the carbon mass balance method, a common implementation of which is shown in Eqs. 5.1 and 5.2 (Box 5.2). The carbon mass balance method assumes all biomass carbon is volatilized as gases and aerosol is measured as excess mixing ratios and included in the sum of Eq. 5.2. In practice, many of the carbonaceous gases produced in combustion are not measured. However, because >90% of the carbon emitted is contained in carbon dioxide (CO_2), carbon monoxide (CO), and methane (CH_4), inclusion of only these gases in Eq. 5.2 results in only a slight overestimate of EFs (Yokelson et al. 1999). Additional assumptions of the carbon mass balance method are uniform mixing of all smoke components and constant background composition.

Box 5.2 Emission Factor by the Carbon Mass Balance Method

$$EFX = F_C \times 1000 \, (g\,kg^{-1}) \times \frac{MM_X}{12} \times \frac{ER_X}{C_T} \quad (5.1)$$

In Eq. 5.1, F_c is the mass fraction of carbon in the dry biomass, MM_X is the molar mass of X (g mole^{-1}), 12 is the molar mass of carbon (g mole^{-1}), ER_X is the emission ratio of X to CO_2, and C_T is given by Eq. 5.2.

$$C_T = \sum_{j=1}^{n} N_j \times \frac{\Delta C_j}{\Delta CO_2} \quad (5.2)$$

In Eq. 5.2, n is the number of carbon-containing species measured, N_j is the number of carbon atoms in species j, and ΔC_j is the excess mixing ratio of species j.

Principal factors that affect combustion, and hence the composition, of fresh wildland fire emissions are the structure and arrangement of fuels—size, shape and packing of fuel particles, and fuel condition—moisture content, growth stage, and soundness of woody material (Chap. 2). Fuel chemistry is also important. Emissions of gases and particles containing trace elements such as nitrogen (N), sulfur (S), and chlorine (Cl) are limited by the amounts of these elements in the fuel. Further, compounds often present in biomass (e.g., terpenoid compounds) can be released through distillation prior to the onset of pyrolysis. Ambient conditions, such as wind and terrain, influence both fire behavior and emissions.

The general relationship among fuel bed properties, combustion processes, and emissions is depicted in Fig. 5.1. Small fuel particles with high surface-to-volume ratio, loosely packed fuels, and low moisture content favor flaming combustion (Chap. 2). Grass, foliage, loosely packed litter, and fine woody debris tend to burn predominantly by flaming combustion, given moderate to low moisture content. Smoldering is an important process in the combustion of large-diameter woody fuels, dominating the burning of duff, organic soil, and peat. The relative amount of smoldering combustion increases with fuel moisture content.

In wildland fires, the combustion processes—preignition/distillation, flaming, smoldering, and glowing/char oxidation—occur simultaneously and often in proximity (Yokelson et al. 1996; Ottmar 2001; Chaps. 2 and 3). The chemical composition of smoke is related to the relative amounts of flaming and smoldering combustion (Chap. 6). Some species are emitted almost exclusively by flaming or smoldering

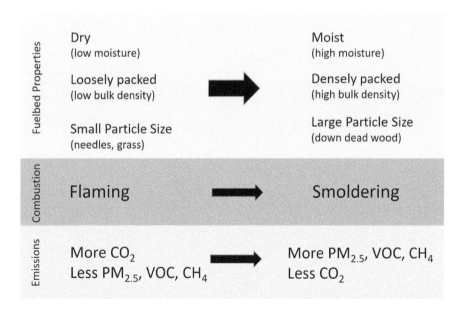

Fig. 5.1 General relationships among fuel bed properties, combustion processes, and emissions. VOC is volatile organic compound

combustion. Flaming combustion produces CO_2, nitrogen oxides (NO_x), hydrogen chloride (HCl), sulfur dioxide (SO_2), nitrous acid (HONO) (Burling et al. 2010), and black carbon (BC) (McMeeking et al. 2009). CO, CH_4, ammonia (NH_3), many non-methane organic gases (NMOG), and organic aerosol (OA) are associated with smoldering combustion (McMeeking et al. 2009; Burling et al. 2010). Several NMOGs are produced during both flaming and smoldering combustion (Burling et al. 2010).

The fraction of combusted fuel carbon emitted as products other than CO_2 increases with the proportion of smoldering combustion. A widely used metric for characterizing burning conditions is modified combustion efficiency, MCE (MCE = $\Delta CO_2/(\Delta CO_2 + \Delta CO)$), an index of the relative amount of flaming and smoldering combustion (Yokelson et al. 1999). Carbonaceous emissions of greatest consequence for air quality (NMOGs and OA) are products of incomplete combustion, and their EFs increase with the proportion of smoldering combustion (Fig. 5.1). The EFs of many NMOGs are negatively correlated with MCE. EFs measured in the laboratory for four NMOGs are plotted versus MCE in Fig. 5.2. The strength of the EF–MCE relationship tends to differ with fuel, being greatest for fine understory forest fuels (litter, woody debris, grass) and weakest for fuels prone to long-term smoldering and glowing combustion such as logs and organic soil.

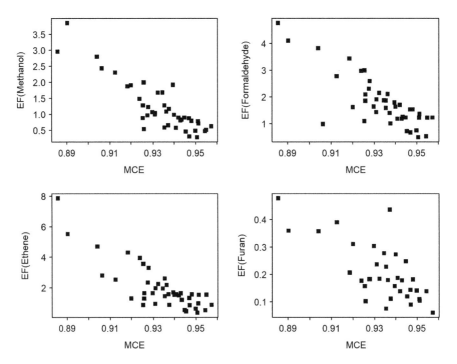

Fig. 5.2 Emission factors for methanol, formaldehyde, ethene, and furan plotted versus modified combustion efficiency (MCE). Data from burning of western US coniferous ecosystem fuels during the FIREX laboratory intensive study (excludes duff and logs) (Selimovic et al. 2018)

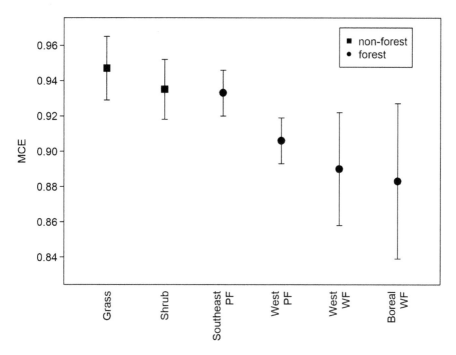

Fig. 5.3 Modified combustion efficiency (MCE) for different fire types. PF = prescribed fire, WF = wildfire. Grass, shrub, and prescribed forest fire based on Urbanski (2014). Wildfire MCE based on Liu et al. (2017), O'Shea et al. (2013), Urbanski (2013), Hornbrook et al. (2011), and Simpson et al. (2011)

The tendency for NMOG and OA EFs to be correlated with MCE provides insight into how emissions of these species differ across fuel types. MCE is highest for fires in herbaceous and shrub fuels and lowest for forest fuels (Fig. 5.3). Forest wildfire MCEs are lower than those for prescribed forest fires. These observed MCEs indicate total NMOG and OA emissions, per unit mass of fuel burned, trend as: herb/shrub < forest prescribed fire < forest wildfire.

5.2.2 Smoke Composition and Emission Factors

The primary emission products of wildland fire are CO_2 and H_2O. However, the minor components of smoke—aerosols, NMOGs, and inorganic gases—are of primary concern to atmospheric scientists, public health officials, air regulators, and land managers. A synthesis by Andreae and Merlet (2001) reported EFs for 92 species. Between 2006 and 2016, a series of laboratory studies at the U.S. Forest Service Missoula Fire Sciences Laboratory brought together over 100 researchers from more than 20 institutions to characterize gaseous and particulate emissions from simulated wildland fires (McMeeking et al. 2009; Burling et al. 2010; Koss et al. 2018). During the same period, several field studies validated laboratory results and developed a framework for extrapolating laboratory-measured EFs to "real fires" in the natural environment. As a result, more than 500 gases have been identified in fresh smoke, and our knowledge regarding the physical characteristics (size and morphology), chemical composition, and optical properties of aerosols has expanded greatly. This section reviews the current state of the science regarding the composition of wildland fire emissions based on recent advances from these laboratory and field studies.

5.2.2.1 Aerosol Emissions

Aerosols are classified by their physical characteristics (size and morphology), chemical composition (inorganic, black carbon, organic species, degree of oxidation, etc.), and/or optical properties (Box 5.1). Of most interest for measuring and modeling impacts of aerosol from smoke are the primary emissions of particles—primarily OA and lesser amounts of BC and inorganic species. In addition, it is important to identify the numerous volatile and semi-volatile organic compounds (SVOCs) that can exist in both the gas phase and particle phase. These SVOC compounds can contribute to secondary organic aerosol (SOA) that is formed by reactions in the atmosphere. SVOC species can also coat BC, which modifies its optical, physical, and chemical properties.

Particulate matter (PM) is the pollutant principally responsible for the detrimental public health impacts and visibility degradation caused by wildland fire smoke (Chap. 7). Although PM air quality has improved across much of the USA over the past 30 years due to reduced anthropogenic emissions, it has deteriorated in regions prone to smoke impacts from wildfires (McClure and Jaffe 2018). Therefore, characterizing the range of EFs for particulate matter (EFPM) for wildfires is critical.

PM produced by wildland fires is dominated by OA with a range of volatilities. In the natural environment, as a fresh smoke plume dilutes and cools, competing condensation/evaporation processes can alter $PM_{2.5}$ mass and hence the measured $EFPM_{2.5}$ (Grieshop et al. 2009). For this reason, extrapolating $EFPM_{2.5}$ measured in laboratory studies, where smoke concentrations are typically very high, to real fires is generally unreliable (May et al., 2014, 2015), so wildfire $EFPM_{2.5}$ are based on limited field observations.

Measurements of EFPM for US wildfires are limited; Liu et al. (2017) reported $EFPM_1$ (aerosol with an aerodynamic diameter <1.0 μm) for only three wildfires. However, Garofalo et al. (2019) reported OA:CO emission ratios ($\Delta OA/\Delta CO$) for 16 western US wildfires. Since wildland fire-produced PM_1 is mostly OA (Box 5.1), this extensive dataset can provide an improved estimate of the average magnitude and range of wildfire $EFPM_1$.

Using methods described below (Sect. 5.4.2), Garofalo et al. (2019) showed that $\Delta OA/\Delta CO$ can be combined with EFCO measured for western wildfires in previous studies to estimate EFOA for a wider range of fires than reported in Liu et al.

(2017). Based on study average EFCO from Liu et al. (2017) (89 g kg^{-1}, $n = 3$) and Urbanski (2013) (135 g kg^{-1}, $n = 9$), the Garofalo et al. (2019) ΔOA/ΔCO (0.26 μg sm^{-3} ppbv^{-1}, $n = 16$) indicates an EFOA range of 26–40 g kg^{-1}.

This exercise suggests EFPM$_1$ for some wildfires may be up to 50% higher than that reported by Liu et al. (2017). The choice of which EF to use in a model can have significant implications for current air quality forecasting and projections of emissions and air quality impacts associated with an anticipated increase in wildfire activity in the western USA (Yue et al. 2013; Liu et al. 2016; Ford et al. 2018; Chap. 1).

Concern has arisen about the health impacts of ultrafine particles (UFPs) or nanoparticles (aerosol with a diameter <100 nm) (Leonard et al. 2007), which may react differently in the body than larger particles (Chap. 7). However, it has been difficult to draw firm conclusions on exposure and health effects of UFPs because of limited field measurements and problems resolving the effects of PM$_{2.5}$ and UFPs in epidemiologic and experimental studies (Baldauf et al. 2016). Nevertheless, it is clear wildland fires release large numbers of UFPs, and their concentration differs with combustion conditions and smoke age. As for other size ranges, UFPs differ with combustion conditions and smoke age. For example, a laboratory study of burning chaparral vegetation found the most numerous particles emitted were in the range of 30–50-nm diameter; the total concentration of particles decreased approximately 100-fold from the flaming to smoldering phase of combustion, while the relative fraction of very fine particles increased (Hosseini et al. 2010).

BC, commonly known as soot, is non-reactive, insoluble, and strongly light absorbing. Globally, biomass burning is the largest single source of BC to the atmosphere (Bond et al. 2013). Terminology for BC is not consistent and generally depends on measurement techniques: thermal–optical methods measure elemental carbon (EC) on filter samples; optical measurements derive BC mass from in situ absorbance and/or scattering data or light attenuation through filter deposit using a mass conversion factor; and laser-induced incandescence (LII) measures refractory BC (rBC) from single-particle incandescence (Petzold et al. 2013). Inconsistencies among measurement techniques and terminology have resulted in uncertainties in EFs, although newer methods (e.g., LII) are beginning to identify relationships between the different methods (May et al. 2014; Li et al. 2019a, b).

Aerosol from biomass burning consists mainly of OA, which typically makes up over 90% of the mass. Almost all BC is produced from flaming phases of combustion, whereas smoldering phases shift emissions toward a greater mass of OA and more particles overall (Bond et al. 2013; May et al. 2014). Jen et al. (2019) found that EFs for EC increase with MCE (flaming), and OC decreases with MCE, with both fitting well to logarithmic functions. Some material is emitted as primary organic aerosol (POA), especially during smoldering phases; other organic compounds are initially emitted as gases, which may condense upon cooling as they move away from the combustion zone. The reverse process also occurs, in which compounds evaporate as the primary particles are diluted in an expanding smoke plume, as much as 80% of POA mass may be lost during this phase (May et al. 2013, 2015). These competing processes will be governed by the temperature and concentration in the plume as it is transported away from the fire. Finally, particles can increase in size through collisions (accumulation mode), growing from a peak count median diameter of ~110 nm at the point of emission to ~250 nm downwind (Janhall et al. 2010). Thus, the size class distribution of particles in an evolving smoke plume is dynamic over seconds to hours after combustion.

Organic gases can be oxidized photochemically or by O_3 as it ages. Oxidation of NMOGs generates SOA. Enhancements of SOA production by up to a factor of two have been observed from burning source materials with different NMOG emissions. A detailed study of the chemistry of particles emitted from laboratory burns of forest and shrubland fuels from the western USA found that 20–65% of the particle emissions (by mass) could be categorized into 12 chemical classes, with the majority of identifiable species being sugars, organic N compounds, and aliphatic or oxy-aliphatic species (Jen et al. 2019). The fraction of emissions that could be classified differed considerably among fuels; decayed logs emitted fewer identifiable substances (~10% classified) than fresher fuels. EFs were approximately log-linear with MCE for both total mass and some of the chemical classes, with $\log(EF) = -a * MCE + b$.

5.2.2.2 Gas Emissions

EFs for the 20 most abundant gases (excluding CO_2, CO, and CH_4) measured in laboratory studies burning common US fuels are shown in Fig. 5.4. The largest EFs for all fuel types are low molecular weight and/or oxygenated species. The NMOGs with the largest EFs common to all fuel types are formaldehyde (HCHO), ethene (C_2H_4), acetic acid (CH_3COOH), and methanol (CH_3OH). The majority of gases emitted are NMOGs with EFs that span >4 orders of magnitude (Yokelson et al. 2013; Koss et al. 2018). The relative magnitude of the NMOGs emitted differs across fuels. Based on laboratory data, southwestern shrubs (e.g., chaparral and mesquite [*Prosopis* spp.]) have the lowest total NMOG emissions (~9 g kg^{-1}), western forest fuels have the highest (~29 g kg^{-1}), and southeastern pine understory fuels have an intermediate value (19 g kg^{-1}) (Yokelson et al. 2013; Koss et al. 2018).

The observed NMOGs can be sorted into structural categories: aromatics (benzene-type compounds), oxygenated aromatics, terpenes, furans, aliphatic hydrocarbons, oxygenated aliphatic hydrocarbons, and compounds containing nitrogen or sulfur. Non-aromatic oxygenated compounds and furans comprise the largest portions of NMOGs (by EF) for western forests, chaparral, and wire grass (*Aristida stricta*) (Fig. 5.5). Terpenoids, a highly reactive class of compounds thought to be important SOA precursors (Chap. 6), are produced and stored in plant resins and can be released when resinous vegetation is heated (Greenberg et al. 2006; Hatch et al. 2019). Because terpenoid emissions result from distillation rather than combustion, they depend strongly on vegetation type (Greenberg et al. 2006; Hatch et al. 2019) and comprise a much larger fraction of western forest fuel emissions compared with non-forest fuels (Fig. 5.5). Total EFNMOG of forest fuels far exceeds that of the non-forest fuels. This stems from a combination of burning conditions and fuel properties.

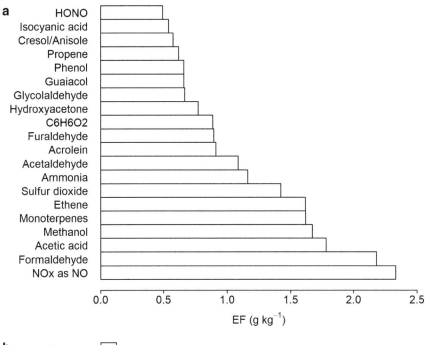
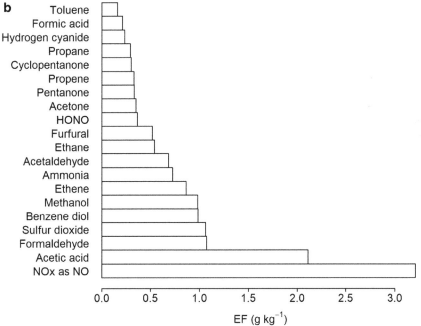

Fig. 5.4 Emission factors (EFs) for the 20 most abundant gas emissions (excluding CO_2, CO, and CH_4) from common US fuel types as reported in laboratory studies (Burling et al. 2010; Gilman et al. 2015; Koss et al. 2018; Selimovic et al. 2018). Panel: **a** western conifer forest, **b** southeastern forest, **c** chaparral

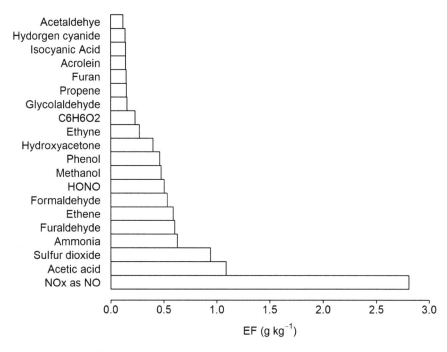

Fig. 5.4 (continued)

The rank in total EFNMOG (western forest > chaparral > wire grass) (Fig. 5.5) is partly a function of burning condition as represented by the MCE of 0.921, 0.955, and 0.971 for western forest, chaparral, and wire grass, respectively.

Photochemical processing of NMOG emissions in the atmosphere can lead to O_3 and SOA formation (see Chap. 6). Quantifying NMOG reactivity with OH identifies which emissions may have the greatest potential to form these secondary pollutants. The variability in OH reactivity of emissions from different fuel types can be considerable due to large differences in the magnitude and relative composition of NMOG emissions. The OH reactivity of NMOG emissions from western forest fuels (~90 s^{-1} [ppb CO]$^{-1}$) is nearly three times that of chaparral fuels (~30 s^{-1} (ppb CO)$^{-1}$), with the reactivity of southeastern understory forest fuels having an intermediate value (Gilman et al. 2015; Koss et al. 2018).

In experiments employing airborne sampling platforms, over 90 gases have been measured in fresh smoke from montane and boreal wildfires and US prescribed fires (Box 5.3). However, emissions have been measured using advanced chemical analysis techniques for relatively few wildfires. There are only three such EF datasets based on in situ airborne measurements in US and Canadian fires (Simpson et al. 2011; Akagi et al. 2013; Liu et al. 2017). Prescribed fire emissions have been more thoroughly studied, in part due to relative ease of logistics and the concerns of land management agencies regarding prescribed burn impacts on air quality.

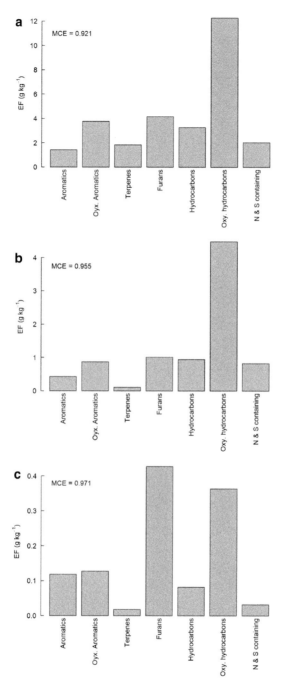

Fig. 5.5 Laboratory-measured non-methane organic gas emission factors (EFs) aggregated by structural class for **a** western forest fuels, **b** chaparral, and **c** wire grass. Based on data from Hatch et al. (2015) and Koss et al. (2018)

Box 5.3 Locations of Airborne Smoke Plume Sampling

Four contemporary peer-review studies have reported detailed NMOG analysis of smoke plumes sampled from airborne platforms: Burling et al. (2011), Simpson et al. (2011), Akagi et al. (2013), and Liu et al. (2017). The most frequently sampled fire types are understory prescribed fires in southeastern forests ($n = 13$).

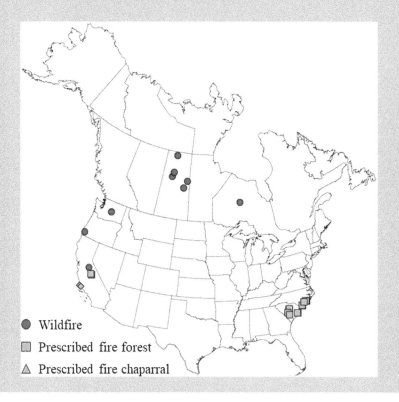

NMOGs for which EFs have been measured in the field comprise 36–58% (by mass) of total NMOG emissions quantified in laboratory studies (Simpson et al. 2011; Yokelson et al. 2013; Liu et al. 2017; Koss et al. 2018). EFs for select compounds measured for prescribed fires in three different fuel types (chaparral, southeastern forest, and western conifer forest) and western wildfires are plotted versus MCE in Fig. 5.6. There is high variability within and across fire types for these chemical species, which are among the most abundant emitted by fires. Large fuel-type differences in NMOG EFs observed in laboratory studies are less pronounced in field data, presumably due to the small sample size and large natural variability in fuels and fire behavior which tend to homogenize the emissions at the point and time of

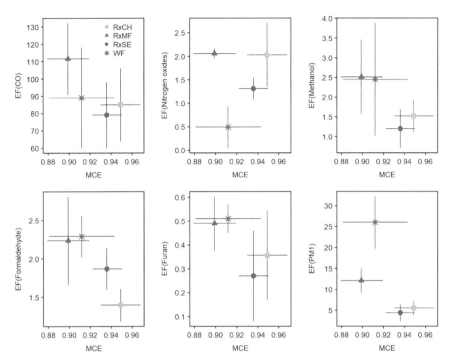

Fig. 5.6 Emission factors (EFs) for select compounds versus modified combustion efficiency (MCE). Data are from airborne measurements of prescribed fires in chaparral [RxCH; Burling et al. (2011)], southeastern conifer forest [RxSE; Akagi et al. (2013)], western conifer forest [RxMF; Burling et al. (2011)], and western wildfires [WF; Liu et al. (2017)]. EF for particulate matter data for prescribed fires is from May et al. (2014). Horizontal and vertical bars are one standard deviation

measurement (Fig. 5.6). The EFs in Fig. 5.6 tend to group according to MCE which is consistent with laboratory findings (see Figs. 5.2 and 5.3).

5.2.2.3 Emissions from Residual Smoldering Combustion

Long-term smoldering combustion that is not influenced by fire-related convection sufficient to loft the smoke above the surface layer is referred to as residual smoldering combustion (RSC; Wade and Lunsford 1989). RSC includes glowing combustion, which is strong smoldering that produces high local temperatures (Santoso et al. 2019) and often does not produce visible smoke. RSC emissions are generated from logs, stumps, duff, and organic soils which are prone to sustained smoldering combustion. Following ignition during flame-front passage, these fuel components can smolder for hours to days (Ottmar 2018). Replicating RSC in the laboratory is challenging for these fuel components, and limited data are available.

Two field studies of prescribed fires in North Carolina and South Carolina pine understories augmented airborne measurements with ground-based sampling of RSC emissions. These studies found EFs of gases associated with smoldering combustion; CO, CH_4, and many NMOGs were much higher for RSC than those measured from airborne platforms (Burling et al. 2011; Akagi et al. 2013). Akagi et al. (2013) measured over 90 NMOGs from airborne and ground-based platforms for three prescribed fires in South Carolina pine understory. They found EFNMOG for RSC (34.18 ± 20.40 g kg^{-1}) was more than twice that measured in the lofted plume (14.56 ± 0.72 g kg^{-1}), with differences between RSC and lofted plume EFs for individual NMOGs being highly variable. Emissions of NO_X, which result from flaming combustion, were negligible from RSC (Burling et al. 2011; Akagi et al. 2013).

Organic soils (peat) and duff burn predominantly by smoldering combustion (Chap. 2), which can persist for days. When wildfires occur in landscapes with deep organic soil layers, such as in the southeast USA and northern boreal ecosystems, smoke production can continue for weeks after fire spread is contained and produce vast quantities of pollutants (Ottmar 2018). Limited field measurements of PM emissions from smoldering organic soil (North Carolina coastal plain) found $EFPM_{2.5} \geq 40$ g kg^{-1} (Geron and Hays 2013). This is more than twice the $EFPM_{2.5}$ observed for the burning of southeastern understory forest fuels with ground-based measurements (Geron and Hays 2013; Urbanski 2014) and considerably larger than $EFPM_1$ measured from aircraft (May et al. 2014) (Fig. 5.6). In situ measurements of gaseous emissions from RSC show EFCO = 200–300 g kg^{-1} and EFVOC ~40 g kg^{-1} (VOC = NMOG + CH_4) (Hao and Babbit 2007; Geron and Hays 2013).

Interpretation and application of RSC EFs are challenging due to the uncertain representativeness and potential sampling biases associated with RSC measurements. A limited comparison of EFs measured for smoldering fuel components and for drift smoke along burn-unit perimeters indicates smoldering, and possibly scattered flaming combustion of other fuel types (e.g., litter and shrubs), may contribute to unlofted emissions (Akagi et al. 2014). Thus, using only EFs based on RSC-prone fuel components may not give an accurate depiction of unit-level emissions, firefighter exposure, or local smoke impacts. Given the scarcity of RSC measurements, extrapolation of data from Geron and Hays (2013) to other ecosystems is needed. In addition, because comprehensive field measurements of EFs for smoldering organic soil and peat are even more limited, laboratory-measured EFs must currently be relied upon to estimate emissions for fires involving these fuel types and associated combustion characteristics.

5.2.3 Emission Calculations

Quantifying EFs of wildland fires is only the starting point for characterizing emissions. Decision support activities (e.g., forecasting smoke impacts) and research (e.g., climate forcing of aerosols) require mass flux estimates (kg m^{-2} s^{-1}) of pollutants released into the atmosphere by wildland fires. Here, we refer to the mass flux of pollutant X as "emissions of X" (E_X) which can be calculated bottom-up or top-down. Bottom-up calculations are based on surface data (fuel loading and

5.2.3.1 Emission Calculations: Bottom-Up Methods

In simplified form, bottom-up emission calculations may be described with Eq. 5.3:

$$E_X = A \times F \times C \times EF_X \tag{5.3}$$

where the mass flux of species X, E_X (kg-X m^{-2} s^{-1}) is the product of area burned (A, m^2), fuel loading (F, kg-fuel m^{-2}), combustion completeness (C), and EF_X (kg-X kg-fuel^{-1}). In practice, this calculation involves several components (Ottmar 2018): (1) fire activity, (2) fuel characteristics, (3) fuel consumption, (4) emission factors, (5) temporal allocation of emissions, (6) vertical allocation of emissions, and (7) the atmosphere (Fig. 5.7).

First, fire activity information is necessary—when and where a fire occurred, and size of the area burned. Availability of fire activity data is constrained by the intended use of the emissions. Smoke forecasting requires rapid data accessibility for recent fire activity (e.g., previous 24 h) as well as predictions of fire activity and resultant emissions for the forecast period (typically 24–48 h). In contrast, research activities can usually use emissions calculated a long period time after the actual fire activity, allowing access to post-fire data products.

Fire activity data collected as part of fire management activities are often available with a timeliness suitable for smoke forecasting. These data include incident management reports for wildfires and burn permit records and agency reporting for prescribed fires. This reporting provides fire location and size, and may include size increase since last report. Prescribed fire data differ widely depending on the agency, jurisdictional reporting requirements, and land ownership. During large wildfire operations, fire perimeter data are commonly obtained from airborne mapping, usually via infrared-based instruments. For both prescribed fire and wildfire, fire size is not necessarily equivalent to the actual area burned. Meddens et al. (2016) determined that approximately 20 percent of the area within a wildfire perimeter was unburned.

Satellite detection of active fires ("hotspots") can provide a large-scale (regional to continental) view of fire activity (Chuvieco et al. 2019a). Satellite fire detection data in the USA and Canada have variable spatial and temporal resolution. The MODIS and VIIRS instruments on polar-orbiting satellites provide data with a nadir (surface point centered directly below the satellite) pixel size of 375 m to 1 km, and a return time of 12 h per satellite. The latest generation National Oceanic and Atmospheric Administration's Geostationary Operational Environmental Satellites (GOES-16 and GOES-17) provide fire detection data with a frequency of 5–15 min and nadir pixel resolution of 2 km. Although widely used, these data have limitations. Clouds, forest canopy cover, and low fire intensity can inhibit satellite fire detection. The data do not provide actual fire size, since detectability depends on many factors including fire intensity (Schroeder et al. 2014; Szpakowski and Jensen 2019).

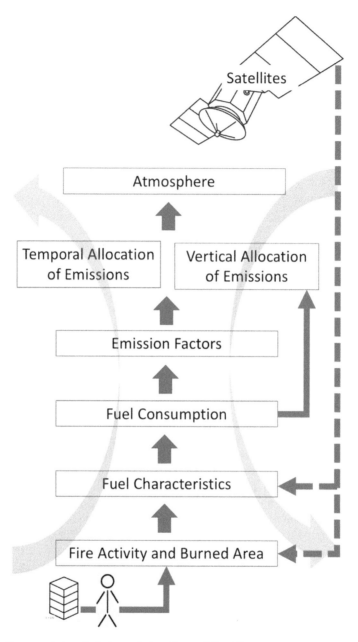

Fig. 5.7 Components in calculating emissions from wildland fire

Emission calculations used in retrospective analyses can leverage fire activity data not available for real-time smoke forecasting. For example, burned area products derived from satellite time series of MODIS and LANDSAT observations (Chuvieco et al. 2019a, b) provide robust burned area mapping. The relaxed time requirements of retrospective analyses also enable use of detailed, vetted databases constructed from multi-agency fire reports such as the Fire Occurrence Database (Short et al. 2020). Combining disparate data sources on fire activity in a consistent dataset optimized for emission calculations is challenging. Tools and efforts described in the Comprehensive Fire Information Reconciled Emissions (CFIRE) Inventory (Larkin et al. 2020) addressed these issues in an attempt to develop a cohesive dataset of fire activity information for a region and time period.

Once a fire is located and its size is estimated, vegetation information is required to infer fuel loading data. Vegetation types, such as Douglas fir (*Pseudotsuga menziesii*) forest or sagebrush (*Artemisia* spp.) shrubland, can be obtained from national-scale mapped datasets such the Fuel Characteristic Classification System (FCCS; Prichard et al. 2013) or on a site-specific basis (Wright et al. 2010b). Fuel classification systems associate vegetation types with an estimate of fuel loading by stratum (duff, litter, woody fuels, etc.). These datasets typically represent the mean for vegetation types whose fuel loading may in reality vary greatly. The high variability of fuel loading is one of the largest contributors of uncertainty in wildland fire emission estimates (Larkin et al. 2014; Chap. 2).

Once burned area and fuel loading are obtained, information on the fraction of fuel consumed across the different fuel strata (combustion completeness) is needed. Fuel consumption (Chaps. 2 and 3) is determined by the combustion process, consisting of four phases: (1) preignition involving distillation and pyrolysis, leading to (2) flaming, (3) smoldering, and (4) glowing (char oxidation) combustion. Fuel properties (type, moisture content, and arrangement), environmental conditions (e.g., wind speed and terrain), and ignition method in the case of prescribed fires can affect the amount of biomass consumed during various combustion stages. CONSUME (Prichard et al. 2014), FOFEM (Lutes 2019), and Pile Calculator (Wright et al. 2010a) are three widely used fuel consumption models.

The composition and relative abundance of emission species produced during fuel consumption are a function of fuel type, combustion process, and atmospheric interactions. The role of these complex processes in determining EFs is discussed in Sect. 5.2.2.

Finally, emissions must be allocated temporally and vertically in the atmosphere. For prescribed fires, temporal allocation of emissions is often conducted using the Fire Emissions Production Simulator (FEPS; Anderson et al. 2004), where soon after ignition, a large spike in flaming emission occurs which then decays exponentially until 6 pm local time, at which time all flaming emissions end and smoldering emissions continue through the evening (Ferguson and Hardy 1994). For wildfires, time profiles based on diurnal cycles derived from a fusion of fire activity observations from geostationary and polar-orbiting satellites (Mu et al. 2011, Li et al. 2019a, b) or from the work of the Western Region Air Partnership (WRAP) are typically applied.

Future work with fire detection data from the GOES-16 and GOES-17 satellites is anticipated to improve temporal profiles for large wildfires.

The vertical distribution of emissions in the atmosphere depends on smoke plume dynamics (Chap. 4). Heat released from the fire is estimated from the consumption model and is often used to estimate the maximum height in the atmosphere under which emissions are distributed, known as plume rise. A Briggs (1976) approach has been historically used in systems such as BlueSky (Larkin et al. 2009). Other plume modeling methods have been used for emissions and smoke modeling (e.g., DAYSMOKE; Achtemeier et al. 2011; Chap. 4).

Concurrent with plume rise is how emissions are distributed underneath the plume top. Typically, smoldering emissions are allocated to the lowest level of the atmospheric model (near the surface). Flaming emissions are usually distributed evenly (vertically) through the atmosphere beneath the nominal plume-rise height. How plume-rise height interacts with mixing height, as well as quantity of flaming versus smoldering emissions, has implications for the quantity of emissions retained near the surface versus lofted and transported long distances.

5.2.3.2 Emission Calculations: Top-Down Methods

Bottom-up emission approaches combine fuel loading maps with estimates of area burned and fuel consumption to derive biomass burned, to which EFs are applied to calculate pollutant emissions (Fig. 5.7; Eq. 5.3). Fuel consumption, the product of fuel loading and combustion completeness, is the largest source of uncertainty in bottom-up emission calculations (French et al. 2011; Urbanski et al. 2011; Leeuwen et al. 2014). Top-down emission methods use satellite observations of fire radiative power (FRP), a measure of the radiant energy release rate from burning vegetation, to estimate fuel consumption, circumventing the need to explicitly consider fuel loading and combustion completeness.

FRP is one of the parameters provided in the active fire products derived from observations of the MODIS and VIIRS sensors (and other satellite-based sensors) (Wooster et al. 2003; Zhang et al. 2017). FRP is based on the fire pixel temperature observed in mid-wavelength infrared, typically around 4 μm (3.96 μm for MODIS) (Wooster et al. 2003). Laboratory and field experiments have shown that (1) FRP is linearly related to the vegetation combustion rate, and (2) fire radiative energy (FRE) (time-integrated FRP) is linearly related to the mass of vegetation combusted (Wooster et al. 2005; Freeborn et al. 2008; Hudak et al. 2016). Most top-down approaches estimate emissions by combining fuel consumption inferred from FRE with biome/land cover-specific EFs (Kaiser et al. 2012; Zhang et al. 2012). A variation of this approach used estimates of atmospheric column PM loading (derived from MODIS aerosol optical depth) to develop land cover-specific PM emission coefficients (kg-PM MJ^{-1}) for predicting PM emissions directly from FRE (kg MJ^{-1}) (Ichoku and Ellison 2014).

Top-down emission inventories typically use FRP retrievals from the MODIS and VIIRS sensors which are on polar-orbiting satellites. In addition to providing

5 Emissions

Table 5.1 Databases, syntheses, and reviews for emission factors (in order of last update)

Emission factor dataset	References	Availability	Last update
Smoke Emissions Repository Application	Prichard et al. (2020)	https://depts.washington.edu/nwfire/sera	2019
Andreae biomass burning emission factors	Andreae (2019)	https://doi.org/10.17617/3.26	2019
Urbanski	Urbanski (2014)	https://www.fs.usda.gov/treesearch/pubs/45727	2014
Wildland fire emissions factors database	Lincoln et al. (2014)	https://www.fs.usda.gov/rds/archive/catalog/RDS-2014-0012	2014
Akagi et al.	Akagi et al. (2011)	http://bai.acom.ucar.edu/Data/fire/	2011
USEPA AP-42	USEPA (1996)	https://www3.epa.gov/ttn/chief/ap42/ch13/index.html	1996

global coverage, these sensors offer a higher spatial resolution (nominal resolution at nadir of 1 km for MODIS and 750 m/375 m for VIIRS) than sensors on geostationary orbiting satellites (e.g., GOES-11/13/15) (nominal 4 km at nadir). However, polar-orbiting satellites offer limited temporal coverage (two observations a day per satellite) compared with geostationary satellites. For example, the GOES imagers provide observations every 5–15 min across the contiguous USA.

Since FRP is an instantaneous indicator of heat flux and does not provide information on fire evolution, the sparse temporal coverage of polar-orbiting satellites is a major limitation of the top-down emission approach. Recent efforts to combine FRP data from polar-orbiting satellites (MODIS/VIIRS) and higher temporal resolution GOES fire products are promising for providing improved spatiotemporal FRP coverage (Li et al. 2019a, b). Application of this approach to the new generation of GOES imagers (GOES-16/17), which have improved spatial resolution (nominal 2 km at nadir for fire products), may be an effective emission inventory method.

5.3 Existing Data, Tools, Models, and Other Technology

5.3.1 Emission Factors

Publicly available EF syntheses and databases are listed in Table 5.1. Andreae (2019) and Akagi (2011) support global emission modeling and provide EFs for broad fire types such as "temperate forest" and "peat fires," as well as other biomass sources (e.g., biofuel use and trash burning). Urbanski (2014) uses more specific fire classifications, designed for US and Canadian fires, such as "prescribed fire southeast conifer forest" and residual smoldering of "stumps and logs."

The Smoke Emission Reference Application (SERA) is an online database that allows users to explore and summarize an extensive repository of EFs for smoke management and emission inventory activities (Prichard et al. 2020). The Lincoln et al. (2014) database compiles EFs from a large body of field and laboratory studies. The SERA and Lincoln et al. databases do not synthesize data to derive "best estimate" EFs. Chapter 13 of "Compilation of Air Pollutant Emissions Factors" (AP-42) (USEPA 1996) provides recommended EFs for a limited number of pollutants for US fire types and was published prior to the advances achieved in the past 15 years in characterizing emissions of wildland fires.

5.3.2 Emission Inventories

An emission inventory is a compilation of data that lists, by source, the amount of air pollutants released into the atmosphere in a defined geographic area during a specific time period. Table 5.2 provides nine wildland fire emission inventories that cover the contiguous USA (CONUS). The domain and temporal coverage differ among the inventories. A number of inventories (GFED, FiNN, QFED, GFAS, FEER, and GBBPx) are global in coverage, and others focus on the USA (WFEIS, MFLEI, and NEI) (Table 5.2). Although the spatial resolution of the inventories uses different metrics (500 m to 0.25°), all provide emissions with a 1-day temporal resolution. Many atmospheric model applications, whether operational forecasts or retrospective analyses, require hourly emissions. High temporal frequency observations of fire activity from geostationary satellites have proven useful for deriving hourly emission profiles from daily estimates (Mu et al. 2011; Li et al. 2018).

Several inventories (FiNN, QFED, GFAS, FEER, and GBBPx) calculate emissions in near-real time for use in atmospheric chemistry forecasting. FiNN and QFED are used in the Whole Atmosphere Community Climate Model (https://www.acom.ucar.edu/waccm/forecast). GFAS is used in Copernicus Atmosphere Monitoring Service (https://atmosphere.copernicus.eu/global-forecast-plots), and GBBEPx is an operational product currently being used by the NGAC v2 aerosol model at the National Center for Environmental Prediction. GFED, WFEIS, MFLEI, and NEI are all retrospective inventories that estimate emissions with a time lag of one to three years. Retrospective inventories have the potential to provide more accurate emission estimates than their real-time counterparts as they can leverage burned area and burn severity geospatial data products that are not available in real time (Urbanski et al. 2018).

Different inventories include different pollutant species. For example, FiNN emissions are speciated for three different atmospheric chemistry model mechanisms; MFLEI provides fuel consumption and emissions of CO_2, CO, CH_4, and $PM_{2.5}$; GFED offers fuel consumption according to fire type, with recommended EFs for over 20 species. Most of the inventories include fuel consumption which can be used to calculate emissions for any species for which EFs are available; this requires information or assumptions regarding fire type and vegetation burned.

Table 5.2 Emission inventories

Inventory	Domain	ΔX	Δt	Active	Access
Bottom-up emission calculations					
Global Fire Emissions Database (GFED)	Global	0.25°	1 day to 1 month	Yes	https://www.globalfiredata.org/
Wildland Fire Emissions Information System (WFEIS)	USA	1 km	1 day	No	https://wfeis.mtri.org/
Fire Inventory from NCAR (FiNN)	Global	1 km	1 day	Yes	https://www2.acom.ucar.edu/modeling/finn-fire-inventory-ncar
Missoula Fire Lab Emission Inventory (MFLEI)	CONUS	500 m	1 day	Yes	https://www.fs.usda.gov/rds/archive/catalog/RDS-2017-0039
National Emission Inventory (NEI)	USA	Variable	1 day	Yes	https://www.epa.gov/air-emissions-inventories/national-emissions-inventory-nei
Top-down emission calculations					
Global Fire Assimilation System (GFAS)	Global	0.1°	1 day	Yes	https://atmosphere.copernicus.eu/global-fire-emissions
Quick Fire Emission Dataset v2.4 (QFED)	Global	0.1°	1 day	Yes	https://www.acom.ucar.edu/waccm/register.shtml
Fire Energetics and Emissions Research v1 (FEER)	Global	0.1°	1 day	Yes	https://feer.gsfc.nasa.gov/projects/emissions/
Blended Global Biomass Burning Emissions Product (GBBEPx V3)	Global	0.1°	1 h	Yes	https://www.ospo.noaa.gov/Products/land/gbbepx/

5.3.2.1 Emission Estimates for CONUS, Canada, and Alaska

A map of annual average PM$_{2.5}$ emissions from 2003 to 2018 estimated by GFED (Werf et al. 2017) is shown in Fig. 5.8. Emission hotspots are concentrated in the boreal regions and, to a lesser extent, in the western USA and southern British Columbia. An emission hotspot is also present on the Georgia–Florida border owing to a series of intense fires in the Okefenokee Swamp region. GFED annual sums of PM$_{2.5}$ emissions for CONUS and Alaska/Canada for 2003–2018 are shown in Fig. 5.9. The GFED-estimated annual PM$_{2.5}$ emissions for Alaska and Canada exceed

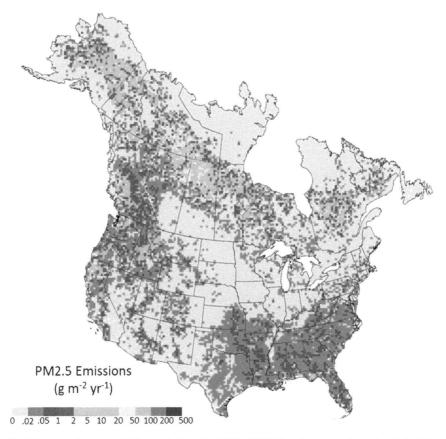

Fig. 5.8 Annual average $PM_{2.5}$ emissions for 2003–2018. Based on data from the global fire emissions database (Werf et al. 2017)

those of CONUS by a factor of 2–20, depending on the year. Interannual variability in emissions is similar for the two regions, with coefficients of variation near 0.5.

Monthly average GFED $PM_{2.5}$ emissions are shown in Fig. 5.10. Across the northern tier, emissions are concentrated in the summer months (90% between June and August). CONUS emissions are spread more broadly across the year, with the peak three months (July–September) accounting for 60% of the annual total. Putting the magnitude of emissions into context, Fig. 5.11 plots summer emissions (July–September) for the western 11 CONUS states with $PM_{2.5}$ emissions from non-fire sources as estimated from the EPA 2014 NEI v2. During the heart of the western USA wildfire season, GFED-estimated $PM_{2.5}$ emissions regularly exceeded anthropogenic sources by a factor of 2–4 during severe fire years (2007, 2012, 2017, 2018).

Annual magnitude, seasonality, and spatial distribution of fire emission across the USA and Canada are summarized in Figs. 5.8, 5.9, 5.10 and 5.11. There is uncertainty in emission inventories, especially at spatiotemporal scales relevant for

5 Emissions

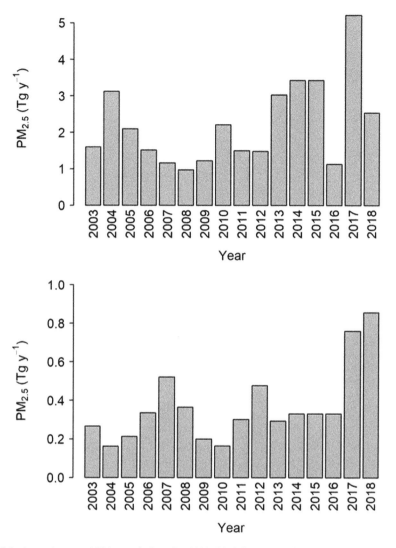

Fig. 5.9 Annual sums of PM$_{2.5}$ emissions for 2003–2018 for Alaska/Canada (top panel) and the CONUS (bottom panel). From the global fire emissions database

understanding and predicting smoke impacts. PM$_{2.5}$ emissions based on four inventories are shown in Fig. 5.12: PM$_{2.5}$ emissions range from 80 to 230% of the ensemble mean. Different data and methods—burned area, fuel-type classification maps, fuel loading and consumption, and EFs—all contribute to this variability.

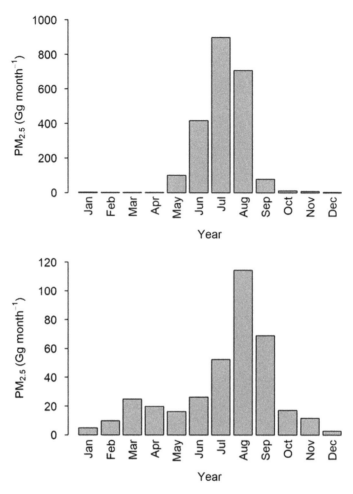

Fig. 5.10 Monthly average PM$_{2.5}$ emissions for 2003–2018 for Alaska/Canada (top panel) and CONUS (bottom panel). From the global fire emissions database

5.3.3 Emission Models for Land Management

Prescribed fire is used to maintain and restore ecosystem function and health and mitigate wildfire risk through reduction of hazardous fuel. Smoke impacts are an important consideration for prescribed burning, and effective smoke management strategies are generally required for successful use of prescribed fire. Emission reduction techniques (ERTs) are central to the basic smoke management practices recommended by the National Wildfire Coordination Group (Peterson et al. 2018). ERTs take into consideration area burned, fuel load, fuel produced, amount of fuel consumed, and combustion efficiency. Smoke emission models designed for land managers and

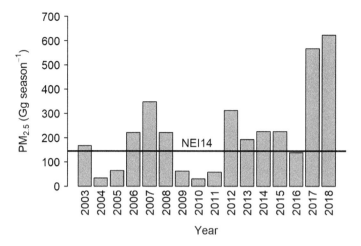

Fig. 5.11 Summer (July–September) $PM_{2.5}$ emissions for the western 11 CONUS states (bars) and $PM_{2.5}$ emissions from non-fire sources as estimated from the USEPA 2014 NEI v2 [solid horizontal line; USEPA (2014)]

prescribed fire practitioners are important tools for implementing ERTs. Smoke emission models commonly used for planning of prescribed fires in the USA (Table 5.3) predict emissions based on fuel loading, fuel moisture, and environmental factors.

A number of models are available for managers to use in prescribed fire planning. The First Order Fire Effects Model (FOFEM) predicts the immediate consequences of wildland fire, including fuel consumption, smoke production, soil heating, and tree mortality. CONSUME is a module within BlueSky, WFEIS, and the Fuel and Fire Tools (FFT) suite that predicts total fuel consumption, emissions, and heat release. FEPS predicts hourly emissions, heat release, and plume-rise values for wildland fires; can import consumption and emission data from CONSUME and FOFEM; and is included in FFT. The software application FFT integrates CONSUME and FEPS with fuel data from the FCCS and Digital Photo Series (Chap. 2) into a single user interface (Ottmar 2014). BlueSky Playground (Larkin 2018) provides interactive access to several models enabled by the BlueSky Framework and allows users to enter basic fire information to simulate fuel consumption and pollutant emissions, as well as model plume rise and smoke dispersion.

5.4 Gaps in Data, Understanding, and Tools/Technology

5.4.1 Emission Factors for Wildfires

The paucity of EF measurements for wildfires is a significant gap in our understanding of emissions. With the exception of prescribed fires in southeastern US forests, most fire types have received limited field investigation. The small number of wildfires

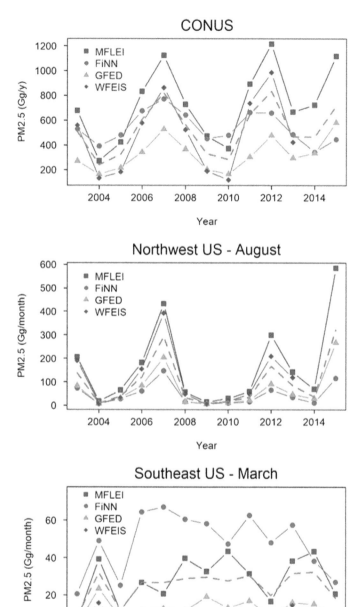

Fig. 5.12 PM$_{2.5}$ emissions based on four different inventories: GFED, FiNN, MFLEI, and WFEIS for three regions and time periods: CONUS-wide—annual (top), northwest USA—August (middle), and southeast USA—March (bottom)

Table 5.3 Emission models for land management

Model	Availability	References
FOFEM	https://www.firelab.org/document/fofem-files/	Lutes (2019)
CONSUME	https://www.fs.fed.us/pnw/fera/research/smoke/consume	Prichard et al. (2020)
FEPS	https://www.fs.fed.us/pnw/fera/feps	Anderson et al. (2004)
FFT	https://www.fs.fed.us/pnw/fera/fft	Ottmar (2014)
BlueSky Playground	https://tools.airfire.org/playground	Larkin (2018)

that have been sampled with detailed chemical speciation does not capture the wide range of fuels and burning conditions that occur across the USA and Canada.

EFs have not been measured from wildfires for most NMOGs known to be present in fresh smoke (based on laboratory studies). Boreal wildfire EFs for the most reactive compounds, which include nearly half the NMOG mass reported, are based on a single fire (Simpson et al. 2011). Similarly, NMOG EFs for western US wildfires are limited to only three fires and may not capture the range of wildfire emissions (Liu et al. 2017). Field studies that did not measure EFs for PM and NMOGs report an MCE range of 0.83–0.95 for 29 western USA and boreal wildfires (Hornbrook et al. 2011; O'Shea et al. 2013; Urbanski 2013). Because EFs for many species are correlated with MCE, the actual range of EFNMOG and EFPM for wildfires may be considerably broader than suggested by Liu et al. and Simpson et al., so applying these data to wildfires may introduce uncertainty in emission estimates.

5.4.2 Connecting Laboratory Studies with Field Observations

Comprehensive emission estimates across the spectrum of relevant fire activity require extrapolating laboratory-measured EFs to real fire conditions. EF extrapolation methods include (e.g., Selimovic et al. 2018; Sekimoto et al. 2018):

- Regression of EF versus MCE to extrapolate field MCE
- Average EF of laboratory burns according to fuels type
- Coupling of laboratory $\Delta X/\Delta CO$ with field EFCO to derive EFX at field conditions
- Pyrolysis profiles based on high- and low-temperature regimes.

These methods may also be used to extrapolate field-measured EFs to fires in different fuel types and burning conditions. A combination of the first three approaches has been used in developing global and regional EF databases that are widely used in emission models and inventories (Akagi et al. 2011; Urbanski 2014; Andreae 2019). However, an extensive evaluation of laboratory-extrapolated EFs has not been published, perhaps due to lack of field data. In a limited evaluation, Sekimoto et al. (2018) found that EFs estimated using high- and low-temperature pyrolysis profiles analyzed with positive matrix factorization can reproduce NMOG EFs from previous field and laboratory burns with reasonable fidelity ($r \geq 0.92$). Additional field data, especially for wildfires, are needed to support a comprehensive evaluation of EF extrapolation methods.

5.4.3 Variability of EFs with Combustion Conditions

Long-term smoldering can result in sustained periods of poor air quality, exacerbation of health conditions among vulnerable residents, and dangerous road conditions due to reduced visibility (Chap. 7). Smoldering that persists into nighttime hours when winds tend to be light and variable, reducing dispersion, can be especially challenging when the shallow nocturnal boundary layer reduces the volume into which smoke is emitted. Nighttime subsidence drainage flows can transport smoke long distances, pooling it in valleys or low-lying areas.

Applying fuel treatments on landscapes with fuels prone to smoldering, while minimizing local smoke impacts, requires models that provide reliable temporal profiles of fuel consumption and pollutant emissions. FOFEM and FEPS are widely used to predict fuel consumption and smoke production (Ottmar 2018). However, the ability of these models to simulate fuel consumption rates for smoldering combustion has not been rigorously evaluated. In addition, the models predict temporal emission profiles using static smoldering-phase EFs.

Consumption and emission rates during long-term smoldering can differ depending on the fuel component (log, stump, basal accumulation, etc.) and fuel condition (Ottmar 2018). Likewise, EFs differ with fuel component type and smoldering characteristics (Hao and Babbit et al. 2007; Reisen et al. 2018). The absence of validated models to predict emission rates from long-term smoldering is a significant obstacle to using prescribed fire.

5.4.4 Validation of Emission Inventories

There are large discrepancies among the various CONUS emission inventories. In a recent study, CONUS-wide average monthly $PM_{2.5}$ emissions estimated by seven inventories, over four years, ranged from 28.2 to 485.6 Gg, with a coefficient of variation of 109% (Li et al. 2019a, b). Comparisons limited to retrospective emission inventories find large differences at fairly coarse scales (Larkin et al. 2014; Koplitz et al. 2018) and increasing variability with decreasing spatiotemporal scale (French et al. 2011; Urbanski et al. 2011).

Improving our ability to forecast smoke events and understand smoke impacts within the USA requires developing and applying thorough validation methods for emission inventories at the fire-event level. Over 10 emission inventories (near-real time and retrospective) include the CONUS, and several comparisons are found in the literature (e.g., Larkin et al. 2014; Koplitz et al. 2018; Urbanski et al. 2018; Li et al. 2019a, b). However, none of the emission inventories has been methodically evaluated using independent data at scales relevant for assessing wildfire smoke impacts on air quality. Validation methods link satellite observations of fire emissions (e.g., aerosol optical depth, CO, NO_2) to fire activity using atmospheric models and meteorological analyses. Although these methods have been used in both forward (Ichoku and Ellison 2014; Petrenko et al. 2017) and inverse (Dubovik et al. 2008; Kopacz et al. 2010) modeling approaches to constrain fire emission inventories at global to regional scales, they have not rigorously validated emission inventories at the fire-event scale.

5.4.5 Forecasting Wildfire Emissions

The lack of reliable near-term (24 h) emission forecasts is a key obstacle to improving forecasts of wildfire smoke impacts on air quality. The main challenge is accurately predicting the growth of many active fires over the next burning period in a timely manner that is compatible with regional-to-continental smoke forecasting systems. Although several fire growth models exist, current operational smoke models use daily persistence in burned area growth to forecast emissions. Daily persistence assumes that the area burned by a given fire in the current day will be that fire's growth the following day. However, given available fuel and variable topography, daily weather plays a major role in the growth of wildfires (Chap. 3).

The sensitivity of wildfire growth to weather is evident in retrospective emission inventories that suggest that the majority of CONUS wildfire emissions occur on a small fraction of days (~5%) (Urbanski et al. 2018). The daily persistence approach will often greatly under-predict these high fire growth/high emission days, which occur during severe fire weather conditions (e.g., Jolly et al. 2019), resulting in a failure to forecast acute smoke episodes. The use of daily persistence can also overestimate fire growth over periods following extreme fire weather days, leading to an overprediction of smoke production. Improving the skill of smoke forecasts will require developing and implementing new methods for predicting short-term (24 h) fire growth and emissions. Methods based on forecast meteorological variables (temperature, relative humidity, wind speed) and fire weather indices have shown promise for improving upon daily persistence in prediction of short-term fire activity and smoke emissions (Peterson et al. 2013; Giuseppe et al. 2017).

5.4.6 Measuring and Modeling $PM_{2.5}$

Inaccurate $PM_{2.5}$ measurements introduce errors in emission models used for air quality modeling and introduce uncertainty in the measurements used to validate these models. Inaccurate ambient $PM_{2.5}$ measurements also may result in public health guidance that is either overly restrictive or not adequately protective.

Due to the semi-volatile and reactive nature of smoke, $PM_{2.5}$ concentration ratios used to calculate EFs can differ depending upon the local conditions at which they are measured. EFs are measured from fresh emissions before significant SOA formation, or other reactions have occurred and altered the chemistry of the emissions. However, at the high concentrations near the fire, the lower-volatility SVOCs will partition to the particle phase, leading to higher $PM_{2.5}$ concentrations than under more dilute conditions (Robinson et al. 2010). These volatility effects may partially explain the wide scatter observed in $EFPM_{2.5}$ across studies (Jolleys et al. 2014; May et al. 2014) and observations that $EFPM_{2.5}$ can be almost twice as high near the fire compared to downwind in a dilute plume (Holder et al. 2016).

The volatility distribution is one way to account for SVOC partitioning and is now being employed in air quality models (Lu et al. 2020). Volatility distributions have been shown to be relatively independent of fuel type and burning conditions (May et al. 2013; Hatch et al. 2018) and can explain up to a 40% loss of PM with 100:1 dilution (Hatch et al. 2018). However, volatility measurements have been limited to laboratory burns, and field measurements are still needed.

Ambient monitoring of smoke also has $PM_{2.5}$ measurement challenges. Air quality information during smoke events is generally derived from Federal Equivalent Method (FEM) monitors that provide hourly measurements. FEMs are validated against 24-h filter-based Federal Reference Measurements (FRMs) at concentrations of 3–200 $\mu g\ m^{-3}$ to ensure broad comparability to FRM $PM_{2.5}$ mass, which is the basis for much of the $PM_{2.5}$ health effect research (USEPA 2020). However, FEM evaluations do not purposefully include smoke-impacted times and do not cover the full range of $PM_{2.5}$ concentrations corresponding to the air quality index range.

Research to identify and resolve FEM monitor measurement accuracy for wildfire smoke is needed. Several FEMs contain measurement artifacts, which may hinder their use for assessing smoke impacts on air quality. Environmental beta attenuation monitors (EBAMs, a near-FEM grade instrument) used in temporary monitoring networks near fires are subject to a high bias at elevated relative humidity, and hourly EBAM $PM_{2.5}$ data at humidity above 40% should be used cautiously (Schweizer et al. 2016). Another FEM (Teledyne T640®) was found to report $PM_{2.5}$ concentrations 40–100% higher than another FEM (MetOne 1020 BAM®), but only when the $PM_{2.5}$ concentrations were elevated (Hassett-Sipple et al. 2020; Landis et al. 2021).

Sensor technologies are increasingly used to monitor wildfire smoke, and work is needed to identify appropriate corrections for $PM_{2.5}$ sensors and methods to ensure high-quality data during extended smoky episodes. Although some $PM_{2.5}$ sensor measurements can report concentrations up to twice as high as nearby FEMs (Mehadi et al. 2020; Holder et al. 2020; Landis et al. 2021), with correction some sensors have been found to report $PM_{2.5}$ with modest error (~20–30%) over a range of conditions (Holder et al. 2020; Barkjohn et al. 2020) and are now displayed as part of the Sensor Data Pilot on the AirNow Fire and Smoke Map (https://fire.airnow.gov).

5.4.7 Emissions of Hazardous Air Pollutants

PM is the major constituent of smoke associated with adverse health effects; however, numerous other hazardous air pollutants are also emitted from fires, such as hydrogen cyanide (HCN), polycyclic aromatic hydrocarbons (PAHs), and other organic compounds (e.g., formaldehyde). The contribution of these gas-phase compounds to health effects is poorly known. O'Dell et al. (2020) identified formaldehyde as the largest gas-phase hazardous air pollutant contributing to cancer risk from wildfire smoke. They also found that acrolein was the major contributor to acute and chronic hazards of young wildfire smoke (<1 day old), whereas HCN was the primary contributor to chronic hazard from aged wildfire smoke. Although O'Dell et al. (2020) estimated that health impacts from gas-phase hazardous air pollutants were small compared to PM, exposure to these pollutants may not be reduced by common actions recommended to reduce smoke exposure (e.g., portable air cleaners and N95 masks). More research is also needed on the impact of gas-phase hazardous air pollutants near fires and human health.

Toxic metals have been measured at trace levels in biomass burning PM (Chen et al. 2007; Alves et al. 2011; Hosseini et al. 2013). As other sources of toxic metal emissions to the atmosphere have decreased through regulations, fires may now be a sizable source of toxic metals to the atmosphere, and some compounds may be the leading source of these emissions to the atmosphere (Reff et al. 2009). Lead is a particular concern because it is a USEPA criteria air pollutant and regulated through the National Ambient Air Quality Standards.

When fires occur in the wildland–urban interface, burning vehicles and structures may emit toxic metals, because the "fuel" in this case may contain high concentrations of these metals (see Sect. 5.4.8). In areas where there has been environmental contamination (e.g., Superfund sites, firing ranges, areas downwind of point sources), lead and other toxic metals deposited in the soil and vegetation can be remobilized as PM, entrained ash, or soil particles (Kristensen and Taylor 2012; Odigie and Flegal 2014; Wu et al. 2017). Radionuclides were remobilized by fires in the Chernobyl Exclusion Zone (Evangeliou and Eckhardt 2020) but had limited long range transport, likely because most of the radionuclides in the fuel partition to ash (Hao et al. 2018).

The toxic metals in PM from wildland fires may be emitted from both the soil and combustion of vegetation. The larger PM size fractions (2.5–10 μm) emitted from fires are enriched in calcium, magnesium, iron, aluminum, and silicon, likely derived from soil particles entrained in the fire plume (Echalar et al. 1995; Alves et al. 2010, 2011; Popovicheva et al. 2016). Several elements (boron, manganese, zinc, copper) are micronutrients that play a vital role in vegetative physiological processes, and some plant species can hyperaccumulate heavy metals (e.g., uptake of lead by the *Brassicaceae* family (mustard family)) (Tangahu et al. 2011).

When present, the higher-volatility metals (e.g., zinc, lead, cadmium, mercury) in vegetation or soils tend to partition to $PM_{2.5}$, whereas the lower-volatility metals (e.g., cobalt, nickel, chromium, vanadium) tend to partition to ash (Narodoslawsky

Table 5.4 USEPA hazardous air pollutant metals in wildland fire smoke

Fire type	Field/lab	Sb	Cd	As	Cr	Co	Pb	Ni	Hg	Se	References
		(μg metal per g particle mass)									
Western conifer	Lab	6.5	2	23			39			2.5	Turn et al. (1997)
Western forest	Field		375		134		1102	78.3			Ward and Hardy (1989)
Southeastern forest	Field	180			29		87			97	Balachandran et al. (2013)
Southeastern forest	Field			1.9			1.5			1.2	Lee et al. (2005)
Southwestern shrub	Field	9.7	9.8	50	220	1.2	22	12	6.5	16	Chow et al. (2004)

and Obernberger 1996). Metal volatility can also be affected by local combustion conditions, such as an oxidizing atmosphere facilitating formation of low-volatility metal oxides, and the presence of other compounds like chlorine that can result in higher-volatility metal chlorides (Linak and Wendt 1993). Other compounds, such as aluminosilicates, may also act as a sorbent for some metals, causing the metals to partition to ash (Linak and Wendt 1993).

There are limited measurements of toxic metal emissions from wildland fires, owing to the lack of real-time measurement methods and the difficulty of obtaining sufficient sample mass for analysis. Table 5.4 provides a summary of field and laboratory measurements of EFs of metals on the USEPA hazardous air pollutant list. The metal contribution to PM mass can vary by one to three orders of magnitude, possibly representing the variation of the metal content in the vegetation that was burned. However, the large variation may be caused in part by analytical uncertainties due to limited sample mass. Accurate emissions for these trace hazardous air pollutants are still needed for many fuel types and regions. Measurements will require large sample masses and sensitive analytical methods to measure EFs above the detection limit.

5.4.8 Emissions from Structure Fires

Wildland fires that occur in the wildland–urban interface have the additional complication of potentially burning different fuels that release toxic emissions when combusted. Research on toxicity of emissions from combustion of building materials and vehicles has shown that numerous toxic compounds are emitted, including hydrogen cyanide, hydrogen fluoride, hydrochloric acid, isocyanates, polycyclic aromatic hydrocarbons, dioxins, NMOGs (e.g., benzene, toluene, xylene, styrene, formaldehyde), and metals (Austin et al. 2001; Lönnermark and Blomqvist 2006; Fabian et al. 2010; Reisen 2011; Stec 2017; Fent et al. 2018). Structural firefighters use a self-contained breathing apparatus to exclude pollutant concentrations that are immediately dangerous to life or health. However, wildland firefighters responding to wildfires where structures are burned do not normally use self-contained breathing apparatus and may be exposed to high concentrations of toxic air pollutants (Chap. 7).

Several studies have quantified EFs for hazardous pollutants from building materials, vehicles, and house fires (Blomqvist et al. 2004; Lönnermark and Blomqvist 2006; Reisen 2011). The range of pollutants measured, as well as the measurement methods, differed among these studies, and substantial gaps remain on the emissions of hazardous air pollutants. In addition, limited information is available on emissions from materials that contain flame retardants or lithium batteries, or that may have highly toxic emissions.

Table 5.5 summarizes studies of emissions from combustion of structures and vehicles, compared with similar measurements from wildland fires. Although EFs for some of the most hazardous compounds are 2–1600 times greater from combustion of vehicles or building materials compared to wildland fuels, total emissions depend on the number and mass of structures or vehicles consumed in the fire. In the 2018 Camp Fire in California, nearly 20,000 structures were consumed, which may have generated sizable emissions compared to those from natural fuels.

As no inventories of emissions from structures burned in the wildland–urban interface exist, they are not included in smoke emission models. For example, in the NEI model, urban areas consumed in wildfires are assigned a default vegetative fuel loading and EFs that likely underestimate the emissions from burning structures and vehicles. Therefore, air quality forecasts estimating fire progression into populated areas may substantially under-predict smoke concentrations.

5.5 Conclusions

Because wildland fires are a major source of gases and aerosols, a thorough understanding of fire emissions is essential for addressing societal and climatic consequences of wildland fire smoke. In recent years, a large body of laboratory and field experiments has led to significant progress in characterizing the composition of fresh smoke. More than 500 gases have been identified, and our knowledge regarding the physical characteristics, chemical composition, and optical properties of aerosols has expanded greatly. Quantifying wildland fire EFs is only the starting point for characterizing emissions.

Decision support and research require emission inventories of pollutants released into the atmosphere by wildland fires. Emission inventory methods for both predictive (e.g., smoke forecasting) and retrospective (e.g., research or air quality regulation) activities have evolved by leveraging scientific advances in smoke composition, fuels and fuel consumption, and satellite remote sensing of fire activity and effects. Several wildland fire emission inventories covering the CONUS are available to support operational forecasts and retrospective analyses.

Despite recent advances, large gaps in smoke emission science remain:

Table 5.5 Emission factors for vehicle and building materials compared with those for western wildland fuels for selected hazardous pollutants

Material	PM (g kg^{-1})	Total PAH[a] (g kg^{-1})	Dioxin PCDD[b] (μg kg^{-1})	Formaldehyde (g kg^{-1})	Benzene (g kg^{-1})	SO$_2$ (g kg^{-1})	HF (g kg^{-1})	HCl (g kg^{-1})	HCN (g kg^{-1})	References
Vehicle						10		170	2.9	Lonnermark and Blomqvist (2006)
Vehicle	46									Reisen (2011)
Vehicle	64	1.1	2.1	1.0	3.0	5.0		13	1.6	Lonnermark and Blomqvist (2006)
Building	50									Reisen (2011)
Building	23	0.08		0.70	2.3					Reisen et al. (2014)
Building	69					27	120	1203	9.0	Blomqvist et al. (2003)
Building						30		300	0.3	Persson and Simonson (1998)
Furniture								313	39.5	Gann et al. (2010)
Living room		1	0.01		0.7	5		0.5	1	Blomqvist et al. (2004)
Urban average	50	0.74	1.1	0.85	2.0	15	120	160	9.1	Prichard et al. (2020)
Wildland fire	20		0.3[a]	2	0.4	1		0.1	0.4	
Urban/wildland fire EF ratio for western wildland fuels[b]	2.5		4.1	0.60	5.0	13		1600	23	

[a] Total dioxin polychlorinated dibenzodioxins (PCDDs) for forest fuels from Gullett et al. (2008)
[b] A unitless ratio

- There is a significant lack of EF measurements for wildfires; however, results from recent field studies may soon address this gap in our understanding of emissions.
- Even with expanded field measurements of EFs, comprehensive emission estimates across the spectrum of relevant fire activity will require extrapolating laboratory-measured EFs to real fire conditions. A thorough evaluation of the different methods used for extrapolating laboratory EFs is needed to identify best practices and quantify uncertainties of derived EF.
- EFs and emission rates from residual smoldering combustion have received only limited research attention. This knowledge shortfall has inhibited the development of reliable models for predicting local smoke impacts from prescribed fire. Field studies characterizing emissions from residual smoldering combustion are needed to provide improved modeling tools to land managers.
- Discrepancies among emission inventories for the CONUS are significant. These discrepancies are further complicated by the natural heterogeneity of wildland systems. Comprehensive evaluation of these emission inventories is needed to quantify their errors and improve their performance across operational and research applications.
- The lack of reliable near-term (24 h) emission forecasts is an obstacle to improving forecasts of smoke impacts on air quality. New methods for predicting short-term fire growth and emissions are needed to improve air quality forecast.
- Toxic metals have been measured in wildland fire PM and may be a large source of toxic metal emissions. Because toxic metal emissions depend on fuel and soil characteristics (e.g., metal content by strata) and fire behavior, understanding how wildland fires may be a source of these hazardous pollutants must be addressed.
- There is a growing need to understand the emmissions from burning structures. Only limited EFs and no emission inventories are available for evaluating potential emission impacts of burned structures on the health of wildland firefighters and nearby communities.

Acknowledgements The authors thank Kelley Barsanti (University of California Riverside), Tom Moore (Western States Air Resources Council) and Talat Odman (Georgia Institute of Technology) for valuable guidance in identifying and developing the topics covered in this chapter. We thank three reviewers for their valuable and constructive comments and suggestions that have improved this chapter and Brian Gullett for providing helpful suggestions.

References

Achtemeier GL, Goodrick SA, Liu YQ et al (2011) Modeling smoke plume-rise and dispersion from Southern United States prescribed burns with daysmoke. Atmosphere 2:358–388

Akagi SK, Yokelson RJ, Wiedinmyer C et al (2011) Emission factors for open and domestic biomass burning for use in atmospheric models. Atmos Chem Phys 11:4039–4072

Akagi SK, Yokelson RJ, Burling IR et al (2013) Measurements of reactive trace gases and variable O-3 formation rates in some South Carolina biomass burning plumes. Atmos Chem Phys 13:1141–1165

Akagi SK, Burling IR, Mendoza A et al (2014) Field measurements of trace gases emitted by prescribed fires in Southeastern US pine forests using an open-path FTIR system. Atmos Chem Phys 14:199–215

Alves CA, Gonçalves C, Pio CA et al (2010) Smoke emissions from biomass burning in a Mediterranean shrubland. Atmos Environ 44:3024–3033

Alves C, Vicente A, Nunes T et al (2011) Summer 2009 wildfires in Portugal: emission of trace gases and aerosol composition. Atmos Environ 45:641–649

Anderson GK, Sandberg DV, Norheim RA (2004) Fire emission production simulator (FEPS) user's guide version 1.0. http://www.fs.fed.us/pnw/fera/feps/index.shtml. 28 Jan 2020

Andreae MO (2019) Emission of trace gases and aerosols from biomass burning—an updated assessment. Atmos Chem Phys 19:8523–8546

Andreae MO, Merlet P (2001) Emission of trace gases and aerosols from biomass burning. Global Biogeochem Cycles 15:955–966

Austin CC, Wang D, Ecobichon DJ et al (2001) Characterization of volatile organic compounds in smoke at municipal structural fires. J Toxicol Environ Health A 63:437–458

Balachandran S, Pachon JE, Lee S et al (2013) Particulate and gas sampling of prescribed fires in South Georgia, USA. Atmos Environ 81:125–135

Baldauf RW, Devlin RB, Gehr P et al (2016) Ultrafine particle metrics and research considerations: review of the 2015 UFP workshop. Int J Environ Res Public Health 13:1054

Barkjohn KK, Gantt B, Clements AL (2020) Development and application of a United States wide correction for $PM_{2.5}$ data collected with the PurpleAir sensor. Atmos Meas Tech Discuss 1–34

Blomqvist P, Hertzberg T, Dalene M et al (2003) Isocyanates, aminoisocyanates and amines from fires—a screening of common materials found in buildings. Fire Mater 27:275–294

Blomqvist P, Rosell L, Simonson M (2004) Emissions from fires part II: simulated room fires. Fire Technol 40:59–73

Bond TC, Doherty SJ, Fahey DW et al (2013) Bounding the role of black carbon in the climate system: a scientific assessment. J Geophys Res: Atmos 118:5380–5552

Briggs GA (1976) National Oceanic and Atmospheric Administration, Oak Ridge, Tenn. (USA). Atmospheric Turbulence and Diffusion Lab; p. 425–478

Brey SJ, Ruminski M, Atwood SA et al (2018) Connecting smoke plumes to sources using hazard mapping system (HMS) smoke and fire location data over North America. Atmos Chem Phys 18:1745–1761

Burling I, Yokelson RJ, Akagi S, Urbanski S, Wold C, Griffith DW et al (2011) Airborne and ground-based measurements of the trace gases and particles emitted by prescribed fires in the United States. Atmos Chem Phys. 11:12197–12216

Burling IR, Yokelson RJ, Griffith DWT et al (2010) Laboratory measurements of trace gas emissions from biomass burning of fuel types from the southeastern and southwestern United States. Atmos Chem Phys 10:11115–11130

Buseck PR, Adachi K, Andras G et al (2014) Ns-Soot: a material-based term for strongly light-absorbing carbonaceous particles. Aerosol Sci Technol 48:777–788

Chen LWA, Moosmuller H, Arnott WP et al (2007) Emissions from laboratory combustion of wildland fuels: emission factors and source profiles. Environ Sci Technol 41:4317–4325

Chuvieco E, Aguado I, Salas J et al (2019a) Satellite remote sensing contributions to wildland fire science and management 6:81–96

Chuvieco E, Mouillot F, van der Werf GR et al (2019b) Historical background and current developments for mapping burned area from satellite earth observation. Remote Sens Environ 225:45–64

Chow JC, Watson JG, Kuhns H et al (2004) Source profiles for industrial, mobile, and area sources in the Big Bend Regional aerosol visibility and observational study. Chemosphere 54:185–208

Di Giuseppe F, Remy S, Pappenberger F et al (2017) Improving forecasts of biomass burning emissions with the fire weather index. J Appl Meteorol Climatol 56:2789–2799

Dubovik O, Lapyonok T, Kaufman YJ et al (2008) Retrieving global aerosol sources from satellites using inverse modeling. Atmos Chem Phys 8:209–250

Echalar F, Gaudichet A, Cachier H et al (1995) Aerosol emissions by tropical forest and savanna biomass burning: characteristic trace elements and fluxes. Geophys Res Lett 22:3039–3042

Evangeliou N, Eckhardt S (2020) Uncovering transport, deposition and impact of radionuclides released after the early spring 2020 wildfires in the Chernobyl exclusion zone. Sci Rep 10:10655

Fabian T, Borgerson JL, Kerber MS et al (2010) Firefighter exposure to smoke particulates. Underwriters Laboratories Inc, Northbrook. https://ulfirefightersafety.org/docs/EMW-2007-FP-02093.pdf. 19 June 2020

Fent KW, Evans DE, Babik K et al (2018) Airborne contaminants during controlled residential fires. J Occup Environ Hyg 15:399–412

Ford B, Martin MV, Zelasky SE et al (2018) Future fire impacts on smoke concentrations, visibility, and health in the contiguous United States. Geohealth 2:229–247

Freeborn PH, Wooster MJ, Hao WM et al (2008) Relationships between energy release, fuel mass loss, and trace gas and aerosol emissions during laboratory biomass fires. J Geophys Res: Atmos 113:D01301

French NHF, de Groot WJ, Jenkins LK et al (2011) Model comparisons for estimating carbon emissions from North American wildland fire. J Geophys Res: Biogeosci 116:G00K05

Ferguson SA, Hardy CC (1994) Modeling smoldering emissions from prescribed broadcast burns in the Pacific-Northwest. Int J Wildland Fire 4:135–142

Gann RG, Averill JD, Johnsson EL et al (2010) Fire effluent component yields from room-scale fire tests. Fire Mater 34:285–314

Garofalo LA, Pothier MA, Levin EJT et al (2019) Emission and evolution of submicron organic aerosol in smoke from wildfires in the western United States. ACS Earth Space Chem 3:1237–1247

Gilardoni S (2017) Advances in organic aerosol characterization: from complex to simple. Aerosol Air Qual Res 17:1447–1451

Geron C, Hays M (2013) Air emissions from organic soil burning on the coastal plain of North Carolina. Atmos Environ 64:192–199

Gilman JB, Lerner BM, Kuster WC et al (2015) Biomass burning emissions and potential air quality impacts of volatile organic compounds and other trace gases from fuels common in the US. Atmos Chem Phys 15:13915–13938

Greenberg JP, Friedli H, Guenther AB et al (2006) Volatile organic emissions from the distillation and pyrolysis of vegetation. Atmos Chem Phys 6:81–91

Grieshop AP, Logue JM, Donahue NM et al (2009) Laboratory investigation of photochemical oxidation of organic aerosol from wood fires 1: measurement and simulation of organic aerosol evolution. Atmos Chem Phys 9:1263–1277

Gullett B, Touati A, Oudejans L (2008) PCDD/F and aromatic emissions from simulated forest and grassland fires. Atmos Environ 42:7997–8006

Hansen J, Nazarenko L (2004) Soot climate forcing via snow and ice albedos. Proc Natl Acad Sci 101:423–428

Hao WM, Babbit RE (2007) Smoke produced from residual combustion (Final report, JFSP-98-1-9-0). U.S. Forest Service, Rocky Mountain Research Station, Missoula. https://www.firescience.gov/projects/98-1-9-01/project/98-1-9-01_final_report.pdf. 10 Feb 2020

Hao WM, Baker S, Lincoln E et al (2018) Cesium emissions from laboratory fires. J Air Waste Manage Assoc 68:1211–1223

Hassett-Sipple B, Hagler G, Vanderpool R, Hanley T (2020) $PM_{2.5}$ temporal trends and instrument performance assessment over 2018–2019 in Sarajevo, BiH. In: 1st Conference on urban planning and regional development, Sarajevo, Bosnia, 30–31 Jan 2020

Hatch LE, Luo W, Pankow JF et al (2015) Identification and quantification of gaseous organic compounds emitted from biomass burning using two-dimensional gas chromatography-time-of-flight mass spectrometry. Atmos Chem Phys 15:1865–1899

Hatch LE, Rivas-Ubach A, Jen CN et al (2018) Measurements of I/SVOCs in biomass-burning smoke using solid-phase extraction disks and two-dimensional gas chromatography. Atmos Chem Phys 18:17801–17817

Hatch LE, Jen CN, Kreisberg NM et al (2019) Highly speciated measurements of terpenoids emitted from laboratory and mixed-conifer forest prescribed fires. Environ Sci Technol 53:9418–9428

Holder AL, Hagler GSW, Aurell J et al (2016) Particulate matter and black carbon optical properties and emission factors from prescribed fires in the southeastern United States. J Geophys Res: Atmos 121:3465–3483

Holder AL, Mebust AK, Maghran LA et al (2020) Field evaluation of low-cost particulate matter sensors for measuring wildfire smoke. Sensors 20:4796

Hornbrook RS, Blake DR, Diskin GS et al (2011) Observations of nonmethane organic compounds during ARCTAS—part 1: biomass burning emissions and plume enhancements. Atmos Chem Phys 11:11103–11130

Hosseini S, Li Q, Cocker D et al (2010) Particle size distributions from laboratory-scale biomass fires using fast response instruments. Atmos Chem Phys 10:8065–8076

Hosseini S, Urbanski SP, Dixit P et al (2013) Laboratory characterization of PM emissions from combustion of wildland biomass fuels. J Geophys Res: Atmos 118:9914–9929

Hudak AT, Dickinson MB, Bright BC et al (2016) Measurements relating fire radiative energy density and surface fuel consumption–RxCADRE 2011 and 2012. Int J Wildland Fire 25:25–37

Ichoku C, Ellison L (2014) Global top-down smoke-aerosol emissions estimation using satellite fire radiative power measurements. Atmos Chem Phys 14:6643–6667

Janhall S, Andreae MO, Poschl U (2010) Biomass burning aerosol emissions from vegetation fires: particle number and mass emission factors and size distributions. Atmos Chem Phys 10:1427–1439

Jen CN, Hatch LE, Selimovic V et al (2019) Speciated and total emission factors of particulate organics from burning western US wildland fuels and their dependence on combustion efficiency. Atmos Chem Phys 19:1013–1026

Jolleys MD, Coe H, Mcfiggans G et al (2014) Organic aerosol emission ratios from the laboratory combustion of biomass fuels. J Geophys Res: Atmos 119:850–12871

Jolly WM, Freeborn PH, Page WG et al (2019) Severe fire danger index: a forecastable metric to inform firefighter and community wildfire risk management. Fire 2:47

Kaiser JW, Heil A, Andreae MO et al (2012) Biomass burning emissions estimated with a global fire assimilation system based on observed fire radiative power. Biogeosciences 9:527–554

Koch D, Del Genio AD (2010) Black carbon semi-direct effects on cloud cover: review and synthesis. Atmos Chem Phys 10:7685–7696

Kopacz M, Jacob DJ, Fisher JA et al (2010) Global estimates of CO sources with high resolution by adjoint inversion of multiple satellite datasets (MOPITT, AIRS, SCIAMACHY, TES). Atmos Chem Phys 10:855–876

Koplitz SN, Nolte CG, Pouliot GA et al (2018) Influence of uncertainties in burned area estimates on modeled wildland fire $PM_{2.5}$ and ozone pollution in the contiguous US. Atmos Environ 191:328–339

Koss AR, Sekimoto K, Gilman JB et al (2018) Non-methane organic gas emissions from biomass burning: identification, quantification, and emission factors from PTR-ToF during the FIREX 2016 laboratory experiment. Atmos Chem Phys 18:3299–3319

Kristensen LJ, Taylor MP (2012) Fields and forests in flames: lead and mercury emissions from wildfire pyrogenic activity. Environ Health Perspect 120:a56–a57

Lack DA, Moosmuller H, McMeeking GR et al (2014) Characterizing elemental, equivalent black, and refractory black carbon aerosol particles: a review of techniques, their limitations and uncertainties. Anal Bioanal Chem 406:99–122

Landis MS, Long RW, Krug J et al (2021) The U.S. EPA wildland fire sensor challenge: performance and evaluation of solver submitted multi-pollutant sensor systems. Atmos Environ 15:118165

Larkin S (2018) BlueSky Playground v3 help. https://sites.google.com/firenet.gov/wfaqrp-airfire-info/playground/playground-v3-help?authuser=0. 28. 22 May 2020

Larkin NK, O'Neill SM, Solomon R et al (2009) The BlueSky smoke modeling framework. Int J Wildland Fire 18:906–920

Larkin NK, Raffuse SM, Strand TM (2014) Wildland fire emissions, carbon, and climate: US emissions inventories. For Ecol Manage 317:61–69

Larkin NK, Raffuse SM, Huang S et al (2020) The comprehensive fire information reconciled emissions (CFIRE) inventory: wildland fire emissions developed for the 2011 and 2014 U.S. National emissions inventory. J Air Waste Manage Assoc 70:1165–1185

Lee S, Baumann K, Schauer JJ et al (2005) Gaseous and particulate emissions from prescribed burning in Georgia. Environ Sci Technol 39:9049–9056

Leonard SS, Castranova V, Chen BT et al (2007) Particle size-dependent radical generation from wildland fire smoke. Toxicology 236:103–113

Li FJ, Zhang XY, Kondragunta S, Roy DP (2018) Investigation of the fire radiative energy biomass combustion coefficient: a comparison of polar and geostationary satellite retrievals over the conterminous United States. J Geophys Res Biogeosci 123:722–739

Li HY, Lamb KD, Schwarz JP et al (2019a) Inter-comparison of black carbon measurement methods for simulated open biomass burning emissions. Atmos Environ 206:156–169

Li FJ, Zhang XY, Roy DP et al (2019b) Estimation of biomass-burning emissions by fusing the fire radiative power retrievals from polar-orbiting and geostationary satellites across the conterminous United States. Atmos Environ 211:274–287

Linak WP, Wendt JOL (1993) Toxic metal emissions from incineration: mechanisms and control. Prog Energy Combust Sci 19:145–185

Lincoln E, Hao WM, Weise DR et al (2014) Wildland fire emission factors database. U.S. Forest Service research data archive, Fort Collins. https://doi.org/10.2737/RDS-2014-0012. 22 May 2020

Liu JC, Mickley LJ, Sulprizio MP et al (2016) Particulate air pollution from wildfires in the western US under climate change. Clim Change 138:655–666

Liu XX, Huey LG, Yokelson RJ et al (2017) Airborne measurements of western US wildfire emissions: comparison with prescribed burning and air quality implications. J Geophys Res: Atmos 122:6108–6129

Lönnermark A, Blomqvist P (2006) Emissions from an automobile fire. Chemosphere 62:1043–1056

Lohmann U, Feichter J (2005) Global indirect aerosol effects: a review. Atmos Chem Phys 5:715–737

Lu Q, Murphy BN, Qin M et al (2020) Simulation of organic aerosol formation during the CalNex study: updated mobile emissions and secondary organic aerosol parameterization for intermediate-volatility organic compounds. Atmos Chem Phys 20:4313–4332

Lu X, Zhang L, Yue X et al (2016) Wildfire influences on the variability and trend of summer surface ozone in the mountainous western United States. Atmos Chem Phys 16:14687–14702

Lutes DC (2019) FOFEM: first order fire effects model v6.5 user guide. http://firelab.org/project/fofem. 22 May 2020

May AA, Levin EJT, Hennigan CJ et al (2013) Gas-particle partitioning of primary organic aerosol emissions: 3. Biomass burning. J Geophys Res: Atmos 118:11327–11338

May AA, McMeeking GR, Lee T et al (2014) Aerosol emissions from prescribed fires in the United States: a synthesis of laboratory and aircraft measurements. J Geophys Res: Atmos 119:11826–11849

May AA, Lee T, McMeeking GR et al (2015) Observations and analysis of organic aerosol evolution in some prescribed fire smoke plumes. Atmos Chem Phys 15:6323–6335

McClure CD, Jaffe DA (2018) US particulate matter air quality improves except in wildfire-prone areas. Proc National Acad Sci USA 115:7901–7906

McMeeking GR, Kreidenweis SM, Baker S et al (2009) Emissions of trace gases and aerosols during the open combustion of biomass in the laboratory. J Geophys Res: Atmos 114:D19210

Meddens AJH, Kolden CA, Lutz JA (2016) Detecting unburned areas within wildfire perimeters using Landsat and ancillary data across the northwestern United States. Remote Sens Environ 186:275–285

Mehadi A, Moosmüller H, Campbell DE et al (2020) Laboratory and field evaluation of real-time and near real-time $PM_{2.5}$ smoke monitors. J Air Waste Manage Assoc 70:158–179

Mu M, Randerson JT, van der Werf GR et al (2011) Daily and 3-hourly variability in global fire emissions and consequences for atmospheric model predictions of carbon monoxide. J Geophys Res: Atmos 116:D24303

Narodoslawsky M, Obernberger I (1996) From waste to raw material—the route from biomass to wood ash for cadmium and other heavy metals. J Hazard Mater 50:157–168

O'Dell K, Hornbrook RS, Permar W et al (2020) Hazardous air pollutants in fresh and aged western US wildfire smoke and implications for long-term exposure. Environ Sci Technol 54:11838–11847

Odigie KO, Flegal AR (2014) Trace metal inventories and lead isotopic composition chronicle a forest fire's remobilization of industrial contaminants deposited in the Angeles national forest. PLoS ONE 9:e107835–e107835

O'Shea SJ, Allen G, Gallagher MW et al (2013) Airborne observations of trace gases over boreal Canada during BORTAS: campaign climatology, air mass analysis and enhancement ratios. Atmos Chem Phys 13:12451–12467

Ottmar RD (2001) Smoke source characteristics. In: Hardy CC, Ottmar RD, Peterson JL et al (eds) Smoke management guide for prescribed and wildland fire: 2001 edition. National Wildfire Coordination Group, Boise, pp 89–106

Ottmar RD (2014) Fuel and fire tools. https://www.fs.usda.gov/pnw/tools/fuel-and-fire-tools-fft. 28 Jan 2020

Ottmar RD (2018) Fuel consumption and smoke production. In: Peterson J, Lahm P, Fitch M et al (eds) NWCG smoke management guide for prescribed fire. National Wildfire Coordinating Group, Boise, pp 110–143

Persson B, Simonson M (1998) Fire emissions into the atmosphere. Fire Technol 34:266–279

Peterson D, Hyer E, Wang J (2013) A short-term predictor of satellite-observed fire activity in the North American boreal forest: toward improving the prediction of smoke emissions. Atmos Environ 71:304–310

Peterson J, Lahm P, Fitch M et al (eds) (2018) NWCG smoke management guide for prescribed fire. National Wildfire Coordination Group, Boise

Petrenko M, Kahn R, Chin M et al (2017) Refined use of satellite aerosol optical depth snapshots to constrain biomass burning emissions in the GOCART Model. J Geophys Res: Atmos 122:10983–11004

Petzold A, Ogren JA, Fiebig M et al (2013) Recommendations for reporting "black carbon" measurements. Atmos Chem Phys 13:8365–8379

Popovicheva OB, Engling G, Diapouli E et al (2016) Impact of smoke intensity on size-resolved aerosol composition and microstructure during the biomass burning season in northwest Vietnam. Aerosol Air Qual Res 16:2635–2654

Prichard SJ, Sandberg DV, Ottmar RD et al (2013) Fuel characteristic classification system version 3.0: technical documentation. General technical report PNW-GTR-887. U.S. Forest Service, Pacific Northwest Research Station, Portland

Prichard SJ, Karau EC, Ottmar RD et al (2014) Evaluation of the CONSUME and FOFEM fuel consumption models in pine and mixed hardwood forests of the eastern United States. Can J Res 44:784–795

Prichard SJ, O'Neill SM, Eagle P et al (2020) Wildland fire emission factors in North America: synthesis of existing data, measurement needs and management applications. Int J Wildland Fire 29:132–147

Reff A, Bhave PV, Simon H et al (2009) Emissions inventory of $PM_{2.5}$ trace elements across the United States. Environ Sci Technol 43:5790–5796

Reid JS, Eck TF, Christopher SA et al (2005a) A review of biomass burning emissions part III: intensive optical properties of biomass burning particles. Atmos Chem Phys 5:827–849

Reid JS, Koppmann R, Eck TF et al (2005b) A review of biomass burning emissions part II: intensive physical properties of biomass burning particles. Atmos Chem Phys 5:799–825

Reisen F (2011) Inventory of major materials present in and around houses and their combustion emission products. Bushfire CRC, Melbourne. https://www.bushfirecrc.com/sites/default/files/managed/resource/inventory.pdf. 22 May 2020

Reisen F, Bhujel M, Leonard J (2014) Particle and volatile organic emissions from the combustion of a range of building and furnishing materials using a cone calorimeter. Fire Saf J 69:76–88

Reisen F, Meyer CP, Weston CJ et al (2018) Ground-based field measurements of $PM_{2.5}$ emission factors from flaming and smoldering combustion in eucalypt forests. J Geophys Res: Atmos 123:8301–8314

Robinson AL, Grieshop AP, Donahue NM et al (2010) Updating the conceptual model for fine particle mass emissions from combustion systems. J Air Waste Manage Assoc 60:1204–1222

Santoso MA, Christensen EG, Yang J et al (2019) Review of the transition from smouldering to flaming combustion in wildfires. Front Mech Eng 5(49):1–20

Schroeder W, Oliva P, Giglio L et al (2014) The new VIIRS 375 m active fire detection data product: algorithm description and initial assessment. Remote Sens Environ 143:85–96

Schweizer D, Cisneros R, Shaw G (2016) A comparative analysis of temporary and permanent beta attenuation monitors: the importance of understanding data and equipment limitations when creating $PM_{2.5}$ air quality health advisories. Atmos Pollut Res 7:865–875

Sekimoto K, Koss AR, Gilman JB et al (2018) High- and low-temperature pyrolysis profiles describe volatile organic compound emissions from western US wildfire fuels. Atmos Chem Phys 18:9263–9281

Selimovic V, Yokelson RJ, Warneke C et al (2018) Aerosol optical properties and trace gas emissions by PAX and OP-FTIR for laboratory-simulated western US wildfires during FIREX. Atmos Chem Phys 18:2929–2948

Simpson IJ, Akagi SK, Barletta B et al (2011) Boreal forest fire emissions in fresh Canadian smoke plumes: C-1-C-10 volatile organic compounds (VOCs), CO_2, CO, NO_2, NO, HCN and CH_3CN. Atmos Chem Phys 11:6445–6463

Short KC, Finney MA, Vogler KC et al (2020) Spatial datasets of probabilistic wildfire risk components for the United States (270 m), 2nd edn. U.S. Forest Service Research Data Archive, Fort Collins. https://doi.org/10.2737/RDS-2016-0034-2. 22 May 2020

Stec AA (2017) Fire toxicity—the elephant in the room? Fire Saf J 91:79–90

Szpakowski DM, Jensen JLR (2019) A review of the applications of remote sensing in fire ecology. Remote Sens 11:2638

Tangahu BV, Sheikh Abdullah SR, Basri H et al (2011) A review on heavy metals (As, Pb, and Hg) uptake by plants through phytoremediation. Int J Chem Eng 2011:939161

Turn SQ, Jenkins BM, Chow JC et al (1997) Elemental characterization of particulate matter emitted from biomass burning: wind tunnel derived source profiles for herbaceous and wood fuels. J Geophys Res: Atmos 102:3683–3699

Urbanski SP (2013) Combustion efficiency and emission factors for wildfire-season fires in mixed conifer forests of the northern rocky mountains, US. Atmos Chem Phys 13:7241–7262

Urbanski S (2014) Wildland fire emissions, carbon, and climate: emission factors. For Ecol Manage 317:51–60

Urbanski SP, Hao WM, Nordgren B (2011) The wildland fire emission inventory: western United States emission estimates and an evaluation of uncertainty. Atmos Chem Phys 11:12973–13000

Urbanski SP, Reeves MC, Corley RE et al (2018) Contiguous United States wildland fire emission estimates during 2003–2015. Earth Syst Sci Data 10:2241–2274

U.S. Environmental Protection Agency (USEPA) (1996) Miscellaneous sources. In: AP 42, 5th edn, vol I, chap 13. Washington, DC. https://www3.epa.gov/ttn/chief/ap42/ch13/index.html. 22 May 2020

U.S. Environmental Protection Agency (USEPA) (2014) 2014 National emissions inventory (NEI) data, tier summaries. Washington, DC. https://www.epa.gov/air-emissions-inventories/2014-national-emissions-inventory-nei-data. 10 April 2020

U.S. Environmental Protection Agency (USEPA) (2020) Integrated Science Assessment (ISA) for particulate matter. (EPA/600/R-19/188). U.S. Environmental Protection Agency, Washington, DC. https://www.epa.gov/isa/integrated-science-assessment-isa-particulate-matter. 03 Feb 2021

van der Werf GR, Randerson JT, Giglio L et al (2017) Global fire emissions estimates during 1997–2016. Earth Syst Sci Data 9:697–720

van Leeuwen TT, van der Werf GR, Hoffmann AA et al (2014) Biomass burning fuel consumption rates: a field measurement database. Biogeosciences 11:7305–7329

Ward DE, Hardy CC (1989) Organic and elemental profiles for smoke from prescribed fires. In: Watson JG (ed) Receptor models in air resources management: Transactions of an international specialty conference of the Air & Waste Management Association. Air and Waste Management Association, Pittsburgh, pp 299–321

Wade DD, Lunsford JD (1989) A guide for prescribed fire in southern forests. Technical Publication R8–TP–11. U.S. Forest Service, Southern Region, Atlanta

Wooster MJ, Zhukov B, Oertel D (2003) Fire radiative energy for quantitative study of biomass burning: derivation from the BIRD experimental satellite and comparison to MODIS fire products. Remote Sens Environ 86:83–107

Wooster MJ, Roberts G, Perry GLW et al (2005) Retrieval of biomass combustion rates and totals from fire radiative power observations: FRP derivation and calibration relationships between biomass consumption and fire radiative energy release. J Geophys Res: Atmos 110:D24311

Wright CS, Balog CS, Kelly JW (2010a) Estimating volume, biomass, and potential emissions of hand-piled fuels. General Technical Report PNW-GTR-805. U.S. Forest Service, Pacific Northwest Research Station, Portland

Wright CS, Eagle PC, Olson DL (2010b) A high-quality fuels database of photos and information. Fire Manage Today 70:27–31

Wu L, Taylor MP, Handley HK (2017) Remobilisation of industrial lead depositions in ash during Australian wildfires. Sci Total Environ 599–600:1233–1240

Yokelson RJ, Griffith DWT, Ward DE (1996) Open-path Fourier transform infrared studies of large-scale laboratory biomass fires. J Geophys Res: Atmos 101:21067–21080

Yokelson RJ, Goode JG, Ward DE et al (1999) Emissions of formaldehyde, acetic acid, methanol, and other trace gases from biomass fires in North Carolina measured by airborne Fourier transform infrared spectroscopy. J Geophys Res: Atmos 104:30109–30125

Yokelson RJ, Burling IR, Gilman JB et al (2013) Coupling field and laboratory measurements to estimate the emission factors of identified and unidentified trace gases for prescribed fires. Atmos Chem Phys 13:89–116

Yue X, Mickley LJ, Logan JA et al (2013) Ensemble projections of wildfire activity and carbonaceous aerosol concentrations over the western United States in the mid-21st century. Atmos Environ 77:767–780

Zhang TR, Wooster MJ, Xu WD (2017) Approaches for synergistically exploiting VIIRS I- and M-Band data in regional active fire detection and FRP assessment: a demonstration with respect to agricultural residue burning in Eastern China. Remote Sens Environ 198:407–424

Zhang XY, Kondragunta S, Ram J et al (2012) Near-real-time global biomass burning emissions product from geostationary satellite constellation. J Geophys Res: Atmos 117:D14201

Open Access This chapter is licensed under the terms of the Creative Commons Attribution 4.0 International License (http://creativecommons.org/licenses/by/4.0/), which permits use, sharing, adaptation, distribution and reproduction in any medium or format, as long as you give appropriate credit to the original author(s) and the source, provide a link to the Creative Commons license and indicate if changes were made.

The images or other third party material in this chapter are included in the chapter's Creative Commons license, unless indicated otherwise in a credit line to the material. If material is not included in the chapter's Creative Commons license and your intended use is not permitted by statutory regulation or exceeds the permitted use, you will need to obtain permission directly from the copyright holder.

Chapter 6
Smoke Chemistry

Matthew J. Alvarado, Kelley C. Barsanti, Serena H. Chung, Daniel A. Jaffe, and Charles T. Moore

Abstract Smoke chemistry (i.e., chemical transformations taking place within smoke plumes) can alter the composition and toxicity of smoke on time scales from minutes to days. Air quality agencies need better information on and better models of smoke chemistry to more accurately characterize the contributions of smoke to ambient ozone and particulate matter, and to better predict good windows for prescribed burning. The ability of these agencies to quantify the contributions of wildland fires to air pollutants and the ability of forest and burn managers to both predict and mitigate these impacts are limited by how current models represent smoke chemistry. This limitation is interconnected with uncertainties in smoke emissions, plume dynamics, and long-range transport. Improving predictive models will require a combination of laboratory, field, and modeling studies focused on enhancing our knowledge of smoke chemistry, including when smoke interacts with anthropogenic emissions and enters indoors.

Keywords Chemical transport models · Ozone · Secondary organic aerosol · Statistical models

---

M. J. Alvarado (✉)
Atmospheric and Environmental Research, Lexington, MA, USA
e-mail: malvarad@aer.com

K. C. Barsanti
Department of Chemical and Environmental Engineering, University of California, Riverside, CA, USA
e-mail: kbarsanti@engr.ucr.edu

S. H. Chung
U.S. EPA Office of Research and Development, Washington, DC, USA
e-mail: chung.serena@epa.gov

D. A. Jaffe
Physical Sciences Division, University of Washington, Bothell, WA, USA
e-mail: djaffe@uw.edu

C. T. Moore
Western States Air Resources Council (WESTAR), Fort Collins, CO, USA
e-mail: tmoore@westar.org

This is a U.S. government work and not under copyright protection in the U.S.; foreign copyright protection may apply 2022
D. L. Peterson et al. (eds.), *Wildland Fire Smoke in the United States*, https://doi.org/10.1007/978-3-030-87045-4_6

6.1 Introduction

Better understanding and improved model representation of the coupled chemical and physical transformations occurring in wildland fire smoke plumes will be critical to improving predictions of the contribution of wildland fires to ozone (O_3), particulate matter <2.5 μm ($PM_{2.5}$), and other air pollutants at local and regional scales. This includes understanding this chemistry as a function of fuel type (Chap. 2) and burn characteristics (Chap. 3). Air quality (AQ) agencies need better information and better models to achieve and maintain compliance with National Ambient Air Quality Standards (NAAQS) (including exceptional events demonstrations; USEPA 2016) and to maintain progress toward the visibility goals set for Class I areas in the Regional Haze Rule. The ability of these AQ agencies to quantify the contributions of wildland fires to air pollutants and the ability of forest and burn managers to predict these impacts are limited by the ability of current models to capture the chemical complexities, which are interconnected to uncertainties in smoke emissions (Chap. 5), plume dynamics, and long-range transport (Chap. 4). Improving predictive models will require a combination of laboratory, field, and modeling studies focused on improving our knowledge of the chemical and physical transformations taking place within smoke plumes, including when smoke interacts with anthropogenic emissions.

6.1.1 Overview and Context of the Issues

The chemically diverse mixture of pollutants in smoke emitted from wildland fires is not inert. Complex and interconnected chemical and physical transformations take place in smoke plumes at time scales from minutes to days (e.g., Hobbs et al. 2003; Baylon et al. 2015), altering the composition of smoke and its effects on human health and climate. As one example, nitrogen oxides ($NO_x = NO + NO_2$) and volatile organic compounds (VOCs) emitted by fires can lead to in-plume production of O_3, a U.S. Environmental Protection Agency (USEPA) criteria pollutant, on a time scale of hours (e.g., Jaffe and Wigder 2012). Smoke can also contribute to O_3 formation at much longer temporal and spatial scales, either through (1) near-source formation of peroxy nitrates (PNs) like peroxyacetyl nitrate (PAN), which can serve as a NO_x reservoir and ultimately a NO_x source (e.g., Alvarado et al. 2010) (Fig. 6.1), or (2) transport of smoke-derived VOCs to NO_x-rich urban areas (e.g., Brey and Fischer 2016). Although these general pathways of O_3 formation from wildland fire smoke are well known, the amount of O_3 formed in each fire event is highly variable (Jaffe and Wigder 2012), and the causes of this variability are not well understood or well represented in current AQ models (Baker et al. 2016).

$PM_{2.5}$ emitted by fires also undergoes chemical and physical transformations, such that the organic component of $PM_{2.5}$ (~80–90% of total $PM_{2.5}$ mass; Akagi et al. 2011) becomes more oxidized during plume transport (e.g., Garofalo et al. 2019). Observed increases in the oxidation state of organic $PM_{2.5}$ can be driven

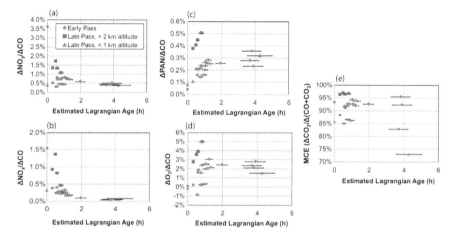

Fig. 6.1 Enhancement ratios of **a** NO_y, **b** NO_x, **c** PAN, and **d** O_3 and **e** modified combustion efficiency (MCE) versus estimated Lagrangian age for a smoke plume from the Lake McKay wildfire (Saskatchewan, Canada; 56.5° N, 106.8° W) sampled during a 1 July 2008 flight of the NASA DC8 during the Arctic Research of the Composition of the Troposphere from Aircraft and Satellites (ARCTAS-B) field campaign. Blue diamonds are for samples taken during the early pass (13:00–15:00 LT), red squares and orange triangles are for samples taken during the late pass (18:20–19:40 LT) above 2 km and below 1 km in altitude, respectively [From Alvarado et al. (2010)]

by physics or chemistry, which also can increase or decrease the total $PM_{2.5}$ mass relative to what was directly emitted. Directly emitted $PM_{2.5}$ can be rapidly diluted as a smoke plume disperses, leading to evaporation of the more volatile organic constituents; this decreases the total $PM_{2.5}$ mass and can lead to an increase in the oxidation state of the remaining organic fraction (May et al. 2013).

At the same time, gas-phase organic compounds (including those partitioned from the particle to the gas phase upon dilution) can be oxidized in plumes. These oxidation reactions can increase functionalization and the tendency of organic compounds to condense (i.e., form secondary organic aerosol [SOA]), increase total $PM_{2.5}$ mass, and increase the oxidation state of the organic $PM_{2.5}$. The extent of SOA formation can differ greatly between laboratory fires, even among similar fuel types (Tkacik et al. 2017), and the relative importance of evaporation versus condensation can vary greatly between smoke plumes (Hodshire et al. 2019a).

The chemical and physical transformations of gases and particles in wildland fire smoke may also change the overall toxicity of smoke, which may also affect human health impacts. For example, air toxics within smoke, such as aldehydes (e.g., formaldehyde and acrolein) and isocyanic acid (HNCO), can be formed or destroyed by in-plume chemistry. In addition, recent studies with mice have indicated that the mutagenicity and toxicity of fresh smoke are a function of the type of fuel burned and burn conditions (Kim et al. 2018, 2019), suggesting that the health impact of smoke differs with the chemical composition of the smoke.

6.1.2 Need for Decision Support

In order to provide better information to the public about ambient AQ and reduce human health impacts of wildland fire, we need better information on the spatial and temporal distributions of the primary (directly emitted) and secondary (formed by chemistry after source emission) pollutants from smoke. Such information can be used by health studies to further elucidate which chemical compounds and/or chemical properties of smoke cause adverse health outcomes and improve the development of relevant multi-pollutant air quality indices (AQIs) for smoke.

State and local agencies need better forecasting of pollutant concentrations from smoke to prepare the appropriate public alerts. Fire and natural resource managers also need better forecast models to inform prescribed burning decisions. Current smoke forecast models have uncertainties related to smoke emissions and transport. In addition, these forecast models either ignore smoke chemistry entirely (Stein et al. 2009), approximate smoke chemistry with simplified (lumped) mechanisms developed for anthropogenic air pollution (Baker et al. 2016), and/or have difficulty handling the changes in the spatial scale of the chemistry as smoke disperses and is transported over long distances (Alvarado et al. 2010). These current approaches are insufficient to accurately estimate the effects of wildfires and prescribed burns on O_3, $PM_{2.5}$, and other air toxics, whose formation and loss rates can differ significantly between smoke plumes. Improved models or statistical approaches are needed to determine whether a specific fire led to non-compliance with an O_3 or $PM_{2.5}$ standard (NAAQS) and to accurately model $PM_{2.5}$ composition and optical properties to determine fire contributions to regional haze.

In addition, we need information on how different prescribed burning methods affect the subsequent chemistry of the smoke. Methods used to start prescribed burns can have significant effects on the plume dynamics, transport, and emissions. These changes will also affect smoke chemistry, but there have been no studies about how ignition methods affect subsequent chemistry in the smoke plume.

6.1.3 Scientific Challenges

6.1.3.1 Ozone Formation in Isolated Plumes

O_3 is a secondary pollutant that is formed from the oxidation of VOCs in the presence of NO_x and UV light. Because fires emit NO_x and VOCs in variable amounts, O_3 may be formed in a smoke plume at varying concentrations depending on the emissions, temperature, UV light, and many complex interactions within the plume. Under warmer conditions, O_3 can form fairly rapidly (hours; Akagi et al. 2013) (Fig. 6.2), whereas in cooler environments, O_3 production takes longer and may not be apparent (Alvarado et al. 2010). An important control on O_3 production is the amount of NO_x emitted and then subsequently removed by chemistry (Mauzerall et al. 1998). NO_x is

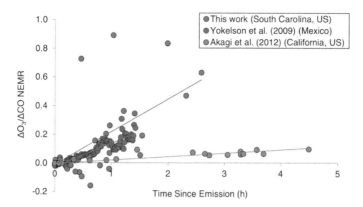

Fig. 6.2 Ozone enhancement ratios ($\Delta O_3/\Delta CO$) versus time since emission from Akagi et al. (2013) (red), Yokelson et al. (2009) (blue), and Akagi et al. (2012) (green) [From Akagi et al. (2013)]

rapidly sequestered as peroxyacetyl nitrate (PAN) in boreal smoke plumes (Alvarado et al. 2010) (Fig. 6.1). A similar result was found for smoke plumes at the Mt. Bachelor Observatory in central Oregon at 2.8 km above sea level (Baylon et al. 2015).

In a review of more than 100 different studies, Jaffe and Wigder (2012) found that O_3 is commonly enhanced downwind from fire plumes, and the production increases with plume age. Tropical and subtropical fires generally produce more O_3 and at a faster rate than temperate and boreal fires, because tropical/subtropical fires emit more NO_x per unit of fuel, and the higher temperatures discourage PAN formation (Jaffe and Wigder 2012). Nonetheless, PAN is only a temporary reservoir; subsequent thermal decomposition will regenerate the original NO_x back and distribute O_3 production further downwind (Val Martin et al. 2006). Rapid O_3 production is likely driven by several sources of oxidants, including OH from HONO (nitrous acid) photolysis. HONO can be either emitted directly (Burling et al. 2011) or produced from heterogeneous reactions (Ye et al. 2017).

6.1.3.2 Ozone Formation When Smoke Mixes in Urban Areas

When a smoke plume enters an urban area, it will mix with all the existing pollutants and change the local photochemical environment. Thus, the presence of smoke can increase urban O_3 either by increasing O_3 production upwind of the city center or by increasing O_3 production in the urban environment. Optimum O_3 production occurs at a VOC/NO_x molar ratio of around 8. Ratios for most urban areas are near this or lower. Fire emissions typically have high VOC/NO_x molar ratios, e.g., ~10–30 (Akagi et al. 2011), so when smoke mixes into an urban area it can facilitate even more O_3 production. Buysse et al. (2019) show that enhanced O_3 in urban areas due to wildland fires is most pronounced at $PM_{2.5}$ concentrations below 60 μg m^{-3}. At higher $PM_{2.5}$ concentrations, O_3 levels appear to be suppressed due to reduced photolysis rates, insufficient reaction times, or heterogeneous chemistry on smoke particles. Photolysis can be complex; there can be multiple scattering influences and photolysis rates will depend on the location within the plume (Alvarado et al. 2015). At moderate smoke levels and with high scattering amounts, photolysis may or may not be significantly reduced inside a smoke plume (Baylon et al. 2018).

Many studies have examined O_3 production in smoke plumes by comparing concentrations in the plume to concentrations outside a plume, defined as the "enhancement." Enhancement of a chemically active compound (e.g., O_3) can be ratioed to a relatively inert compound (e.g., CO_2 or CO) to give an enhancement ratio that shows the chemical production or loss of the species after accounting for plume dilution. Lindaas et al. (2017) documented enhancements in O_3 associated with transported smoke plumes of around 15 ppb in Colorado.

Significant impacts on surface O_3 via intercontinental transport of wildfire emissions can also occur, such as from Siberian smoke reaching the western USA (Teakles et al. 2017; Jaffe et al. 2004) or Alaskan smoke reaching the North Atlantic (Real et al. 2007). Canadian wildfires have been shown to enhance O_3 in the southeastern USA (McKeen et al. 2002), Maryland (Dreessen et al. 2016), and New England (DeBell et al. 2004). Smoke from wildland fires raised the maximum daily 8-h average (MDA8) O_3 levels by 3–6 ppb on average, with a maximum enhancement of up to 40 ppb for six cities in the western USA (Gong et al. 2017). During an especially smoky summer in Boise, Idaho, smoke increased the O_3 MDA8 by an average of ~15 ppb and significantly increased the number of days over the 70 ppb MDA8 AQ threshold (McClure and Jaffe 2018).

The details of nighttime chemistry are poorly understood and are therefore not well represented in models. Although O_3 production is driven by UV photolysis in the daytime, chemical processing can still occur at night, though few studies on this nocturnal chemistry of smoke have been done. From other (non-smoke) studies, we know that NO_2 and O_3 will react to form the NO_3 radical, which can oxidize many organic species and further react to form N_2O_5. Several nighttime reactions, mostly with the NO_3 radical, can significantly modify the reactivity of smoke, aerosols, and O_3 (Finewax et al. 2018; Decker et al. 2019). Nighttime chemical processing in smoke generates both N_2O_5 and $ClNO_2$, both of which regenerate NO_2 through photolysis (Ahern et al. 2018). $ClNO_2$ can also generate reactive Cl radicals that can increase oxidation of VOCs.

6.1.3.3 Secondary Organic Aerosol Formation and Organic Aerosol Dilution/Oxidation

Organic aerosol (OA) composes a significant fraction of the total $PM_{2.5}$ measured in smoke plumes and in smoke-impacted regions. OA includes compounds that are directly emitted, as well as compounds that are formed in-plume during chemical transformations of the smoke. The OA formed by chemical transformations of gas-phase compounds is known as secondary organic aerosol (SOA). Although the fundamental understanding of SOA formation is well developed, there are uncertainties about the gas-phase and particle-phase chemistry of SOA precursors emitted from wildland fires. This limits accurate representation of SOA formation from wildland fires in models and predictions of the contribution of wildland fires to SOA and $PM_{2.5}$.

Field measurements and laboratory experiments have been conducted to understand the aging of PM mass and composition in biomass-burning plumes, as well as to provide bottom-up estimates of biomass-burning PM (Hodshire et al. 2019b). Markers of OA oxidation (e.g., O:C ratios) increase with smoke aging, indicating that chemical reactions are taking place within the smoke plumes and/or there is preferential loss of more volatile, lower O:C containing compounds. However, this aging does not always lead to a net increase in $PM_{2.5}$ mass, and there is significant variability in the observed change of $PM_{2.5}$ mass among studies. In the field, increases in $PM_{2.5}$ mass have occurred at shorter transport ages (<5 h) and have been relatively small (mean increases <10%). In the laboratory, larger increases in $PM_{2.5}$ mass (mean increases of ~70%) have been observed, but these estimates are sensitive to methods used to correct for particle loss on smog chamber walls. Variability within and between field and laboratory studies currently inhibits a unified framework for predicting the mass and composition of wildland fire-derived $PM_{2.5}$ as a function of aging.

6.1.3.4 Modeling Challenges

The chemical mechanisms used in most three-dimensional (3D) photochemical transport models were derived to simulate atmospheric chemistry in polluted urban regions, and as a result the representation of organic species within these models is focused on species common in emissions from cars, trucks, power plants, and other anthropogenic sources. This means that the models are potentially missing or misrepresenting the chemistry of the NMOCs from wildland fires, which can lead to errors in the predicted chemical formation of O_3 and $PM_{2.5}$ from wildland fire smoke. For example, furan compounds (containing a 4-carbon, 1-oxygen aromatic ring) are a significant part of the NMOC emissions from biomass burning, accounting for ~25% of the OH reactivity of fresh smoke (Koss et al. 2018). However, few atmospheric chemistry models account for the distinctive chemistry of these aromatic compounds, either ignoring them or lumping them with other aromatic or alkane species. For many of the recently identified compounds that are potential SOA precursors, few data exist on reaction mechanisms and SOA formation potentials (Hatch et al. 2017).

The chemistry of smoke plumes presents scale challenges as they can start out with a small horizontal extent (~1 km or less) but can rapidly disperse to the continental scale. Most 3D AQ models have a minimum horizontal resolution of 4–12 km for regional models and hundreds of km for global models. They are not able to correctly represent the near-source chemistry of smoke plumes, as they automatically dilute the plumes into large-scale grid boxes, which when combined with the nonlinear chemical processes, can lead to significant errors in model predictions (Sakamoto et al. 2016). For example, the 3D Eulerian model CMAQ overestimates the effect of biomass burning on individual hourly O_3 measurements at CASTNET monitoring sites near fires by up to 40 ppb, underestimating the impact downwind by up to 20 ppb (Baker 2015), possibly due to mistreatment of photolysis rates and peroxy acyl nitrate (PAN) formation.

Plume-scale process models allow examination of chemical and physical transformations of trace gases and aerosols within smoke plumes (Chap. 4) and are used to develop parameterizations for the aging process in coarser grid-scale models (Alvarado et al. 2015). For example, McDonald-Buller et al. (2015) used a subset of the parameterization of Lonsdale et al. (2017) to adjust the chemistry of biomass burning in 3D Eulerian model CAMx, concluding that this approach reduced the median impact of biomass burning on MDA8 O_3 by 0.3 ppb, or 15%.

6.2 Current State of the Science

6.2.1 Well-Understood Aspects of Smoke Chemistry

O_3 production from wildfires is a complex process involving numerous variables including fire emissions, chemical and photochemical reactions, aerosol effects on chemistry and radiation, and local and downwind meteorological patterns (Jaffe and Wigder 2012). Although O_3 production requires both NO_x and NMOCs, wildfire smoke tends to be NO_x-limited. Nitrogen emissions from wildfires are a function of fuel nitrogen content (0.2–4.0%) and the efficiency of combustion, which correlates with NO_x emissions. Modified combustion efficiency (MCE), a unitless ratio, can be used to quantify combustion efficiency; typical MCE values for individual fires range between 0.80 and 1.00. Chemical and photochemical reactions also significantly affect O_3 production in wildfire plumes. Thermal decomposition of downwind PAN, up to two weeks after being emitted, has been linked to O_3 production. The colder temperatures of high latitudes favor sequestration of NO_x in the form of PAN, although tropical wildfire plumes also show evidence of this.

Wildfires generate substantial emissions of O_3 precursors (Akagi et al. 2011), and most observations suggest some degree of O_3 production from wildfires (e.g., Jaffe and Wigder 2012). However, a smaller number of studies, mainly in boreal regions, have found that O_3 is minimally enhanced or even depleted downwind of some smoke plumes (Alvarado et al. 2010). The low O_3 production in these plumes was likely caused by low mixing ratios of NO_x due to low emissions, sequestration of NO_x as PAN, and/or a reduction in photochemical reactions (especially due to aerosol effects on radiation).

Most studies show that O_3 is produced in wildfire smoke, and that typical O_3-to-CO enhancement ratios ($\Delta O_3/\Delta CO$) range from nearly 0 to 0.7 (Jaffe and Wigder 2012). The $\Delta O_3/\Delta CO$ ratio increases with plume age and is higher in tropical regions, which is consistent with higher NO_x emissions per unit of fuel consumed, likely due to higher fuel N content and combustion efficiency. Species that produce HO_x radicals by photolysis (e.g., HONO and formaldehyde) can accelerate O_3 production in smoke plumes.

Mixing of wildfire plumes with air masses containing natural and anthropogenic O_3 precursors has been shown to enhance O_3 production. Smoke-influenced O_3 mixing ratios are highest in locations with the highest emissions of NO_x (Brey and Fischer 2016) (Fig. 6.3). The northeastern US corridor, Dallas, Houston, Atlanta,

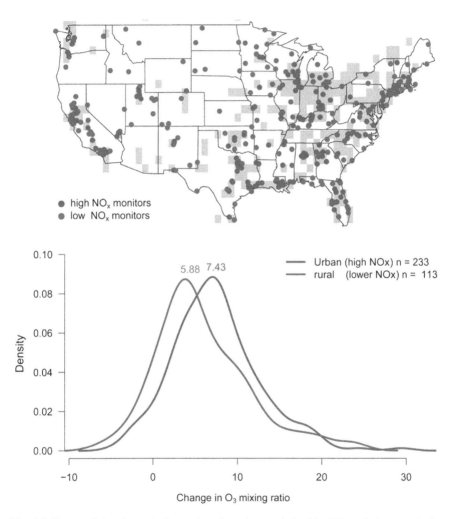

Fig. 6.3 Top panel: locations of urban and rural monitors as defined by NO_x emission rates in the 2008 National Emissions Inventory. Bottom panel: probability distribution of the difference between mean smoke-impacted MDA8 value and smoke-free MDA8 value for urban and rural monitors. Smoke-free days are required to have temperatures that exceed the mean -0.5 standard deviation of the smoke-impacted days temperature values to reduce the impact of varying temperatures on the comparison [From Brey and Fischer (2016)]

Birmingham, and Kansas City stand out as having smoke present 10–20% of the days when 8-h average O_3 mixing ratios exceed 75 ppbv.

The mass, composition, and properties of biomass-burning vapors and particles evolve as smoke ages due to complex chemical and physical processes (Hodshire et al. 2019a). The vapors and particles are composed of thousands of chemical compounds with a range in volatility (vapor pressures), reactivity, and other properties (Shiraiwa et al. 2014). Volatility governs the partitioning of compounds between the gas and particle phases (Pankow 1994) and is often reported as the effective saturation concentration in $\mu g\ m^{-3}$ (Donahue et al. 2006). Organic compounds are often grouped into five volatility categories (Murphy et al. 2014):

- Volatile organic compounds (VOCs; $C^* \geq \sim 10^7$),
- Intermediate-volatility organic compounds (IVOCs, $C^* \sim 10^3-10^6$),
- Semivolatile organic compounds (SVOCs, $C^* \sim 10^0-10^2$),
- Low-volatility organic compounds (LVOCs, $C^* \sim 10^{-3}-10^{-1}$),
- Extremely low-volatility organic compounds (ELVOCs, $C \leq \sim 10^{-4}$).

The higher the OA concentrations, the larger the portion of SVOCs that is partitioned in the particle phase. IVOCs are in the gas phase under dilute OA conditions, but some will partition in the particle phase under very high OA concentrations. In general, higher molecular weight and more oxygenated molecules have lower volatility.

During chemical aging within plumes, the volatility of compounds may both decrease—primarily through functionalization or oligomerization reactions—and increase, primarily through fragmentation (Kroll et al. 2009). Gas-phase compounds that decrease in volatility may partition to the particle phase through condensation, forming SOA and thereby adding to aerosol mass. Conversely, semivolatile condensed compounds may evaporate upon dilution of the smoke plume, thereby decreasing aerosol mass. These evaporated compounds may act as SOA precursors that can undergo volatility-lowering reactions and then condense with continued aging of the plume. The time scales for evaporation may differ based on the particle-phase state, and evaporation may also be modified by particle-phase and surface reactions (Zaveri et al. 2014). The net amount of OA or PM that exists as a function of time represents the balance of primary PM mass lost due to evaporation and secondary PM formation.

6.2.2 Existing Data, Tools, Models, and Other Technology

6.2.2.1 Improved Emission Speciation Data

The most comprehensive speciation data have come from laboratory studies. Until recently, IVOCs emitted from biomass burning were poorly characterized (and nearly entirely unaccounted for in models). Hatch et al. (2017) combined FLAME-IV data collected using multiple measurement techniques into the most comprehensive biomass-burning gas-phase organic emissions inventory to date. Over 500 compounds were identified using multiple measurement techniques, and ~90% of the gaseous NMOC mass was accounted for; 6–11% of the gaseous NMOC mass was associated with IVOCs. These compounds are largely missing in existing emissions inventories, with only a few in the USEPA SPECIATE database.

Sekimoto et al. (2018) showed that only two emission profiles (each being a mass spectral representation of the relative abundance of emitted VOCs) explained 85% of the VOC emissions across various fuel mixtures representative of the western USA. The two profiles are related to low- and high-temperature pyrolysis and are quantitatively similar between different fuel types. The two profiles can represent previously reported VOC data for laboratory and field burns ($r \geq 0.92$), suggesting that these two profiles could be used to determine emissions of VOCs not measured in other studies.

6.2.2.2 Ozone Data

Current data on wildland fire effects on O_3 come from: laboratory studies where smoke from small fires is oxidized within smog chambers (e.g., Bian et al. 2017), surface monitoring (e.g., Brey and Fischer 2016), and field campaigns (e.g., Hobbs et al. 2003). These studies show that O_3 production in smoke is generally NO_x-limited, and thus boreal forest smoke on average forms less O_3 than savannah and grassland smoke on average (Jaffe and Wigder 2012). These studies also show high variability in O_3 production between smoke plumes, for which the causes are not fully understood.

6.2.2.3 SOA Data

There are no direct long-term measurements of SOA. Long-term measurement networks such as IMPROVE and CSN measure the mass of total organic carbon in $PM_{2.5}$. These total carbon measurements do not provide enough chemical composition information to distinguish between primary and secondary components. Although some studies have approximated secondary versus primary components of organics based on emission ratios of particulate OA to black (or elemental) carbon (Yu et al. 2017), this methodology does not apply if there is evaporative loss of organic compounds from the aerosol after emissions (Sect. 6.2.1).

Data on wildland fire effects on secondary formation of OA are mostly from laboratory (Cubison et al. 2011) or intensive field studies (Garofalo et al. 2019). Quantifying the contribution of chemistry to OA is challenging due to complex chemistry and gas–particle partitioning processes, resulting in variability among studies in contribution of SOA to PM mass (Hodshire et al. 2019b).

Enhancement in total OA relative to CO (i.e., increase in normalized excess mixing ratio $\Delta OA/\Delta CO$) provides evidence of SOA formation. However, the SOA amount cannot be quantified simply from $\Delta OA/\Delta CO$, because ΔOA includes changes in OA concentrations due to evaporation of more semivolatile constituents of primary OA as well as SOA formation. Field studies show OA becoming more oxidized with plume age, suggesting formation of SOA; but again quantification of SOA is potentially complicated by evaporative loss of semivolatile organic compounds with low O:C ratio. Zhou et al. (2017) applied positive matrix factorization (PMF) to HR-AMS mass spectra data measured at Mt. Bachelor Observatory (Oregon) in 2013, identifying three biomass-burning aerosol factors, two of which represent SOA.

6.2.2.4 Smoke Plume Models

Accurate modeling of O_3 and SOA production from wildland fire emissions is needed to understand chemical processing and impacts on human health (Brown et al. 2014; Chap. 7). Here we describe three different approaches that have been used to model O_3 and SOA production: Eulerian gridded chemical transport models (CTMs), Lagrangian plume or box models, and statistical methods.

Chemical Transport Models

Chemical transport models (CTMs) characterize the chemical environment in three dimensions plus time. Modeling O_3 and SOA production in a CTM depends on accurately knowing the flux, timing, and location of the primary emissions (PM, NO_x, CO, VOCs, etc.). Modeling the resulting concentrations requires spatial and temporal knowledge of injection heights, 3D wind fields, and other meteorological parameters (temperature, relative humidity, etc.) (Koplitz et al. 2018). Modeling O_3 and SOA also requires a detailed reaction mechanism and accurate UV radiation fields.

Grid resolution is a key component in CTMs. Smaller grid size means greater spatial resolution but increases computational demands. Grid size is especially important for understanding wildfire O_3 production, since this is nonlinear with NO_x and VOCs (Jaffe and Wigder 2012). For a primary pollutant, even if the spatial distribution is not well described, the integrated flux downstream will still reflect the emission flux, assuming no loss or additional production. This is not true for O_3 due to the nonlinearity in production rates. Wildland fires have large emissions of the PAN precursor acetaldehyde, which can result in rapid sequestration of NO_x into PAN (as described above). The degree to which a model captures this process depends on its spatial resolution and accuracy of emissions. Models that overpredict NO_x emissions and/or underpredict acetaldehyde will probably overpredict O_3 near fires (Zhang et al. 2014; Baker et al. 2016).

An additional challenge for CTMs is the large number of VOCs and oxygenated VOCs that are emitted by wildland fires, most of which are not included in standard chemical mechanisms. For example, furans (5-carbon aromatic compounds) can be responsible for 10% of the O_3 production in smoke plumes (Coggon et al. 2019), but these reactions are not included in most chemical mechanisms. Given that over 500 VOCs have been identified in biomass-burning plumes (Hatch et al. 2017), it is necessary to simplify the reaction scheme for CTMs. Despite the challenges in modeling O_3 from wildland fires, an advantage of CTMs is that all sources (multiple fires, industrial emissions, etc.) can be modeled simultaneously for all receptor locations, and the contribution from each source can be determined from the results (e.g., Baker et al. 2016).

Many studies have also examined the production of SOA from biomass-burning emissions using CTMs (Shrivastava et al. 2015). However, there is disagreement about the global production of SOA from biomass-burning emissions due to high uncertainty in precursor emissions and chemistry.

Lagrangian Plume Models

Box models have been used to overcome the challenges of grid resolution and accurately simulate transport (Wolfe et al. 2016). In this approach, a hypothetical box (or air mass) is identified, whereby detailed chemistry is simulated in the box as it moves downwind (with the prevailing wind) in a Lagrangian framework. The concentrations in the box can be initialized with observations, emission factors, and dilution rates. There are several variations in this approach, but it generally outperforms CTMs in simulating O_3 production (Alvarado et al. 2015) (Fig. 6.4).

Box models can allow for a complex chemical scheme, as only one grid cell needs to be simulated. By simulating emissions from a single fire plume, more accurate representation of the emissions can be incorporated, and transport is removed as an uncertainty (the box follows the prevailing plume direction). In the future, box models for individual plumes could be embedded in CTMs to carry out higher-resolution chemistry simulations.

Statistical Modeling

Statistical models attempt to model or "predict" O_3 concentrations (hourly or 8-h average) using a variety of meteorological indicators (daily maximum temperature, vector winds, 24-h backward trajectories, relative humidity, 500-mb geopotential height, etc.). Typical approaches use either multiple linear regression (Jaffe et al. 2013) or generalized additive models (Gong et al. 2017). The data are split into a non-smoke training dataset, an evaluation or cross-validation dataset, and a smoke dataset. The difference between the prediction from the non-smoke training set and the actual observation indicates the contribution to O_3 due to fire emissions. In practice, these models can predict the O_3 MDA8 for non-smoke days with R^2 values of 0.5–0.8. This suggests that for urban environments, the average contribution on smoke days to the MDA8 is 3–10 ppb, with a maximum contribution in extreme

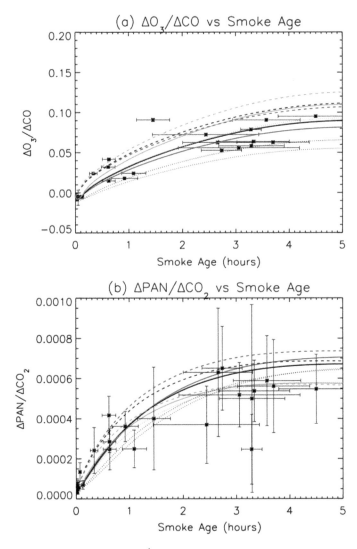

Fig. 6.4 Enhancement ratios (mol mol^{-1}) of **a** O_3 to CO and **b** PAN to CO_2 versus estimated smoke age, when the chemistry of the unidentified SVOCs is included in the model. Asterisks are measured mixing ratios; horizontal error bars show uncertainty in estimated age; vertical error bars show uncertainty in measurement. Red, black, and green are Aerosol Simulation Program results for the slow, best fit (medium), and fast plume dilution rates. Dashed lines are for above-plume photolysis rates; solid lines are for the middle of the plume; dotted lines are for the bottom of the plume [From Alvarado et al. (2015)]

6 Smoke Chemistry

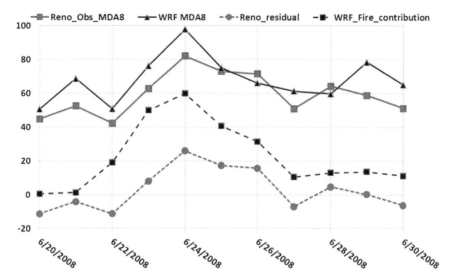

Fig. 6.5 Upper two lines: observed MDA8 (solid red line and squares) and calculated MDA8 from the WRF-Chem model (solid black line and triangles). Bottom two lines: the residual (difference of observed value from predicted value) from the statistical model for Reno, Nevada (dashed red line and circles) and the WRF-Chem model calculated contribution from fires (dashed black line and squares) [From Jaffe et al. (2013)]

cases of up to 50 ppb (Fig. 6.5). These models are simpler to apply than the CTM approach and give statistically robust predictions. However, a statistical approach does not clearly indicate cause and effect.

6.3 Gaps in Data, Understanding, and Tools/Technology

6.3.1 Ozone Data Gaps

More data from laboratory studies, field studies, and long-term monitoring are needed to improve our understanding of the effects of wildland fire smoke on O_3. Although several laboratory studies have examined the chemistry of biomass-burning in smog chambers (Bian et al. 2017), these studies lack measurements needed to close the reactive nitrogen budget and understand how variation in emissions of organics affects O_3 and PAN formation. Thus, smog chamber studies of wildland fire smoke need a more complete set of observations, especially for highly reactive VOCs (alkenes and aldehydes), oxidized nitrogen species (NO_x, HONO, PAN, etc.), HO_x radicals, and measurements of total OH reactivity.

Until the recent Western Wildfire Experiment for Cloud Chemistry, Aerosol Absorption, and Nitrogen (WE-CAN) study, most field experiments did not include enough measurements to close the reactive nitrogen budget. More field observations of this type are needed to understand the sources of the variability in O_3 production in smoke plumes. Better information is also needed on how smoke aerosol radiative effects alter O_3 and SOA chemistry within optically thick smoke plumes, as well as how this chemistry varies horizontally and vertically within smoke plumes.

There are few O_3 and $PM_{2.5}$ monitors in rural areas, so insufficient data are available to help communities in these areas predict and respond to O_3 and $PM_{2.5}$ produced by smoke. Although many urban areas have long-term monitoring of O_3 and $PM_{2.5}$, few have measurements of biomass-burning tracer species (e.g., CO, HCN) that allow easy identification of smoke-influenced days. Multi-channel aethalometers that can help separate $PM_{2.5}$ contributions from OA and BC can also be used to identify smoke. This limits our ability to confidently attribute the formed O_3 to smoke, or to understand how the interaction of smoke with urban emissions affects urban AQ.

6.3.2 Secondary Organic Aerosol Data Gaps

A large number of IVOCs from the FLAME-IV laboratory experiments have been identified as potential SOA precursors, using reactivity with OH and carbon number scaling approach (Hatch et al. 2017). However, many of these newly identified compounds are not represented in current models, and few data exist on their reaction mechanisms and SOA formation potentials. Fifty-five to 77% of the reactive carbon was associated with compounds for which SOA yields are unknown or understudied. They also identified best-candidate compounds (furan derivatives and polyunsaturated aliphatic hydrocarbons) for future smog chamber experiments. However, the technique used by Hatch et al. (2017) may not identify all potential SOA precursors, and other studies have suggested that biogenic compounds are an important precursor (Ahern et al. 2019).

There is significant variability among studies regarding the enhancement of OA (or PM) mass due to aging. Differences in prescribed burning ignition procedure, fuel characteristics (species, composition, moisture), and burn conditions (e.g., combustion efficiency) lead to differences in absolute and relative emission rates of various gaseous organic compounds that are SOA precursors. These precursors produce chemical reaction products with different volatility distributions and SOA yields. Oxidant concentrations (affected by the amount NO_x emitted from burning), amount of sunlight, and relative humidity also affect the distribution of reaction products and SOA yields. Differences in dilution/entrainment, losses in experimental chambers and lines, and differences in the baseline from which changes are estimated also contribute to the variability among studies (Hodshire et al. 2019b).

Nighttime oxidation of wildland fire emissions by NO_3 radicals and O_3 is expected to lead to SOA formation. For example, Decker et al. (2019) reported that, during the 2013 Southeast Nexus (SENEX) campaign, nighttime oxidation for rice straw and ponderosa pine (*Pinus ponderosa*) emissions was dominated by NO_3 (72% and 53%, respectively), but O_3 oxidation (25% and 43%) was also significant. Vakkari et al. (2014) observed lower production of OA in nighttime biomass-burning plumes compared to daytime plumes in South Africa. Vakkari et al. (2018) observed no net increase in total PM_1 mass concentrations at night but did see net increases in aged mass during the day. In a study at Mt. Bachelor Observatory in Oregon, wildfire plumes transported primarily at night versus day showed little difference in mass enhancement, but the OA was less oxidized in the nighttime plumes (Zhou et al. 2017).

6.3.3 Model Gaps

Although understanding O_3 production in smoke plumes is critical for policy and health studies, in situations where wildland fires cause ambient concentrations to exceed the NAAQS, the Clean Air Act allows for exclusion of these data based on the "exceptional event" rule (USEPA 2016). Statistical models have been used to support exceptional-event cases, showing concentrations that would be expected for the prevailing meteorological conditions (USEPA 2016).

There have been few comparisons of the three modeling methods (CTM, Lagrangian plume or box, and statistical) used to quantify O_3 from fire emissions. A statistical and CTM approach was compared to estimate the O_3 contribution due to transported smoke into Salt Lake City in August 2012 (Jaffe et al. 2013) (Fig. 6.5). Significant differences found between the two approaches were large enough to complicate any regulatory decisions based on this analysis. Therefore, further comparisons of O_3 production using multiple models is a key research need.

More work is needed on combining models based on different spatial scales to better simulate the long-range impacts of wildland fire smoke. Regional and global AQ models are needed to examine these impacts and account for the interactions of smoke with anthropogenic pollution, but these models cannot represent the near-source chemistry of smoke plumes as they are still at subgrid scale. Thus, plume-scale models need to be used to develop parameterizations for these subgrid processes (Chap. 4). These parameterizations can be incorporated into regional and global AQ models to correctly represent this near-source chemistry. Some progress has been made (Sakamoto et al. 2016), although more work is needed to develop and test these parameterizations in large-scale models.

Until recently, most plume-scale modeling studies have been limited to single box model studies, which cannot account for the horizontal and vertical variations in concentrations, photolysis rates, and other parameters in smoke plumes. Recent work has tried to address this by representing smoke plumes as 2D Lagrangian "walls" to look at vertical and horizontal gradients within some plumes (Fig. 6.6). For example, the SAM-TOMAS model was used in this fashion to study coagulation of aerosols in smoke plumes (Sakamoto et al. 2015, 2016). A new variant, the SAM-ASP model (Lonsdale et al. 2020), can examine the effects of plume gradients on gas-phase chemistry, aerosol condensation/evaporation, and coagulation within smoke plumes. However, as instrument time response and accuracy have improved, data have become

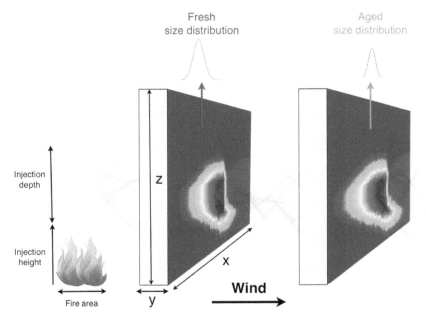

Fig. 6.6 Schematic of a 2D Lagrangian wall SAM-TOMAS simulation [From Sakamoto et al. (2016)]

available to examine these models to allow multiple data points per plume transect (Garofalo et al. 2019). These models require further development and testing to see if they can be used to help parameterize near-source chemistry for regional and global AQ models.

6.4 Vision for Improving Our Understanding of Smoke Chemistry

As wildland fire emissions in the USA are increasing (e.g., McClure and Jaffe 2018; Jaffe et al. 2020) at the same time that most anthropogenic emissions of O_3 and $PM_{2.5}$ precursors are decreasing, due to emission controls (Chan et al. 2018), there is a greater need to understand the chemistry of wildland fire smoke. Here, we discuss opportunities for significant, near-term contributions to the understanding of smoke chemistry (Sect. 6.4.1). We then suggest an approach for addressing and prioritizing research on smoke chemistry knowledge gaps to ensure timely improvement (Sect. 6.4.2).

6.4.1 Near-Term Opportunities

Recent improvements in instrumentation can help identify the organic species emitted by biomass burning (Jen et al. 2019), greatly improving our capability of identifying emitted compounds and understanding their chemistry. Laboratory studies on the OH, O_3, and NO_3 oxidation of newly identified compounds in wildland fire smoke will provide the data needed to improve atmospheric chemical mechanisms for these compounds and predict their influence on O_3 and SOA. Laboratory oxidation studies, including updated instrumentation, are needed for this work.

There is also a near-term opportunity to use existing laboratory and field data to develop simplified, empirical models of smoke $PM_{2.5}$ oxidation and aging to better understand the evolution of organic aerosol within smoke plumes, by accounting for the oxidation, condensation, and evaporation of the organic compounds. These simplified mechanisms could then be incorporated into plume-scale, regional, and global AQ models to better understand and forecast the effects of these chemical changes on the $PM_{2.5}$ impacts of wildland fire smoke.

Recent field campaigns (WE-CAN, FIREX-AQ) have provided new data on the organic species, nitrogen budget, and photolysis rates within wildland fire smoke plumes, including how these parameters vary within the smoke plume (Garofalo et al. 2019). Further analysis of these datasets will clarify the effects of aerosol shading and plume gradients on smoke plume chemistry. These data can be used in box and 2D Lagrangian plume-scale modeling to better understand O_3 and SOA chemistry within these plumes.

Finally, both physical and statistical models of the effects of biomass burning on O_3 have advanced significantly in recent years, but comparisons between these models and observed O_3 have not been conducted. Intercomparisons among statistical, plume-scale, and 3D regional air quality models will help understand the strengths and limitations of each approach and identify priorities for further development of each approach.

6.4.2 Long-Term Priorities for Improving Smoke Chemistry Knowledge

Priorities for smoke chemistry studies should be set according to how much the study will provide near-term, concrete improvements in the information needed to minimize effects of wildfire smoke on human health and other values. We recommend a focus on studies of the impact of wildland fire smoke on exposure to O_3 and $PM_{2.5}$ in rural and urban populations. In addition, as common public health advice during smoke exposure events is to stay indoors and turn on air conditioning and particulate filters, additional study is needed of the indoor chemistry of smoke and how this affects exposure.

6.4.2.1 Exposure of Rural Populations

For rural populations, the key need is for more low-cost and federal equivalent monitors to better assess the near-source chemistry of smoke plumes that are not interacting with significant anthropogenic emissions. Our understanding of the exposure of rural populations is limited by a lack of long-term data on atmospheric chemistry in rural areas, because AQ monitors are primarily near large urban populations. Traditional Federal Reference Method (FRM) and Federal Equivalent Method (FEM) monitoring techniques will likely be too expensive to provide the needed coverage. However, there have been recent advances in low-cost monitoring instruments for air quality, focusing mainly on $PM_{2.5}$, O_3, NO_x, and CO. These instruments are usually connected to Web portals that provide real-time AQ data, allowing fire managers, AQ managers, and the public to quickly assess the impacts of a given fire. These low-cost sensors require the use of more expensive reference methods for calibration. This could be provided by a small number of FRM/FEM sites, supplemented by mobile laboratories with more detailed instrumentation (Yacovitch et al. 2015) that can be deployed to areas where smoke is present.

Long-term data on wildland smoke concentrations at multiple rural locations can be used to develop improved plume-scale, statistical, and 3D models of smoke chemistry by providing tests of new smoke chemical mechanisms and modeling approaches. These data can also be used in a data assimilation framework to improve smoke chemistry forecasts, ensuring that they are consistent with the latest observations of the smoke plume and chemistry, thus addressing inherent variability in smoke emissions and chemistry.

6.4.2.2 Exposure of Urban Populations

The high levels of anthropogenic emissions of O_3 and $PM_{2.5}$ precursors in urban areas increase the complexity of wildland fire smoke chemistry. Although many urban areas have long-term monitoring of O_3 and $PM_{2.5}$, few have measurements of biomass-burning tracer species (e.g., CO, HCN) that allow easy identification of smoke-influenced days by determining days with concentrations of these species above the normal variation in the urban area. Previous studies looking at the interaction of smoke and urban pollution have used satellite observations of smoke to determine smoke-influenced days (Brey and Fischer 2016), but these observations of smoke throughout the vertical column are not necessarily good proxies for surface-level smoke impacts.

Progress on mitigating the urban impacts of wildland fire smoke will require additional monitoring sites with O_3, $PM_{2.5}$, and wildland fire tracers. In addition, a "rapid response" observation unit for urban ground measurements, using mobile labs, could be deployed to provide more detailed observations during smoke events. These data would provide more accurate identification of smoke-influenced days and data on how smoke alters urban air quality. They could also be used to train statistical models of air quality on non-smoke-influenced days, which can be used to determine effective smoke contribution to O_3 and $PM_{2.5}$ on smoke-influenced days. These data would also help refine the treatment of smoke–urban pollution interactions in typical regional air quality models like CMAQ and CAMx. Although plume-scale modeling is unlikely to be directly relevant to studies of urban chemistry, representation of near-source smoke chemistry in these 3D air quality models can be improved, especially downwind interactions with urban areas.

6.4.2.3 Indoor Smoke Chemistry

Although elevated concentrations of ambient $PM_{2.5}$ and O_3 due to fire emissions are well documented, little is known about smoke exposure indoors. This is especially important given that community health warnings during smoke events suggest that residents stay indoors (Chap. 7). The assumption is that indoor air will be significantly cleaner than outdoor air when smoke is present. However, the degree that this assumption is accurate is unclear.

More information is needed to understand indoor smoke concentrations across a range of building types and conditions, how the chemistry and toxicity of wildland fire smoke are altered as it is transported indoors, and the effectiveness of particle filtration for reducing the health impacts of smoke. Building environments are heterogeneous; buildings have wide variability in compound concentrations, temperature, relative humidity, and air exchange (Morrison 2015). Combined with the large number of reactants, products, secondary reactants, and side reactions, each microenvironment has a discrete chemical signature.

Wildland smoke $PM_{2.5}$ can enter buildings, where it can be inhaled or deposited on indoor surfaces. Wheeler et al. (2020) showed that roof-cavity dust could be analyzed for smoke compounds to estimate exposure. The rate of smoke infiltration depends on the rate of air exchange and ventilation in the building. Particulate filters (stand-alone or as part of an HVAC system) can reduce indoor $PM_{2.5}$ concentrations, but older buildings without central HVAC can have indoor $PM_{2.5}$ levels like those outside during smoke events (Kirk et al. 2018).

O_3 from wildland fire smoke can react with indoor surfaces to produce additional pollutants that cannot be removed by particulate filters. If oils or other organics are present on these surfaces, the reaction with O_3 can produce SOA and other toxic species such as carbonyls and carboxylic acids. O_3 can also react with terpenes emitted by wood, cleaners, air fresheners, and personal care products (Hodgson et al. 2002), producing similar toxic species. Ozone will also react with tobacco smoke residuals on indoor surfaces and generate volatile products such as cotinine (Petrick et al. 2011). In addition, HONO can be produced indoors from surface reactions of NO_x and will react with tobacco smoke residue on surfaces and produce carcinogenic tobacco-specific nitrosamines (Sleiman et al. 2010). These reactions may also take place within homes where wildland fire smoke residue is present on the walls, ceilings, and floors.

Other toxic species in wildland fire smoke (e.g., aldehydes, isocyanic acid) are also unlikely to be removed by particulate filters. If the observed health effects of smoke are partially related to these other species, and not just $PM_{2.5}$ mass, filtering out particles may not fully protect the public from the health impacts of smoke.

In one of the few relevant studies on smoke intrusion into homes, Kirk et al. (2018) examined two homes in the Pacific Northwest during the summer of 2015, a year with high fire occurrence and frequent high $PM_{2.5}$ days. They found high correlations between indoor (I) and outdoor (O) concentrations, but relatively low I/O ratios of 0.10 to 0.26 for $PM_{2.5}$. Data from another study show a wider range for the I/O ratio (0.16–0.91) during smoke events for several homes and commercial buildings in the Pacific Northwest during 2018 and 2019 smoke events (D.A. Jaffe, unpublished data). In some cases, homes had indoor $PM_{2.5}$ concentrations of up to 100 $\mu g\ m^{-3}$ (I/O ratio of 0.91). Therefore, the assumption that indoor air is "safe" during smoke events is not uniformly correct.

Further work is thus needed to examine the infiltration of wildland fire smoke into different types of buildings, and how the smoke is transformed by reactions inside the building. As a first step, collaboration is needed between smoke chemists and indoor air quality specialists to identify potential studies that can provide initial data on these indoor transformations.

6.5 Emerging Issues

6.5.1 Higher Particulate Matter, Ozone, and Hazardous Air Pollutants from Fires in Western States

$PM_{2.5}$ emissions from wildland fires in 2017 and 2018 were the highest ever observed, at that point in time, at US regulatory monitors (Laing and Jaffe 2019; Jaffe et al. 2020):

- San Francisco, CA: Highest daily $PM_{2.5}$ ever recorded was 177 $\mu g\ m^{-3}$ (November 16, 2018).
- Seattle, WA: Highest daily $PM_{2.5}$ ever recorded was 110 $\mu g\ m^{-3}$ (August 21, 2018).
- Medford OR: 8 days of $PM_{2.5}$ over 100 $\mu g\ m^{-3}$ in 2017, highest daily $PM_{2.5}$ of 268 $\mu g\ m^{-3}$ (September 6, 2017).
- Seeley Lake, MT: $PM_{2.5}$ of 642 $\mu g\ m^{-3}$, highest ever observed in the USA, (September 6, 2017).

Similar extremes have been observed for O_3. For example, during the major smoke events of 2017 and 2018, numerous locations experienced O_3 MDA8 values over 100 ppb: 103 ppb at Enumclaw, WA (August 3, 2017), 116 ppb at Carus, OR (August 3, 2017), and 115 ppb at Auburn, CA (August 1, 2018). These extremes occurred due to a large number of concurrent wildfires in the greater Pacific Northwest region and British Columbia (McClure and Jaffe 2018).

6.5.2 How Prescribed Burning Affects Smoke Chemistry

There have been no studies about how ignition methods affect chemistry in the smoke plume, so there is little guidance available for fire managers on the consequences of different methods on local and regional air quality. The effects of fuel conditions (moisture content, nitrogen content, 3D structure) on smoke emissions and chemistry have been studied in a few cases, but not with the goal of providing information to fire managers on how to plan prescribed burning to minimize impacts on human exposure to O_3, $PM_{2.5}$, and other air toxics.

Different ignition methods and fuel conditions can lead to variations in smoke emissions and plume dynamics, including the initial plume size, injection height, and plume chemistry (Chap. 4). For example, a more concentrated plume will have lower photolysis rates within and below the core of the plume, potentially delaying the photolysis of radical sources like HONO and HCHO, delaying the loss of NO_x, and delaying or reducing the formation of O_3 and the oxidation of the NMOCs. A more concentrated plume will also have more of the semivolatile and intermediate-volatility organic compounds (S/IVOCs) present in the particle phase, potentially altering downwind chemistry. A higher plume injection height will also alter the chemistry taking place within the smoke plume. Higher altitudes tend to have colder temperatures, which favors the formation of PAN and condensation of organic vapors into $PM_{2.5}$. Smoke at higher altitudes is also more likely to be transported long distances in the troposphere before being transported back to the surface where it can affect O_3 and $PM_{2.5}$ downwind of the fire source.

However, plume-scale LES models like SAM-TOMAS and SAM-ASP could be used to investigate how focused field studies could clarify how different prescribed burn ignition methods and fuel conditions affect downwind air quality impacts. The chemistry of smoke plumes could be sampled downwind using small aircraft and/or surface mobile laboratories to characterize the formation of O_3 and $PM_{2.5}$. These studies can then be used to revise plume-scale models to better account for the effects of ignition methods and provide guidance to fire manager efforts to minimize the formation of O_3 and $PM_{2.5}$.

6.5.3 Clarifying Specific Health Effects

Little work has been done to examine how the aging of wildland fire smoke $PM_{2.5}$ affects human health (Chap. 7). Toxicological studies could help to determine if aged smoke has similar impacts as fresh smoke, building on the work of Gilmour et al. (2015). However, epidemiological studies are likely to be difficult to perform. As aged smoke tends to be more dilute than fresh smoke, the public may take fewer precautions to reduce exposure to aged smoke than they do for concentrated fresh smoke, confounding the analysis of the relative health effects.

There is also a need to distinguish between the health effects due to multiple air toxics within smoke (e.g., carbonyls, isocyanic acid, aromatics) versus $PM_{2.5}$. Even if $PM_{2.5}$ is not a major concern, the remaining air toxics may represent substantial health impacts. However, measurements of these additional air toxics in wildland fire smoke are rare outside of dedicated field campaigns, so estimating exposure to these species is difficult.

6.6 Links with Other Components of the Smoke Assessment

Many of the challenges in other components of understanding smoke (Chaps. 2, 3, 4, 5 and 7) have links to understanding wildland fire smoke chemistry. We note a few of these links below.

6.6.1 Fire Behavior and Plume Dynamics

As noted above, injection height and initial plume concentrations of different pollutants are critical to understanding the subsequent chemistry of the smoke plume. Injection height affects the temperature where the initial plume chemistry is occurring and affects the dilution rate of smoke, both of which alter the chemical formation of O_3 and evolution of $PM_{2.5}$ within the smoke plume. In addition, the size of the plume will have significant effects on the dilution and chemistry taking place within the smoke plume. Thus, a better understanding of fire behavior and near-source plume dynamics is critical for better predictions of the effects of wildland fires on local and regional O_3 and $PM_{2.5}$.

6.6.2 Fuel Characterization

The size, chemical composition, and 3D structure of fuels can directly affect the chemical composition of emissions, and thus downwind chemistry. For example, because O_3 formation in smoke plumes is generally NO_x-limited (Jaffe and Wigder 2012), understanding the emissions of NO_x from wildland fires is critical to understanding smoke plume chemistry. Emissions of NO_x from wildland fires are in turn a function of the combustion efficiency (itself a function of fuel size, structure, and moisture content) and fuel nitrogen content, which can vary from 0.2 to 4.0% between fuel types and can vary between seasons for a single ecosystem (Mebust and Cohen 2013). Different combustion efficiencies and fuel composition can also lead to changes in emissions of organic $PM_{2.5}$ and other air toxics (Jen et al. 2019). An assessment of the typical nitrogen content of wildland fuels would benefit studies of smoke chemistry by reducing uncertainty about the initial NO_x emissions and chemistry.

6.6.3 Smoke Emissions

Understanding the chemical transformations of smoke requires understanding the speciated emission of trace gases and aerosols within smoke and how these mixtures of species depend on fuel type, fuel structure, fuel moisture, and fire behavior (e.g., rate of spread). This requires more than an estimate of the total PM emissions from a fire. Information is needed on emissions of individual gases, including air toxics such as formaldehyde and HNCO, as well as information on the organic and inorganic species within PM (e.g., volatility, hygroscopicity, and chemical reactivity). Detailed measurements are needed to understand the chemical composition of fresh smoke, extending it to subsequent chemical reactions as the smoke moves downwind.

6.6.4 Effects on People, Health, Transportation, and Commerce

Understanding the impacts of smoke on health requires an understanding of the chemical transformations taking place within smoke, as not all people exposed to wildfire smoke are exposed to the same mixture of compounds. Fresh and aged smoke differ chemically, and these differences may alter impacts on human health. For example, understanding the chemistry of smoke plumes is important for understanding the impact of smoke on O_3 exposure. Fresh smoke is usually depleted in O_3, whereas aged smoke can be enhanced in O_3 and thus have increased health effects. Understanding how smoke chemistry increases or decreases the amount of OA and $PM_{2.5}$ present in smoke also is needed to accurately estimate health impacts. In addition, individual air toxics within smoke have different chemical lifetimes and may increase downwind (e.g., formaldehyde), so understanding rates of change of individual air toxics will provide a better picture of which components of smoke are responsible for health impacts.

6.7 Conclusions

6.7.1 Key Research Needs and Priorities

Initial smoke emissions are transformed through chemical processes at time scales from a few minutes to days and at spatial scales from meters to thousands of kilometers. These transformations alter the effects of smoke on human health and climate. Critical reviews, laboratory campaigns, field studies, and model development efforts in the last decade have improved our knowledge of the organic species in wildland fire smoke and their subsequent chemistry. However, current knowledge is insufficient to explain the wide variability in smoke chemistry from different wildland fires. Key scientific challenges in smoke chemistry include: (1) understanding O_3 formation in isolated and mixed smoke plumes, as well as when smoke mixes with urban emissions, (2) understanding SOA formation in smoke plumes and the aging of OA in smoke plumes, and (3) developing improved statistical and physical models of smoke chemistry.

Several emerging issues in smoke chemistry also merit research attention including the role of increasing PM, O_3, and hazardous air pollutants from fires in Western states, how different prescribed burn approaches can affect the chemistry of the smoke, and how the indoor environment alters the chemistry of wildland fire smoke.

Critical research needs include:

- Better information on the spatial and temporal distributions of the primary (directly emitted) and secondary (formed by chemistry after emission) pollutants formed from wildland fire smoke,
- Better forecast models to inform prescribed burning decisions,
- Information on how different prescribed burning methods affect the subsequent chemistry of the smoke.

Near-term opportunities for improving our understanding of smoke chemistry include:

- Examining the chemistry of newly identified smoke species,
- Developing empirical models of smoke fine particulate matter ($PM_{2.5}$) aging,
- Investigating the impact of plume gradients,
- Performing physical and statistical model intercomparisons.

Long-term priorities include increasing the data available on the effects of wildland fire smoke on O_3 and SOA in rural and urban areas, and examining how wildland fire smoke affects indoor AQ. More research is needed to understand how chemical properties affect the toxicity of smoke as it is chemically transformed in the atmosphere. This would inform how the complexity of chemistry may be simplified in models to predict the chemical properties most relevant to informing decisions on public health.

Additional data on smoke chemistry from laboratory and field studies are needed to:

- Improve forecast and impact models and close the nitrogen budget,
- Determine the chemistry of recently identified organic compounds, and
- Examine how aerosol shading and other effects result in horizontal and vertical gradients in the smoke plume chemistry.

Model development is needed to:

- Improve the chemical mechanisms within air quality models to represent smoke species,
- Develop parameterizations of the near-source chemistry of fires for inclusion in regional and global models, and
- In form development of statistical models of the impact of fires on air quality.

6.7.2 Opportunities for Shared Stewardship to Improve Smoke Science and Management

There are several opportunities for federal agencies, states, tribes, and the research community to collaboratively to address the research priorities above. First, coordination among the US Forest Service, USEPA, USDOD, and other scientific agencies could facilitate laboratory studies of the oxidation of newly identified organic compounds in wildland fire smoke. This would improve atmospheric chemical mechanisms for these compounds and predict their impacts on O_3 and SOA. These studies will require coordinating large numbers of investigators and instruments, including significant data analysis and modeling work after laboratory studies are completed. In addition, fully mining the WE-CAN and FIREX-AQ datasets to improve air quality models would be facilitated by a coordinated approach among agencies.

Increasing the amount of monitoring data on smoke-related O_3 and $PM_{2.5}$ would also be easier with coordination between agencies and other interested parties. This will require expertise in wildland fire, air quality, and local conditions to develop the best monitoring plan for each region. In addition, improving smoke forecast and air quality models would benefit from coordination among federal agencies, including the US Forest Service, USEPA, and NOAA.

References

Ahern AT, Goldberger L, Jahl L et al (2018) Production of N_2O_5 and $ClNO_2$ through nocturnal processing of biomass-burning aerosol. Environ Sci Technol 52:550–555

Ahern AT, Robinson ES, Tkacik DS et al (2019) Production of secondary organic aerosol during aging of biomass burning smoke from fresh fuels and its relationship to VOC precursors. J Geophys Res Atmos 124:3583–3606

Akagi SK, Yokelson RJ, Wiedinmyer C et al (2011) Emission factors for open and domestic biomass burning for use in atmospheric models. Atmos Chem Phys 11:4039

Akagi SK, Craven JS, Taylor JW et al (2012) Evolution of trace gases and particles emitted by a chaparral fire in California. Atmos Chem Phys 12:1397–1421

Akagi SK, Yokelson RJ, Burling IR et al (2013) Measurements of reactive trace gases and variable O3 formation rates in some South Carolina biomass burning plumes. Atmos Chem Phys 13:1141–1165

Alvarado MJ, Logan JA, Mao J et al (2010) Nitrogen oxides and PAN in plumes from boreal fires during ARCTAS-B and their impact on ozone: an integrated analysis of aircraft and satellite observations. Atmos Chem Phys 10:9739–9760

Alvarado MJ, Lonsdale CR, Yokelson RJ et al (2015) Investigating the links between ozone and organic aerosol chemistry in a biomass burning plume from a prescribed fire in California chaparral. Atmos Chem Phys 15:6667–6688

Baker KR, Woody MC, Tonnesen GS et al (2016) Contribution of regional-scale fire events to ozone and $PM_{2.5}$ air quality estimated by photochemical modeling approaches. Atmos Environ 140:539–554

Baker K (2015) Simulating fire event impacts on regional O_3 and $PM_{2.5}$ and looking forward toward evaluation. Paper presented at the 14 Annual CMAS Conference, UNC-Chapel Hill, 5–7 Oct 2015

Baylon P, Jaffe DA, Wigder NL et al (2015) Ozone enhancement in western US wildfire plumes at the Mt. Bachelor Observatory: the role of NO_x. Atmos Environ 109:297–304

Baylon P, Jaffe DA, Hall SR et al (2018) Impact of biomass burning plumes on photolysis rates and ozone formation at the mount bachelor observatory. J Geophys Res Atmos 123:2272–2284

Bian Q, Jathar SH, Kodros JK et al (2017) Secondary organic aerosol formation in biomass-burning plumes: theoretical analysis of lab studies and ambient plumes. Atmos Chem Phys 17:5459–5475

Brey SJ, Fischer EV (2016) Smoke in the city: How often and where does smoke impact summertime ozone in the United States? Environ Sci Technol 50:1288–1294

Brown T, Clements C, Larkin N et al (2014) Validating the next generation of wildland fire and smoke models for operational and research use—A national plan (Final report, Project 13-S-01-01. U.S. Joint Fire Science Program, Boise. https://www.firescience.gov/projects/13-S-01-01/project/13-S-01-01_final_report.pdf. 22 June 2020

Burling IR, Yokelson RJ, Akagi SK et al (2011) Airborne and ground-based measurements of the trace gases and particles emitted by prescribed fires in the United States. Atmos Chem Phys 11:12197–12216

Buysse CE, Kaulfus A, Nair U, Jaffe DA (2019) Relationships between particulate matter, ozone, and nitrogen oxides during urban smoke events in the Western US. Environ Sci Technol 53:12519–12528

Chan EA, Gantt B, McDow S (2018) The reduction of summer sulfate and switch from summertime to wintertime PM2.5 concentration maxima in the United States. Atmos Environ 175:25–32

Coggon MM, Lim CY, Koss AR et al (2019) OH chemistry of non-methane organic gases (NMOGs) emitted from laboratory and ambient biomass burning smoke: evaluating the influence of furans and oxygenated aromatics on ozone and secondary NMOG formation. Atmos Chem Phys 19:14875–14899

Cubison MJ, Ortega AM, Hayes PL et al (2011) Effects of aging on organic aerosol from open biomass burning smoke in aircraft and laboratory studies. Atmos Chem Phys 11:12049–12064

DeBell LJ, Talbot RW, Dibb JE et al (2004) A major regional air pollution event in the northeastern United States caused by extensive forest fires in Quebec, Canada. J Geophys Res Atmos 109:D19305

Decker ZC, Zarzana KJ, Coggon M et al (2019) Nighttime chemical transformation in biomass burning plumes: a box model analysis initialized with aircraft observations. Environ Sci Technol 53:2529–2538

Donahue NM, Robinson AL, Stanier CO, Pandi SN (2006) Coupled partitioning, dilution, and chemical aging of semivolatile organics. Environ Sci Technol 40:2635–2643

Dreessen J, Sullivan J, Delgado R (2016) Observations and impacts of transported Canadian wildfire smoke on ozone and aerosol air quality in the Maryland region on June 9–12, 2015. J Air Waste Manag Assoc 66:842–862

Finewax Z, de Gouw JA, Ziemann PJ (2018) Identification and quantification of 4-nitrocatechol formed from OH and NO_3 radical-initiated reactions of catechol in air in the presence of NO x: implications for secondary organic aerosol formation from biomass burning. Environ Sci Technol 52:1981–1989

Garofalo LA, Pothier MA, Levin EJ et al (2019) Emission and evolution of submicron organic aerosol in smoke from wildfires in the western United States. ACS Earth Space Chem 3:1237–1247

Gilmour MI, Kim YH, Hays MD (2015) Comparative chemistry and toxicity of diesel and biomass combustion emissions. Anal Bioanal Chem 407:5869–5875

Gong X, Kaulfus A, Nair U, Jaffe DA (2017) Quantifying O_3 impacts in urban areas due to wildfires using a generalized additive model. Environ Sci Technol 51:13216–13223

Hatch LE, Yokelson RJ, Stockwell CE et al (2017) Multi-instrument comparison and compilation of non-methane organic gas emissions from biomass burning and implications for smoke-derived secondary organic aerosol precursors. Atmos Chem Phys 17:1471–1489

Hobbs PV, Sinha P, Yokelson RJ et al (2003) Evolution of gases and particles from a savanna fire in South Africa. J Geophys Res Atmos 108:8485

Hodgson AT, Beal D, McIlvaine JER (2002) Sources of formaldehyde, other aldehydes and terpenes in a new manufactured house. Indoor Air 12:235–242

Hodshire AL, Bian Q, Ramnarine E et al (2019a) More than emissions and chemistry: Fire size, dilution, and background aerosol also greatly influence near-field biomass burning aerosol aging. J Geophys Res Atmos 124:5589–5611

Hodshire AL, Akherati A, Alvarado MJ et al (2019b) Aging effects on biomass burning aerosol mass and composition: a critical review of field and laboratory studies. Environ Sci Technol 17:10007–10022

Jaffe DA, Wigder NL (2012) Ozone production from wildfires: a critical review. Atmos Environ 51:1–10

Jaffe D, Bertschi I, Jaeglé L et al (2004) Long-range transport of Siberian biomass burning emissions and impact on surface ozone in western North America. Geophys Res Lett 31:L16106

Jaffe DA, Wigder N, Downey N et al (2013) Impact of wildfires on ozone exceptional events in the western US. Environ Sci Technol 47:11065–11072

Jaffe DA, O'Neill SM, Larkin NK et al (2020) Wildfire and prescribed burning impacts on air quality in the United States. J Air Waste Manag Assoc 70:583–615

Jen CN, Hatch LE, Selimovic V et al (2019) Speciated and total emission factors of particulate organics from burning western US wildland fuels and their dependence on combustion efficiency. Atmos Chem Phys 19:1013–1026

Kim YH, King C, Krantz Q et al (2019) The role of fuel type and combustion phase on the toxicity of biomass smoke following inhalation exposure in mice. Arch Toxicol 93:1501–1513

Kim YH, Warren SH, Krantz QT et al (2018) Mutagenicity and lung toxicity of smoldering vs. flaming emissions from various biomass fuels: Implications for health effects from wildland fires. Environ Health Perspect 126:017011

Kirk WM, Fuchs M, Huangfu Y et al (2018) Indoor air quality and wildfire smoke impacts in the Pacific Northwest. Sci Technol Built Environ 24:149–159

Koplitz SN, Nolte CG, Pouliot GA et al (2018) Influence of uncertainties in burned area estimates on modeled wildland fire PM2.5 and ozone pollution in the contiguous US. Atmos Environ 191:328–339

Koss AR, Sekimoto K, Gilman JB et al (2018) Non-methane organic gas emissions from biomass burning: Identification, quantification, and emission factors from PTR-ToF during the FIREX 2016 laboratory experiment. Atmos Chem Phys 18:3299–3319

Kroll JH, Smith JD, Che DL et al (2009) Measurement of fragmentation and functionalization pathways in the heterogeneous oxidation of oxidized organic aerosol. Phys Chem Chem Phys 11:8005–8014

Laing JR, Jaffe D (2019) Wildfires are causing extreme PM concentrations in the Western United States. EM Magazine (June). Air and Waste Management Association, Pittsburgh. http://pubs.awma.org/flip/EM-June-2019/jaffe.pdf. 22 June 2020

Lindaas J, Farmer DK, Pollack IB et al (2017) Changes in ozone and precursors during two aged wildfire smoke events in the Colorado Front Range in summer 2015. Atmos Chem Phys 17:10691–10707

Lonsdale CR, Alvarado MJ, Hodshire AL et al (2020) Simulating forest fire plume dispersion, chemistry, and aerosol formation using SAM-ASP version 1.0. Geoscientific Model Dev Discuss 13:4579–4593

Lonsdale CR, Brodowski CM, Alvarado MJ (2017) Improving the modeling of wildfire impacts on ozone and particulate matter for Texas air quality *planning* (Final report, Project 16–024, August). Texas Air Quality Research Program, Austin. http://aqrp.ceer.utexas.edu/viewprojects FY16-17.cfm?Prop_Num=17-024. 22 June 2020

Mauzerall DL, Logan JA, Jacob DJ et al (1998) Photochemistry in biomass burning plumes and implications for tropospheric ozone over the tropical South Atlantic. J Geophys Res Atmos 103:8401–8423

May AA, Levin EJ, Hennigan CJ et al (2013) Gas-particle partitioning of primary organic aerosol emissions: 3. Biomass burning. J Geophys Res Atmos 118:11327–11338

McClure CD, Jaffe DA (2018) US particulate matter air quality improves except in wildfire-prone areas. Proc National Acad Sci USA 115:7901–7906

McDonald-Buller E, Kimura Y, Wiedinmyer C et al (2015) Targeted improvements in the fire inventory form NCAR (FINN) model for Texas air quality planning (Final report, Project 14–011, December). Texas Air Quality Research Program, Austin. http://aqrp.ceer.utexas.edu/projec tinfoFY14_15%5C14-011%5C14-011%20Final%20Report.pdf. 22 June 2020

McKeen SA, Wotawa G, Parrish DD et al (2002) Ozone production from Canadian wildfires during June and July of 1995. J Geophys Res Atmos 107:4192

Mebust AK, Cohen RC (2013) Observations of a seasonal cycle in NOx emissions from fires in African woody savannas. Geophys Res Lett 40:1451–1455

Morrison G (2015) Recent advances in indoor chemistry. Curr Sustain/renew Energy Rep 2:33–40

Murphy BN, Donahue NM, Robinson AL, Pandis SN (2014) A naming convention for atmospheric organic aerosol. Atmos Chem Phys 14:5825–5839

Pankow JF (1994) An absorption model of the gas/aerosol partitioning involved in the formation of secondary organic aerosol. Atmos Environ 28:189–193

Petrick LM, Sleiman M, Dubowski Y et al (2011) Tobacco smoke aging in the presence of ozone: a room-sized chamber study. Atmos Environ 45:959–965

Real E, Law KS, Weinzierl B et al (2007) Processes influencing ozone levels in Alaskan forest fire plumes during long-range transport over the North Atlantic. J Geophys Res Atmos 112:D10S41

Sakamoto KM, Allan JD, Coe H et al (2015) Aged boreal biomass-burning aerosol size distributions from BORTAS 2011. Atmos Chem Phys 15:1633

Sakamoto KM, Laing JR, Stevens R et al (2016) The evolution of biomass-burning aerosol size distributions due to coagulation: dependence on fire and meteorological details and parameterization. Atmos Chem Phys 16:7709–7724

Sekimoto K, Koss AR, Gilman JB et al (2018) High- and low-temperature pyrolysis profiles describe volatile organic compound emissions from western US wildfire fuels. Atmos Chem Phys 18:9263–9281

Shiraiwa M, Berkemeier T, Schilling-Fahnestock KA et al (2014) Molecular corridors and kinetic regimes in the multiphase chemical evolution of secondary organic aerosol. Atmos Chem Phys 14:8323–8341

Shrivastava M, Easter RC, Liu X et al (2015) Global transformation and fate of SOA: Implications of low-volatility SOA and gas-phase fragmentation reactions. J Geophys Res Atmos 120:4169–4195

Sleiman M, Gundel LA, Pankow JF et al (2010) Formation of carcinogens indoors by surface-mediated reactions of nicotine with nitrous acid, leading to potential thirdhand smoke hazards. Proc Nat Acad Sci USA 107:6576–8651

Stein AF, Rolph GD, Draxler RR et al (2009) Verification of the NOAA smoke forecasting system: model sensitivity to the injection height. Weather Forecast 24:379–394

Teakles AD, So R, Ainslie B et al (2017) Impacts of the July 2012 Siberian fire plume on air quality in the Pacific Northwest. Atmos Chem Phys 17:2593

Tkacik DS, Robinson ES, Ahern A et al (2017) A dual-chamber method for quantifying the effects of atmospheric perturbations on secondary organic aerosol formation from biomass burning emissions. J Geophys Res Atmos 122:6043–6058

U.S. Environmental Protection Agency (USEPA) (2016) Final guidance on the preparation of exceptional events demonstrations for wildfire events that may influence ozone concentrations. U.S. Environmental Protection Agency, Office of Air Quality Planning and Standards, Research Triangle Park. https://www.epa.gov/sites/production/files/2018-10/documents/exceptional_events_guidance_9-16-16_final.pdf. 22 June 2020

Vakkari V, Kerminen VM, Beukes JP et al (2014) Rapid changes in biomass burning aerosols by atmospheric oxidation. Geophys Res Lett 41:2644–2651

Vakkari V, Beukes JP, Dal Maso M et al (2018) Major secondary aerosol formation in southern African open biomass burning plumes. Nat Geosci 11:580–583

Val Martin M, Honrath RE, Owen RC et al (2006) Significant enhancements of nitrogen oxides, black carbon, and ozone in the North Atlantic lower free troposphere resulting from North American boreal wildfires. J Geophys Res Atmos 111:D23S60

Wheeler AJ, Jones PJ, Reisen F et al (2020) Roof cavity dust as an exposure proxy for extreme air pollution events. Chemosphere 244:125537

Wolfe GM, Marvin MR, Roberts SJ et al (2016) The framework for 0-D atmospheric modeling (F0AM) v3. 1. Geoscientific Model Dev 9:3309–3319

Yacovitch TI, Herndon SC, Pétron G et al (2015) Mobile laboratory observations of methane emissions in the Barnett Shale region. Environ Sci Technol 49:7889–7895

Ye C, Zhang N, Gao H, Zhou X (2017) Photolysis of particulate nitrate as a source of HONO and NOx. Environ Sci Technol 51:6849–6856

Yokelson RJ, Crounse JD, DeCarlo PF et al (2009) Emissions from biomass burning in the Yucatan. Atmos Chem Phys 9:5785–5812

Yu S, Bhave PV, Dennis RL, Mathur R (2017) Seasonal and regional variations of primary and secondary organic aerosol over the continental United States: Semi-empirical estimates and model evaluation. Environ Sci Technol 41:4690–4697

Zaveri RA, Easter RC, Shilling JE, Seinfeld JH (2014) Modeling kinetic partitioning of secondary organic aerosol and size distribution dynamics: representing effects of volatility, phase state, and particle-phase reaction. Atmos Chem Phys 14:5153–5181

Zhang L, Jacob DJ, Yue X et al (2014) Sources contributing to background surface ozone in the US Intermountain West. Atmos Chem Phys 14:5295–5309

Zhou S, Collier S, Jaffe DA et al (2017) Regional influence of wildfires on aerosol chemistry in the western US and insights into atmospheric aging of biomass burning organic aerosol. Atmos Chem Phys 17:2477–2493

Open Access This chapter is licensed under the terms of the Creative Commons Attribution 4.0 International License (http://creativecommons.org/licenses/by/4.0/), which permits use, sharing, adaptation, distribution and reproduction in any medium or format, as long as you give appropriate credit to the original author(s) and the source, provide a link to the Creative Commons license and indicate if changes were made.

The images or other third party material in this chapter are included in the chapter's Creative Commons license, unless indicated otherwise in a credit line to the material. If material is not included in the chapter's Creative Commons license and your intended use is not permitted by statutory regulation or exceeds the permitted use, you will need to obtain permission directly from the copyright holder.

Chapter 7
Social Considerations: Health, Economics, and Risk Communication

Sarah M. McCaffrey, Ana G. Rappold, Mary Clare Hano, Kathleen M. Navarro, Tanya F. Phillips, Jeffrey P. Prestemon, Ambarish Vaidyanathan, Karen L. Abt, Colleen E. Reid, and Jason D. Sacks

Abstract At a fundamental level, smoke from wildland fire is of scientific concern because of its potential adverse effects on human health and social well-being. Although many impacts (e.g., evacuations, property loss) occur primarily in proximity to the actual fire, smoke can end up having a significant social impact far from the source. This dynamic, combined with lengthening fire seasons, suggests

---

The views and opinions expressed in this article are those of the individual authors, and do not necessarily reflect the views and opinions of the co-authors or the official policies and positions of the U.S. Government, the Department of Health and Human Services (CDC/ATSDR), the U.S. Forest Service, the U.S. Environmental Protection Agency, or other participating institutions.

S. M. McCaffrey (✉)
U.S. Forest Service, Rocky Mountain Research Station, Fort Collins, CO, USA
e-mail: sarah.m.mccaffrey@usda.gov

A. G. Rappold
Center for Public Health and Environmental Assessment, U.S. Environmental Protection Agency, Research Triangle Park, Washington D.C., NC, USA
e-mail: rappold.ana@epa.gov

M. C. Hano
Office of Research and Development, U.S. Environmental Protection Agency, Chapel Hill, NC, USA
e-mail: hano.maryclare@epa.gov

K. M. Navarro
Centers for Disease Control and Prevention, National Institute for Occupational Safety and Health, Denver, CO, USA
e-mail: knavarro@cdc.gov

T. F. Phillips
Jackson County Public Health, Medford, OR, USA
e-mail: phillitf@jacksoncounty.org

J. P. Prestemon · K. L. Abt
U.S. Forest Service, Southern Research Station, Research Triangle Park, Asheville, NC, USA
e-mail: jeff.prestemon@usda.gov

K. L. Abt
e-mail: karen.abt@usda.gov

This is a U.S. government work and not under copyright protection in the U.S.; foreign copyright protection may apply 2022
D. L. Peterson et al. (eds.), *Wildland Fire Smoke in the United States*,
https://doi.org/10.1007/978-3-030-87045-4_7

that understanding how wildland fire smoke affects diverse social values will be increasingly critical. This chapter reviews the existing scientific knowledge related to wildland fire smoke with respect to four topic areas: human health, economics, social acceptability, and risk communication. The broadest existing knowledge base, regarding the health effects attributed to wildland fire smoke exposure, stems from decades of research on the health effects of exposures to ambient fine particulate matter ($PM_{2.5}$). Despite the potential consequences, scientific knowledge about chronic health effects, economic impacts, and effectiveness of protective actions in response to wildfire smoke risk communication is fairly limited. The chapter concludes with identification of (1) key areas where the need for more empirical information is most critical, and (2) challenges that inhibit an improved scientific understanding.

Keywords Health effects · Economic impacts · Firefighter exposure · Risk communication · Social acceptability

7.1 Introduction

At a fundamental level, smoke from wildland fire is of scientific concern because of its potential adverse effects on an array of social values (e.g., health, economic, cultural). Compared to impacts that tend to occur in proximity to the actual fire, such as evacuations and property loss, smoke can have a significant social impact far from the source. In the fall of 2019, smoke from Australian bushfires greatly affected New Zealand air quality, and across the USA wildland fire smoke has been observed to account for a disproportionate number of poor air quality days (Liu et al. 2015; Larsen et al. 2018). As wildland fire seasons grow longer (Jolly et al. 2015) and wildfires have greater air quality impacts (McClure and Jaffe 2018; O'Dell et al. 2019), understanding potential social impacts from wildland fire smoke becomes increasingly critical. Despite this need, research on the social impacts of smoke is limited compared to other areas of smoke science.

This chapter reviews existing scientific knowledge related to the effects of wildland fire smoke on different social values. The first section summarizes research

---

A. Vaidyanathan
Centers for Disease Control and Prevention, National Center for Environmental Health, Atlanta, GA, USA
e-mail: rishv@cdc.gov

C. E. Reid
Department of Geography, University of Colorado Boulder, Boulder, CO, USA
e-mail: colleen.reid@colorado.edu

J. D. Sacks
Office of Research and Development, U.S. Environmental Protection Agency, Research Triangle Park, Washington D.C., NC, USA
e-mail: sacks.jason@epa.gov

studies which focus on acute health effects of smoke, including occupational exposure. The remaining sections discuss what is known regarding economic impacts, social acceptance, and risk communication specific to wildland fire smoke. The chapter ends with a summary of key findings and research needs.

7.2 Health Effects Attributed to Wildland Fire Smoke

Scientific evidence examining health effects attributed to wildland fire smoke exposure has grown significantly in the last decade in response to the increased frequency of large fires, the need to understand their public health impacts, and the desire to develop effective response plans. The growth in research efforts is reflected in the increasing number of systematic and critical reviews identifying the potential health effects of smoke exposure (Naeher et al. 2007; Youssouf et al. 2014; Liu et al. 2015; Adetona et al. 2016; Reid et al. 2016a; Black et al. 2017; Cascio 2018; Kondo et al. 2019). In addition, collaborations among federal, state, tribal, local, and territorial governments, as well as nongovernmental organizations have led to development of guidance documents and training materials to inform public health officials, medical professionals, and fire managers of the potential health risks of wildland fire smoke exposure.

In this section, we provide background information on air pollutants in wildland fire smoke and summarize the current scientific evidence on health effects and risk factors that may increase the likelihood of experiencing adverse health effects. The information draws heavily on published documents that have culminated from several interagency collaborations, including "Wildfire Smoke: A Guide for Public Health Officials" (USEPA 2019c) and the continuing medical education course Wildfire Smoke and Your Patients' Health (USEPA 2019b), and review articles (Naeher et al. 2007; Youssouf et al. 2014; Liu et al. 2015; Adetona et al. 2016; Reid et al. 2016a; Black et al. 2017; Cascio 2018; Kondo et al. 2019). The evidence presented within this section is a broad overview of the current state of science with respect to health effects attributed to wildland fire smoke exposure and is not intended to be a comprehensive systematic review.

7.2.1 Wildland Fire Smoke Exposure

Wildland fire smoke contains a number of air pollutants that are known to be harmful to health, including particulate matter (PM), nitrogen oxides, carbon dioxide, carbon monoxide (CO), hydrocarbons, and other organic chemicals. Of these, particulate matter (PM) is the most significant concern to public health due to widespread exposure and known health effects. PM is largely a by-product of combustion, with the fuel and conditions of combustion being important predictors of the size of particles produced (Chap. 2). Particle pollution is categorized most often by size fraction.

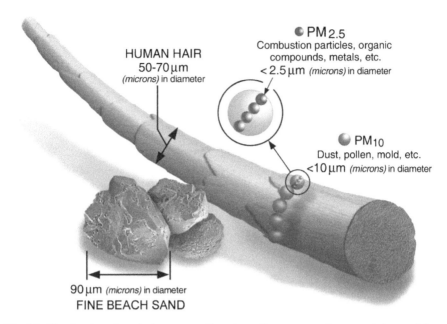

Fig. 7.1 Size fraction of particulate matter. From https://www.epa.gov/pmcourse/what-particle-pollution

Particles that are ≤2.5 μm in aerodynamic diameter (PM$_{2.5}$, fine particles) comprise >90% of total particle mass emitted from wildland fires (Groß et al. 2013) and are of primary interest when considering health impacts based on scientific evidence from short- and long-term ambient PM$_{2.5}$ exposures.

Although particles having an aerodynamic diameter of ≤10 μm (PM$_{10}$) are able to enter the respiratory tract (Fig. 7.1), PM$_{2.5}$ particles are of the greatest risk to health and are associated with both less severe (e.g., eye and respiratory tract irritation, wheezing, difficult breathing, persistent coughing, excessive phlegm) and more serious health effects (e.g., exacerbation of asthma and heart failure, premature death) (Box 7.1). Therefore, the focus of the health effects discussion in this section is based mostly on PM$_{2.5}$; potential effects of exposure to other pollutants found in wildland fire smoke are discussed in Sect. 7.2.3.

Once inhaled, PM$_{2.5}$ can cause serious health effects because of its ability to pass through the nose and throat and enter the lungs, thus affecting the lungs and heart. Inhaled particles cause systemic inflammation and oxidative stress that can exacerbate respiratory and cardiovascular disease. Particles can also lead to autonomic dysfunction and central nervous system activation, increasing heart rate, blood pressure, coagulation, restriction of blood vessels, and heart rhythm abnormalities leading to adverse cardiovascular outcomes. Some smaller particles (e.g., particles with an aerodynamic diameter <0.1 μm) may be able to translocate from the lung to the circulatory system, contributing to effects in other organ systems (Brook et al. 2010; USEPA 2019a).

Box 7.1 Prominent Impacts of Wildland Fire Smoke on Human Health

Asthma is a common chronic respiratory disease that affects all age and sociodemographic groups. It is characterized by chronic inflammation of the bronchi and smaller airways, with intermittent airway constriction, causing shortness of breath, wheezing, chest tightness, and coughing, sometimes accompanied by excess mucus production. During an asthma attack, the muscles tighten around the airways, and the lining of the airways become inflamed and swollen, constricting the flow of air. Symptoms are commonly triggered by exposure to air pollution or allergens and are usually worse at night and in the early morning. Physical exertion and cold air also trigger asthma symptoms.

A significant fraction of the population may have airway hyperresponsiveness; an exaggerated tendency of the large and small airways (bronchi and bronchioles, respectively) to constrict in response to respiratory irritants including cold air, dry air, and other stimuli, as well as wildfire smoke. Although airway hyperresponsiveness is considered a hallmark of asthma, this tendency may also be found in individuals without asthma, for example, during and following a lower respiratory tract infection. In such individuals, smoke exposure may cause asthma-like symptoms and bronchitis.

Chronic obstructive pulmonary disease (COPD), which is generally considered to encompass emphysema and chronic bronchitis, is a chronic respiratory disease characterized by irreversible breathing problems and restricted air flow (USEPA 2019a). COPD is also related to presence of other respiratory and heart conditions including heart failure, leading to highly compromised lung and heart capacity. In addition, COPD patients often may experience asthma-like symptoms. However, because their lung capacity has typically been seriously compromised, additional constriction of the airways in individuals with COPD may result in symptoms requiring medical attention. In addition, cigarette smoke is the primary cause COPD and individuals with this condition may also have heart and vascular disease and are potentially at risk of health effects due to smoke exposure from both conditions (General 2014; Morgan et al. 2018).

Cardiovascular diseases are the leading cause of mortality in the USA, comprising 30–40% of all deaths each year (NHBLI 2012). Most of these deaths occur in people over 65 years of age. Diseases of the circulatory system include high blood pressure, heart failure, vascular diseases such as coronary artery disease, and cerebrovascular conditions. These chronic conditions can render individuals susceptible to attacks triggered by air pollutants, such as wildfire smoke, including angina pectoris (transient chest pain), heart attacks, and sudden death due to cardiac arrhythmia, heart failure, or stroke.

In response to exposure to particulate matter, people with chronic heart disease may experience one or more of the following symptoms: shortness of

> breath; chest tightness; pain in the chest, neck, shoulder or arm; palpitations; or unusual fatigue or lightheadedness. Chemical messengers released into the blood because of particle-related lung inflammation may increase the risk of blood clot formation, angina episodes, heart attacks, and strokes.

PM may also contribute to respiratory infections by impairing physiological processes that remove inhaled viruses and bacteria and prevent them from entering the lungs and circulation. Even in healthy people, exposures to $PM_{2.5}$ can lead to respiratory effects, including reduced lung function and pulmonary inflammation, but these effects generally are considered transient. A review of biological mechanisms by which PM found in smoke can affect the human body is found in Neaher et al. (2007) and Adetona et al. (2016). Additional resources are in Brook et al. (2010) and "Integrated Science Assessment for Particulate Matter" (USEPA 2019b).

Coarse particles, also referred to as $PM_{10\text{-}2.5}$ (i.e., particles ≥ 2.5 μm to <10 μm in aerodynamic diameter), are generated primarily from mechanical operations rather than directly from wildland fires or formed downwind. Larger particles (aerodynamic diameter >10 μm) are generally of less concern because they usually do not enter the lower respiratory tract; however, they can irritate the eyes, nose, and throat.

The chemical composition of particles, particularly $PM_{2.5}$, in smoke can vary geographically (Chap. 5), and multiple toxicological studies have shown that some individual components (e.g., black carbon, metals) within the $PM_{2.5}$ mixture may be more toxic than others. As a result, research efforts have attempted to identify whether health effects are more consistently attributed to individual components or specific sources of $PM_{2.5}$. Evaluation of this evidence has found that, although many components and sources have been linked with health effects, the evidence does not indicate that any specific individual component or source is related more strongly to health effects than $PM_{2.5}$ (USEPA 2019a). More recent epidemiologic studies focusing on the health effects of wildland fire-specific $PM_{2.5}$ support this conclusion by reporting associations similar in magnitude between cardiovascular effects and $PM_{2.5}$ generated on days affected by smoke and days not affected by smoke (DeFlorio-Barker et al. 2019). However, some recent studies indicate that associations between respiratory effects, including asthma exacerbations, and wildfire-specific $PM_{2.5}$ may be larger in magnitude compared to associations reported in studies of ambient $PM_{2.5}$, creating a need for additional exploration into these potential differences (Borchers Arriagada et al. 2019; Kiser et al. 2020; DeFlorio-Barker et al. 2019).

7.2.2 Epidemiologic Evidence—Wildfire Smoke and $PM_{2.5}$

The primary body of evidence that forms the basis of the understanding about the health effects of wildland fire smoke stems from decades of research on the health effects of ambient particle pollution, specifically $PM_{2.5}$, conducted primarily in urban settings. Those research efforts, which have generally supported a linear concentration-response relationship, provide extensive information on health risks and biological mechanisms by which exposure to particle pollution can lead to health effects. Based on a comprehensive evaluation of the scientific evidence in support of National Ambient Air Quality Standards (NAAQS), the U.S. Environmental Protection Agency (USEPA) has concluded that, for both short- (days to weeks) and long-term (months to years) $PM_{2.5}$ exposure, there is a "causal relationship" for cardiovascular effects and mortality and "likely to be a causal relationship" for respiratory effects (USEPA 2019a).

In recent years, the number of studies examining the health effects specifically of wildfire smoke exposure has grown as well. However, many of these studies are conducted in different geographic locations and use various exposure metrics (e.g., monitoring data or modeled estimates of $PM_{2.5}$ or PM_{10}, smoke versus no-smoke days), which complicates the quantitative comparison of risk estimates across studies. There have been recent advancements in approaches used to estimate smoke exposure by blending chemical transport model predictions and satellite data with ground-based measurements through machine learning and data fusion methods. These new methods have the potential to reduce uncertainty and facilitate future quantitative comparisons of health effects across studies.

Some studies are beginning to assess whether there are differences in health effects between smoke from prescribed fire and wildfire. For example, a recent study by Prunicki et al. (2019) provided initial evidence of differences in markers of immune function, DNA methylation, and worsened respiratory outcomes in school-aged children exposed to wildfire smoke compared to prescribed fire smoke. However, it is unclear if these differences are primarily due to the difference in smoke concentrations between prescribed fires and wildfires.

Although the biological mechanisms that can lead to adverse health outcomes are similar between particles emitted from different sources, there are several considerations when generalizing health effects from ambient air pollution to effects from wildland fires. During wildfire events, populations are exposed to a complex mixture as with ambient air pollution, but at much higher concentrations of particles and gases that may have different effects (e.g., synergistic or additive effects) not present during exposure to non-wildfire ambient air pollution. And although much is known about the shape of the $PM_{2.5}$ concentration response function in ambient settings, less is known about how the response function may differ at hazardous smoke exposure levels, at repeated exposure to hazardous levels, or at peak exposures at the sub-daily level. Some wildfires burn infrastructure in addition to vegetation, likely shifting the mixture from those created solely through the combustion of natural fuels to emissions from the burning of often more toxic, man-made materials (Chap. 6). Finally, population exposure patterns and behavior modification patterns, as well as health impacts during lower intensity events such as prescribed burning, are largely unknown. The following sections characterize the current state of science with respect to health effects of wildland smoke exposure by integrating evidence from studies of ambient $PM_{2.5}$ and smoke exposure, specifically from wildfires spanning a variety of exposure assignment approaches. Except with firefighters, epidemiologic studies have largely not focused on the population-level health effects from exposure to prescribed fire smoke.

7.2.2.1 Health Outcomes

Respiratory Morbidity

Short-term exposure to $PM_{2.5}$ during smoke episodes can lead to breathing difficulties, especially for people with chronic lung diseases, such as asthma, chronic obstructive pulmonary disease (COPD), and other reactive airway diseases (Box 7.1). More than 24 million people in the USA, including more than 5 million children, experience chronic asthma, and 16 million experience COPD (CDC 2017, 2019). Epidemiologic studies on exposure to $PM_{2.5}$ demonstrate increased risk of emergency department visits and hospital admissions related to these outcomes (USEPA 2019a).

During smoke episodes, increased rates of respiratory-related physician visits (Mott et al. 2002; Moore et al. 2006; Lee et al. 2009; Henderson et al. 2011), emergency department (ED) visits (Tham et al. 2009; Rappold et al. 2011; Thelen et al. 2013; Johnston et al. 2014), and hospitalizations (Mott et al. 2005; Cançado et al. 2006; Chen et al. 2006; Delfino et al. 2009; Ignotti et al. 2010; Morgan et al. 2010; Henderson et al. 2011; Martin et al. 2013; DeFlorio-Barker et al., 2019) are reported with consistency. Among these, asthma- and COPD-related exacerbations and increased use of related medications are among the most commonly reported outcomes (Liu et al. 2015; Adetona et al. 2016; Reid et al. 2016a; Black et al. 2017; Cascio 2018; Gan et al. 2020). An analysis of asthma healthcare utilizations during the 2013 wildfire season in Oregon (Fig. 7.2) found a positive association between wildfire smoke $PM_{2.5}$ and various asthma morbidity measures, including ED visits, ambulatory care provided in an office setting, outpatient hospital visits, and asthma-rescue-inhaler medication prescriptions filled (Short Acting Beta-2 Agonists [SABA]) (Gan et al. 2020). This suggests that, in communities impacted by smoke, asthma-related healthcare utilizations could increase significantly by people seeking medical countermeasures and treatment in diverse health care facilities (e.g., primary care physician office, clinics, emergency rooms, hospitals).

Cardiovascular Morbidity

Epidemiologic and experimental studies have linked $PM_{2.5}$ exposure to increased risks of heart attacks, heart failure, cardiac arrhythmias, and other adverse effects in those with cardiovascular disease (USEPA 2019a). As a result, $PM_{2.5}$ is a concern for those with chronic heart diseases. Although fewer studies have examined the relationship between smoke exposure and cardiovascular outcomes, evidence is increasing concurrently with the increased frequency of large wildfires (Delfino et al. 2009; Henderson et al. 2011; Rappold et al. 2011, 2012; Dennekamp et al. 2015; Gan et al. 2017; Wettstein et al. 2018; DeFlorio-Barker et al. 2019; Yao et al. 2020). At the time of the last critical review of this literature completed in August 2015, the

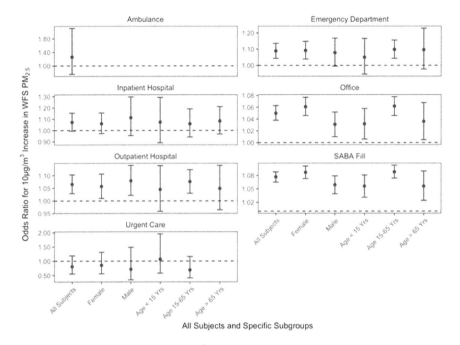

Fig. 7.2 Association between a 10 μg m^{-3} increase in wildland fire smoke-related PM$_{2.5}$ concentration and risk for asthma-related healthcare utilization events (Gan et al. 2020). Asthma-related health risks differed across subpopulations by age and gender, although those differences were not statistically significant for the 2013 wildfire season in Oregon. SABA stands for short acting beta-2 agonists, an asthma-rescue-inhaler medication

evidence of a relationship between smoke exposure and cardiovascular outcomes was considered less consistent compared with studies examining ambient PM$_{2.5}$ exposure (Reid et al. 2016a). However, several recent studies have reported elevated risks of specific cardiovascular outcomes—such as ischemic heart disease, heart failure, and dysrhythmia (Dennekamp et al. 2015; Yao et al. 2016; Wettstein et al. 2018; DeFlorio-Barker et al. 2019; Jones et al. 2020)—with the magnitude of association consistent with those previously reported in studies of ambient PM exposure (USEPA 2019a).

A number of factors may explain the inconsistent results reported in earlier reviews of evidence related to wildland fire smoke exposure and cardiovascular outcomes as compared to those examining ambient PM$_{2.5}$ effects. These factors include differences in the exposure metric used across studies (e.g., PM$_{10}$ versus PM$_{2.5}$ and smoke day versus no-smoke day) and differences in the ability to accurately assess exposure to ambient PM$_{2.5}$ versus exposure to wildland fire smoke (Liu et al. 2015; Fann et al. 2018). In addition, it has been hypothesized that the difference in results for cardiovascular effects could reflect individuals taking protective action to address acute respiratory effects, which may reduce the risk of other severe outcomes, including cardiovascular, that might result in hospitalizations (DeFlorio-Barker et al. 2019).

Mortality

Extensive epidemiologic evidence from studies conducted across the USA and elsewhere has demonstrated a relationship between short-term $PM_{2.5}$ exposure and mortality (USEPA 2019a). The limited number of studies examining wildland fire smoke exposure and mortality provide evidence consistent with the larger body of evidence examining ambient $PM_{2.5}$ exposure (Morgan et al. 2010; Johnston et al. 2011; Analitis et al. 2012; Faustini et al. 2015; Linares et al. 2015, 2018; Kollanus et al. 2016; Yao et al. 2019; Doubleday et al. 2020).

Epidemiologic studies examining cause-specific mortality and wildland fire smoke exposure are limited in number and have only reported evidence of positive associations with cardiovascular-related mortality, although there is extensive evidence indicating a relationship between short-term ambient $PM_{2.5}$ exposure and respiratory- and cardiovascular-related mortality (Johnston et al. 2011; Analitis et al. 2012; Faustini et al. 2015; Kollanus et al. 2016). This difference in results between wildfire smoke and ambient $PM_{2.5}$ exposure could be attributed to a number of factors including that cardiovascular mortality accounts for a larger fraction of total non-accidental mortality (~33%) in comparison with respiratory mortality (~7%); in combination with wildland fire events being of short duration (a few days to a few months), the statistical power of a study to observe an association is reduced. Although it is worthwhile to speculate on why there is a difference in results between studies examining mortality associated with short-term wildland fire smoke exposures compared to ambient $PM_{2.5}$ exposures, it is important to reiterate the extensive evidence demonstrating positive associations between short-term $PM_{2.5}$ exposure and mortality in studies conducted using different exposure assessment methodologies, in different geographic locations, and in populations with different demographic characteristics (USEPA 2019a).

Other Health Outcomes

In addition to respiratory and cardiovascular effects and mortality, new evidence exists on other potential health effects from both short- and long-term $PM_{2.5}$ exposure, including metabolic effects and effects on the nervous systems, which may also occur in response to wildland fire smoke exposure. There is some evidence indicating that short-term ambient $PM_{2.5}$ exposure may lead to altered metabolic function, such as changes in glucose and insulin homeostasis, whereas long-term ambient $PM_{2.5}$ exposures may lead to the development of metabolic syndrome and diabetes (USEPA 2019a). One study has reported that short-term exposure to wildfire $PM_{2.5}$ was associated with calls related to diabetes in ambulance dispatches and physician's assessments but not hospital diagnosis in British Columbia (Yao et al. 2020). Recent studies of ambient $PM_{2.5}$ exposures also provide evidence of relationships between long-term exposure and nervous system effects in adults, including cognitive declines and altered brain volume (USEPA 2019a). Evidence is more limited for associations between $PM_{2.5}$ exposures and other outcomes, such as developmental effects, including autism spectrum disorder and cognitive development (USEPA 2019a). Evidence is also limited for nervous system effects in relation to short-term PM$_{2.5}$ exposures (USEPA 2019a). Overall, there is limited research on the effects of short-term wildfire smoke exposure on metabolic effects and the effects on the nervous system.

Smoke also may have substantial effects on the mental health and emotional stress of communities. Although mental health effects have been studied to a limited degree for general wildfire contexts, the literature on mental health impacts from smoke exposure is even more sparse. When individuals in communities in the Northwest Territories (Canada) were told to stay indoors for most of the summer of 2014 because of prolonged air pollution from nearby wildfires, residents reported decreased physical activity and community engagement, both of which were associated with adverse mental health impacts (Dodd et al. 2018).

7.2.2.2 Life Stages and Populations Potentially at Risk of Smoke-Related Health Effects

Most healthy adults and children may experience transient health effects from smoke exposure without long-term consequences. However, some individuals may experience more severe effects. Although our understanding of the long-term health implications of wildfire smoke exposure is minimal, there is extensive evidence indicating that long-term exposure to ambient PM$_{2.5}$ can lead to a range of health effects (USEPA 2019a). The concentration and duration of exposure, individual susceptibility (including the presence of preexisting lung [e.g., asthma, COPD] or heart disease), and other factors play significant roles in determining whether someone will experience smoke-related health effects. Beyond those with preexisting health conditions, specific life stages and populations potentially at greater risk of experiencing an adverse health outcome include children under 18 years of age, pregnant people, developing fetuses, older adults, those of lower socioeconomic position (SEP), and outdoor workers.

Evidence of the particular life stages and populations potentially at increased risk of health effects from wildland fire smoke exposure stems from the large number of epidemiologic studies examining PM$_{2.5}$ in urban settings, which indicate that the risk of health effects attributed to PM$_{2.5}$ exposures differs based on life stage (children, older adults), health status, and SEP. Risk factors that influence whether a population or individual is at increased risk of health effects from smoke are similar to those for ambient PM$_{2.5}$ (Naeher et al. 2007; Liu et al. 2015; Adetona et al., 2016; Reid et al. 2016a).

Children

All children are considered at risk for experiencing a health effect because of air pollution and wildland fire smoke, regardless of whether they have a preexisting health condition. Compared with adults, children inhale more air per kilogram of body weight, spend more time outside, and may engage in more vigorous activity, all of which can contribute to increased exposure to PM$_{2.5}$, and ultimately, affect developing lungs (Sacks et al. 2011).

Short-term exposure to $PM_{2.5}$ can lead to increased respiratory symptoms, asthma exacerbations, and decreased lung function in children (USEPA 2019a). Similar respiratory effects have been reported in studies of smoke exposure on children, which have demonstrated increased coughing, wheezing, difficulty breathing, and chest tightness, resulting in school absences and declines in lung function (Jacobson et al. 2012, 2014). In addition, an experimental study conducted in infant monkeys has provided initial evidence indicating that smoke exposure during infancy may lead to altered lung and immune function in adolescence (Black et al. 2017).

In the USA, more than 6 million children have chronic lung diseases, such as asthma (CDC 2017, 2019). Higher rates of asthma ED visits and hospital admissions for children, especially infants and very young children, have been observed during and after wildland fires (Hutchinson et al. 2018). However, children without asthma can also experience respiratory symptoms in response to smoke exposure (Jalaludin et al. 2000; Jacobson et al. 2012, 2014), resulting in school absences and other limitations of normal childhood activities.

Although respiratory effects represent the primary adverse health outcome in children in response to smoke exposure, children also may be more likely to be exposed to ash from the fire itself, by engaging in outdoor activities and cleanup after a wildfire, or simply by their proximity to the ground. Fire ash can contain high concentrations of chemicals harmful to health.

Pregnant People and Fetuses

Individuals who are pregnant may be at increased risk of adverse health effects from wildland fire smoke because of the numerous physiologic changes that occur during pregnancy, such as increased blood and plasma volumes and increased respiratory rates, all of which can increase vulnerability to environmental exposures (USEPA 2019c). Developing fetuses also may be at risk when mothers are exposed to smoke during critical phases of human development. For example, Miller et al. (2019) showed that pregnant monkeys exposed to wildfire smoke could pass immune dysregulation on to the next generation.

Although only a few studies have examined the health effects of smoke exposure on pregnancy outcomes, studies on other combustion-related air pollutants have provided some evidence of adverse health effects. Specifically, there is substantial evidence of low birth weight caused by repeated exposures to cigarette smoke, including both active smoking and passive exposure (Martin and Bracken 1986; Windham et al. 2000; Wang et al. 2002).

Studies examining chronic maternal exposure to $PM_{2.5}$ (USEPA 2019a; DiCicca et al. 2020 Inoue et al. 2020; Li et al. 2021;) and indoor biomass smoke from wood-fire home heating devices have provided some evidence of adverse birth and obstetrical outcomes (e.g., decreased infant birth weight, preterm birth, birth defects) and perinatal mortality (Lakshmi et al. 2013; Amegah et al. 2014; Gehring et al. 2014; Weber et al. 2020). Smoke may also affect the developing fetus, resulting in lower birth weight in children that were in utero when wildland fire smoke was present (Holstius et al. 2012; Candido da Silva et al. 2014). For example, lower birth weight and preterm birth were found to be associated with wildfire smoke exposure in a study conducted across the state of Colorado (Abdo et al. 2019). In addition, there is some evidence for an increase in risk of congenital heart birth defects (atrial and ventricular septal defects) in relation to $PM_{2.5}$ exposure (Hu et al. 2020). Results from studies examining birth outcomes and smoke exposure are similar to the large number of studies examining $PM_{2.5}$ exposure and birth outcomes in terms of inconsistency in results. Some studies provide evidence of an association and others do not, potentially a result of inconsistencies in the exposure window (i.e., exposure over the entire pregnancy or specific trimesters) in which associations are observed (USEPA 2019a).

Older Adults

Epidemiologic studies of smoke exposure have demonstrated increased risk of health effects in older populations (often defined as people > 65 years of age). This is of particular concern because the number of US adults 65 years of age and older is expected to double by 2030 (Ortman et al. 2014). Older adults often are at increased risk because of higher prevalence of preexisting lung and heart diseases, reduced lung capacity, and a decline in physiologic processes (e.g., defense mechanisms) that occur with age (Sacks et al. 2011). Wettstein et al. (2018) reported a positive association between wildland fire smoke $PM_{2.5}$ density and ED visits for cardiovascular, cerebrovascular, and respiratory disease, with the greatest impact observed among adults 65 years and older, in contrast to no change in risk for those under 65 (Fig. 7.3). However, other studies examining smoke exposure did not find evidence

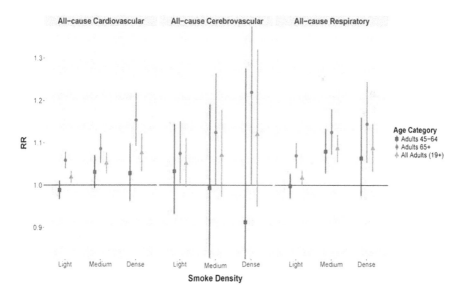

Fig. 7.3 Relative risk and 95% confidence intervals for all-cause cardiovascular outcomes relative to smoke-free days. Data are cumulative 0–4 days following exposure (lag days 0 to lag days 4), by age groups for eight California Air Basins (1 May–30 September 30 2015). From Wettstein et al. (2018)

that adults over 65 years of age are at increased risk; in fact, some studies have found higher risk for working-age adults (Henderson et al. 2011; Reid et al. 2016b), and others have found no clear differences by age category (Alman et al. 2016). Differences in findings across studies could result from examining different health endpoints, different exposure assessment methods (Gan et al. 2017), or different underlying health of the population.

Low Socioeconomic Position (SEP)

Metrics of SEP are used to characterize access to resources, both social and economic (Duncan et al. 2002; Galobardes et al. 2006). These indicators, which include education, employment, income, access to health care, and housing, may be considered at different scales, including individual, family, or community levels. SEP is associated with differential exposures to air pollution; individuals who have lower SEP or live in communities with lower SEP profiles are often exposed to higher concentrations of ambient air pollutants (USEPA 2019c). However, a recent study found that the non-Hispanic white populations in the USA, on average, live in communities with higher wildfire smoke exposure (Burke et al. 2021). When considering exposures to wildland fire smoke, individuals of lower SEP may have limited access to exposure-reducing resources, such as in-home filtration or portable air purifiers, as well as healthcare (USEPA 2019c).

Epidemiologic studies examining short-term $PM_{2.5}$ exposure demonstrate increased risk of health effects for individuals with lower SEP profiles. However, different studies use different metrics to represent low SEP (e.g., educational attainment, percent below poverty line). The few studies that have examined smoke and the role of SEP as a modifier of risk have similar results. Reid et al. (2016b) reported an inverse relationship between ZIP code-level higher median income and the risk of asthma, COPD, pneumonia, and all-cause respiratory ED visits during a wildland fire event (Fig. 7.4). Rappold et al. (2012) reported higher rates of ED visits in counties with lower SEP status compared with those with higher SEP status during smoke events. Conversely, Liu et al. (2017) found no difference in the likelihood of respiratory hospital admissions by educational attainment in an elderly cohort during periods more affected by smoke.

7.2.3 Other Smoke Pollutants Associated with Health Risks

Although particle pollution is of greatest concern to public health, wildland fire smoke is a complex mixture of pollutants that, individually on their own, also have been associated with health effects. Other pollutants found in smoke that are related to various health effects include tropospheric ozone, CO, and hazardous air pollutants (HAP).

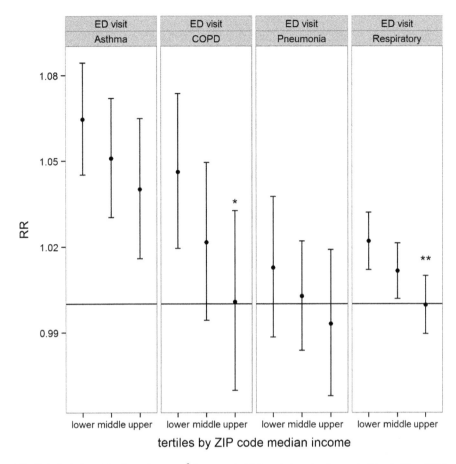

Fig. 7.4 Relative risks for a 5 μg m^{-3} increase in PM$_{2.5}$ during the fire period by tertile of ZIP code-level median income. **denotes p < 0.01, * denotes p < 0.05, and + denotes p < 0.10 compared with the lower tertile. From Reid et al. (2016a, b)

7.2.3.1 Ozone

Ground-level or tropospheric ozone is a widespread pollutant formed by the photochemical reaction of volatile organic compounds and nitrogen oxides in the presence of sunlight (Chaps. 5 and 6). Wildland fires emit large quantities of volatile organic compounds that can be transported in the atmosphere over large distances and enhance ozone production downwind, particularly over urban areas rich in nitrogen oxides from other sources (Brey and Fischer 2016; Larsen et al. 2018). Both epidemiologic and experimental studies have demonstrated that ground-level ozone exposure can result in adverse health effects, such as reduced lung function, inflammation of the airways, chest pain, coughing, wheezing, and shortness of breath—even in healthy people. These effects can be more serious in people with asthma and other lung diseases (USEPA 2020). Respiratory effects attributed to ozone exposure may lead to increased use of medication, school absences, respiratory-related hospital admissions, and ED visits for asthma and COPD. Evidence is more limited for the effects of ozone exposure on the cardiovascular system. However, short-term ozone exposures may lead to premature mortality (USEPA 2020).

Few studies have examined the role of ozone during wildland fire events on health outcomes. Reid et al. (2019) found that ozone downwind during a fire was associated with increased ED visits for respiratory symptoms, but that the associations were not apparent when the analysis was adjusted for $PM_{2.5}$, which was more strongly associated with respiratory health endpoints than was ozone. However, the study investigated only one fire and, given variability in the timing and location of impacts of wildland fire plumes on ozone production (Buysse et al. 2019), more research is needed on the health impacts of ozone during fire events.

7.2.3.2 Carbon Monoxide

CO is a highly toxic and odorless gas that impairs delivery of oxygen to the body's organs. Wildland fire smoke can contain high CO concentrations, but this generally does not pose a significant risk except for nearby at-risk populations or individuals (e.g., firefighting personnel). Exposure to CO poisoning is dangerous to all individuals, but people with cardiovascular disease may experience health effects, such as chest pain or cardiac arrhythmias, at lower levels of CO than do healthy people. At higher levels (such as those that occur in major structural fires), CO exposure can cause headache, weakness, dizziness, confusion, nausea, disorientation, visual impairment, coma, and death, even in otherwise healthy individuals (USEPA 2010).

7.2.3.3 Hazardous Air Pollutants

In addition to $PM_{2.5}$, ozone, and CO, wildland fire smoke also contains additional pollutants characterized as HAPs or toxic air contaminants (Reinhardt and Ottmar 2004). In the extensive list of HAPs, acetaldehyde, acrolein, formaldehyde, and benzene are among those of greatest concern (Chap. 6). These pollutants can be harmful to infants, children, pregnant people and their fetuses, older adults, persons engaging in physical activity, and those with existing lung, heart, or liver diseases. More information on concerns related to HAP exposures can be found in the publication "Wildfire Smoke: A Guide for Public Health Officials" (USEPA 2019c). However, not many epidemiologic studies examine these pollutants because they are not widely measured. Truly understanding the health effects of other pollutants from wildfire smoke would require more monitoring, especially in non-urban areas. Fully assessing population exposures to these additional pollutants is growing in importance as the number of structures burned during wildfire events increases.

7.2.4 Occupational/Cumulative and Chronic Exposures

Although workers in a range of outdoor occupations (e.g., agriculture, construction, landscaping, utility work) can be exposed to smoke, to date the only occupational smoke exposure research has been on wildland firefighters, and this work is relatively limited. Research on wildland firefighters is also the main source of the limited data related to the health effects of cumulative or chronic smoke exposure. While these findings provide insight into potential occupational health effects, caution should be used in applying them broadly as firefighters perform a variety of tasks, whether suppressing wildfires or implementing a prescribed burn, that can have different levels of smoke exposure (Box 7.2).

> **Box 7.2 Potential Smoke Exposure Incurred by Wildland Firefighters and Other Occupations**
> During the peak of the 2018 wildfire season, approximately 30,000 wildland firefighters were mobilized across the USA to suppress wildfires. When working on a large wildfire, firefighters sleep and eat at a base camp (incident command post) that is often near the fire, experiencing exposure to smoke, emissions from vehicles and generators (diesel exhaust), and road dust. Firefighters also perform a variety of tasks—operating a fire engine, constructing fireline, holding, mop up, and firing operations—all of which can have distinct smoke exposure.
>
> **Engine operators** work as a part of an engine crew, operating diesel pumps that provide water to crews working near the fire. **Fireline construction** involves clearing vegetation and digging or scraping down to mineral soil with hand tools to create a break in burnable vegetation to stop the spread of a fire. **Holding** refers to activities in which firefighters engage to ensure that the active fire has not crossed the fireline. After the fire has been controlled, crews **mop up** the area by extinguishing any burning material by digging out the material or applying water to stop smoldering material from re-igniting a fire. **Firing operations** involve setting an intentional fire, typically using torches filled with a 3:2 diesel/unleaded gasoline mixture, to reduce the available flammable material for the wildfire to consume.
>
> Workers who have outdoor occupations, such as agriculture, construction, landscaping, utility operations, and maintenance can also be exposed to wildfire smoke. There is little research measuring smoke exposure and associated adverse health effects for outdoor workers. In 2020, California adopted an emergency regulation (Regulation 5141.1, Protection from Wildfire Smoke under the California Code of Regulations, Title 8, Division 1, Chapter 4, of the General Industry Safety Orders) to protect outdoor workers from wildfire smoke, using $PM_{2.5}$ as an indicator for exposure to smoke. The regulation requires employers to determine the Air Quality Index (AQI) for $PM_{2.5}$

throughout a work shift, communicate and train employees about the hazards of smoke, and reduce exposures when AQI >151 for $PM_{2.5}$ (0.055 mg m^{-3}).

To reduce exposures at AQI <151, employers can implement engineering or administrative controls, such as providing enclosed spaces with filtered air, changing work schedules, reducing work intensity, or providing more rest breaks. At AQI values between 151 and 500, the regulation requires that respirators (approved by the National Institute for Occupational Safety and Health) be provided to employees for voluntary use, which does not require fit testing or medical evaluations. If AQI >500, employers are required to provide respirators and follow requirements under the respiratory protection regulation, reducing worker exposure to $PM_{2.5}$ <0.055 mg m^{-3}.

7.2.4.1 Acute Occupational/Firefighter Exposure

Assessments of the health effects of smoke exposure in wildland firefighters have focused mainly on acute effects across individual shifts and entire fire seasons. Wildland firefighters suppressing fires work long hours performing physically demanding work and can be exposed to high levels of smoke. Measuring exposures to smoke can be challenging because of the extreme environment in which wildland firefighters operate. Currently, wildland firefighters do not have respiratory protection available that both meets the demands of the arduous work performed and protects against all potentially hazardous exposures (Domitrovich et al. 2017). In addition, according to the National Wildfire Coordinating Group, only respirators approved by the National Institute for Occupational Safety and Health (NIOSH) shall be used on the fireline. While respiratory-type products (such as bandanas) are marketed to wildland firefighters, they are not NIOSH-approved (NIFC 2020).

Although smoke exposure in firefighters has been studied for decades (e.g., Reinhardt and Ottmar 2000), recent studies have focused primarily on exposure to CO, $PM_{2.5}$, and PM_4 (Navarro 2020). The permissible occupational exposure limit (OEL) standard set by the Occupational Safety and Health Administration (OSHA) for CO is 35 ppm, and 5 mg m^{-3} for respirable particles not otherwise regulated (PM_4) (OSHA 2017). Although field studies have measured different size fractions of PM, the particle size of combustion-generated particles for wood smoke is within a similar size range, which makes $PM_{2.5}$ and PM_4 comparable across wildfire smoke studies (Navarro et al. 2019). Across all field studies conducted since 2009, no measured exposure exceeded the OSHA OELs for CO or PM_4. However, these exposure limits do not consider the extended hours that wildland firefighters often work, nor does the PM_4 OEL account for the toxicity of wildland fire smoke from various compounds that make up or adsorb to the airborne particulates from smoke. In addition, most field studies collect data only at specific fire incidents, which makes their measurements limited to certain fire conditions and fuel types.

Reinhardt and Broyles (2019) collected field data across many prescribed fires and wildfires in the continental USA, providing smoke data that captures the variability of exposures at different fires. They reported that 22% (at wildfires) and 20% (at prescribed fires) of the measured PM_4 exceeded OELs that had been derived specifically for wildland firefighters to account for longer work shift, arduous work demands, and the exposure to multiple chemicals in smoke. They also examined factors in the wildfire environment that may predict exposures, finding that work task, time spent performing the work task, wind position, and type of wildfire crew contributed to exposure. Using the same data as Reinhardt and Broyles, Henn et al. (2019) found that fuel quantity, relative humidity, type of suppression strategy, and wind speed were significantly associated with elevated levels of CO exposure.

Recent field studies measuring smoke exposure found higher concentrations of PM and CO at prescribed fires compared with wildfires. This exposure difference could be due to the job tasks performed on prescribed fires. Past field studies report that wildland firefighters performing holding and firing, the two main job tasks performed on prescribed fires, can be exposed to higher concentrations of PM (Adetona et al. 2017; Reinhardt and Broyles 2019). At prescribed fires, Neitzel et al. (2009) reported the highest mean concentration for $PM_{2.5}$ (1.2 mg m^{-3}), and Reinhardt and Broyles (2019) reported the highest mean concentration for CO (4.4 ppm). At wildfires, the highest concentrations of measured PM_4 (0.51 mg m^{-3}) and CO (1.93 ppm) were reported for wildland firefighters performing mop up and fireline construction (Box 7.2), respectively (Gaughan et al. 2014). Highlighting the complexity of assessing smoke impacts, when looking across different smoke studies in the USA and Australia the highest firefighter exposures to $PM_{2.5}$ and CO were seen on prescribed fires in the southeast USA and wildfires in Colorado (Navarro 2020).

In addition to examining lung function, certain biomarkers have been measured in wildland firefighters to understand systemic inflammation (Swiston et al. 2008; Adetona et al. 2011a, b; Hejl et al. 2013). Gaughan et al. (2014) reported that wildland firefighters had a significant decline in lung function associated with high exposure to levoglucosan (a tracer for wood smoke) across work shifts. Adetona et al. (2017) demonstrated that, during a prescribed burn operation, firefighters engaged in lighting operations had elevated measurements for C-reactive protein, serum amyloid, and interleukin-8 compared with firefighters involved in holding activities (Box 7.2), an often smoky task where firefighters ensure that the fire does not cross control lines. The researchers hypothesized that in addition to smoke, exposure to combustion of diesel and gasoline during lighting operations could have led to this increase in inflammatory markers. Such elevated measures can be a sign of increased inflammation throughout the body, which can be associated with cardiovascular and other chronic diseases, and possibly the development of cancer. The same researchers reported positive association between creatinine-adjusted urinary mutagenicity and measures of exposure (polycyclic aromatic hydrocarbon metabolites, malondialdehyde, and light-absorbing carbon) across a work shift (Adetona et al. 2019). Results from this study indicate that exposure to smoke may include mutagens that can alter DNA and may cause cancer.

7.2.4.2 Chronic Exposure

While short-term smoke exposure is measured over days to weeks, cumulative or chronic exposures need to be examined over weeks to months, or even across multiple fire seasons. Extensive evidence among the general population indicates relationships between long-term $PM_{2.5}$ exposure and human health, including respiratory effects (e.g., changes in lung function), cardiovascular effects (e.g., development of atherosclerosis), and premature mortality (USEPA 2019a). However, evidence of health effects from chronic or long-term exposures specific to wildland fire smoke is limited, in part due to significant methodological challenges, such as identifying populations that experience wildfire smoke exposures over many years. The limited research on this topic to date has focused on wildland firefighters; caution should be used in comparing their occupational exposure and health effects to those experienced by the general population (Adetona et al. 2016).

Initial evidence from studies of wildland firefighters indicates that continuous occupational smoke exposure over multiple days or multiple consecutive fire seasons may have a cumulative effect on lung function, with some studies observing a progressive decline in lung function during burn seasons (Liu et al 1992; Gaughan et al. 2014). However, it is unclear if this decline persists across multiple fire seasons or if lung function is recovered in the winter off-season. A survey of firefighters found an association between the duration of their careers and self-reported health outcomes, including associations between ever being diagnosed with two cardiovascular measures (hypertension and/or heart arrhythmia) and the number of years worked as a firefighter (Semmens et al. 2016). Another study estimated that firefighters could have an increased risk of lung cancer (8–43% excess risk) and cardiovascular disease (16 to 30% excess risk) mortality over careers that ranged from 5 to 25 years (Navarro et al. 2019). Although this study was unable to adjust for individual factors (i.e., smoking, diet), the researchers used field measurements of PM_4 and heart rate (to calculate breathing rate) collected on wildfires across the USA in a dose–response model to estimate these risks.

7.3 Economic Costs and Losses from Smoke

Smoke from both wildfires and prescribed fires can affect local economies by altering production and consumption of economic goods and services, including transportation, manufacturing, agriculture, health services, recreation, and tourism. These impacts will differ by distance from the fire, and over time as the fire produces different quantities of smoke at different times.

This section assesses the state of science in measuring the economic impacts of smoke from wildland fires. Because research on economic impacts specific to smoke is limited, this section begins with an overview of potential ways that smoke may affect economies and discusses impacts that could result.

7.3.1 Theoretical Costs and Losses

Economists measure the costs and losses of a natural hazard (such as smoke production from a wildfire) or from an externality (such as smoke production from a prescribed fire) as changes in economic welfare, which includes gains and losses to both producers and consumers in each market, as well as true losses to the economy. Losses can be both direct and indirect. For wildfire, loss of life from smoke-induced illnesses or traffic fatalities due to decreased visibility are both direct losses. Indirect losses related to wildfire and smoke can include (1) business losses because of damage to capital, labor, and supply chains; (2) premature mortality resulting from damage to human health caused by wildfire emissions (Rittmaster et al. 2006; Fann et al. 2018; Borgschulte et al. 2019); and (3) losses due to altered perception of a community that might lead to changes in trade and investment (OECD 2016; World Bank 2016; Wouter Botzen 2019).

Economic costs can also result from actions taken to avoid or mitigate impact, also known as "averting behavior" (e.g., recreation shifts, evacuation, smoke avoidance) (Kochi et al. 2010). Averting behaviors, along with redirecting of recreation, tourism, investment, and trade, create losses in the affected area but may create gains elsewhere. Although there is no empirical evidence that intermittent smoke events have affected air quality sufficiently to alter investments or migration, economic theory would imply that changes in air quality would affect long-term investments and migration, potentially resulting in either spatial or temporal shifts in behavior. An overview of the costs of wildfires in Oregon noted the possibility of investment and migration losses, but did not attempt to quantify such losses (Lehner 2018). Box 7.3 summarizes the types of economic losses that could be enumerated from a smoke event.

Few studies have examined how wildland fire smoke affects any of the above potential economic losses. This is likely due, in part, to challenges in distinguishing between smoke-specific losses versus losses from the fire more broadly, although some work has addressed valuation of economic losses by measuring willingness to pay for a smoke-free day (World Bank 2016). Fann et al. (2018) applied the fire-attributable PM effects from Delfino (2009) in a national assessment of health-related economic losses associated with wildfire smoke using BenMAP-CE (USEPA 2014). However, the potential smoke-related costs and losses described in Box 7.3 have not been comprehensively assessed in any empirical studies. The following sections discuss three areas where limited research is available: (1) direct and indirect health costs and losses, (2) evacuation as an averting behavior, and (3) displaced recreation/tourism.

> **Box 7.3 Classification of Potential Damages, Costs, and Losses Caused by Wildland Fire Smoke**
>
> *Direct (occurs during the smoke event)*
> - Damage to productive capital and inventories
> - Damage to housing stock
> - Damage to infrastructure
> - Damage to human capital—includes direct fatalities and injuries from smoke, and fatalities and injuries from traffic accidents due to reduced visibility during the event
>
> *Indirect (can be during or after the smoke event)*
> - Damage to human capital (manifested after the event, including premature mortality and long-term morbidity)
> - Business losses because of supply shifts resulting from
>   - Damage to capital (human, infrastructure, manufacturing and business, agricultural)
>   - Disruption to supply chains (including transportation disruption)
> - Business losses because of demand shifts resulting from
>   - Averting behaviors affecting consumption (recreation, evacuation, and transportation)
>   - Increased health expenditures that reduce expenditures on other things
> - Losses because of perceptions
>   - Trade: shutdowns, cancellations, and transportation difficulties can make trade partners less confident, reducing trade in the short and long run—will trade elsewhere
>   - Investment: averting behavior—will invest elsewhere

7.3.2 Health Costs and Losses

A smoke event, whether from wildfire or prescribed fire, can lead to direct welfare losses through damage to human capital (fatalities and injuries), and indirect welfare losses from longer-term impacts on human health. Except in rare conditions, air pollution has not tended to result in hazardous visibility conditions (see 1952 London fog for a counter example; Wang et al. 2016). However, recent years have seen growing concern about smoke impacts on transportation, particularly in relation to formation of superfog (see Chap. 4). Here, we assess what is known about economic costs and losses resulting from wildland fire smoke.

Similar to health effects research, studies on the economic impacts of air pollution focus on the role of $PM_{2.5}$. Two main methods have been used to value morbidity impacts: (1) cost of illness (COI), and (2) contingent valuation-based willingness to pay (WTP). The two methods would be expected to generate different values, because the WTP measures also might include consideration of averting costs incurred by victims, as well as the direct costs of suffering an air-quality-based illness.

The economic value of disease varies by type of illness. Kochi et al. (2010) found that the ratio of WTP to COI is 1.3–2.4 for asthma, cataracts, and angina symptoms, with typical ratios ranging from 1.5 to 2.0. They also report that the USEPA estimates that WTP of each case during a wildfire smoke event ranges from $8 (2007 dollars) to avoid shortness of breath, to $10,971 to avoid respiratory hospital admissions, and $15,105 for a cardiovascular hospital admission. Nationwide annual hospital admissions that could be attributed to wildfire smoke from 2008 to 2012 ranged from 3900 to 8500 respiratory admissions and 1700–2800 cardiovascular admissions (Fann et al. 2018). Based on the economic value of those admissions, annual nationwide costs would be $43–93 million for respiratory illness and $26–42 million for cardiovascular illnesses caused by wildfires.

Annual economic losses from mortality effects of wildfire smoke have been estimated as nearly 1000 times higher than the economic losses attributed to morbidity from wildfire smoke in the USA. Kochi et al. (2010) identified a range of $2–14 million annually for economic loss per human life lost, conceived as the value of statistical life (VSL). A central estimate was established as $7.6 million (in 2007 dollars), which is the figure used by the USEPA in its impact assessments for air quality, and by Fann et al. (2018) in their nationwide assessment of the economic impacts of wildfire smoke emissions in the USA. Fann et al. (2018) summarized economic losses based on VSL for mortality from wildfire smoke events from 2008 to 2012, finding a median estimate of $11–20 billion per year for short-term exposures and hospital admissions and $76–130 billion per year for long-term effects of smoke exposure and respiratory hospital admissions. Therefore, the total value of the lives lost ranged from $88 billion in 2009 to $142 billion in 2008, with interannual variation depending on the area burned and location of fires relative to human populations. Overall, for the five years of the study, losses from premature mortality and hospital visits related to smoke averaged $103 billion per year in the USA.

7.3.3 Evacuation as an Averting Behavior

Evacuation is an economically disruptive activity, resulting in time off work, longer commutes, increased expenditures on housing and food, loss of social networks, stress, and discomfort. However, evacuation specifically because of smoke is not recommended; rather, sheltering in place is generally the preferred public health and safety redress (CARPA 2014; USEPA 2019c). A number of recommended public health actions can result in additional expenditures for residents and homeowners (e.g., creating a "clean room" at home, or missing work).

The accepted "correct" methodology for evaluating averting costs and losses resulting from a smoke event is to query residents on their willingness to pay to avoid a smoky day. A recent study found that residents were willing to pay an average of $129 per person per day to avoid a smoky day (Jones 2018). The author suggested this amount might account for the costs of averting behaviors, including evacuation and the need to clean indoor air or filter outdoor air. It may reflect expected expenditures on hospital visits to the extent that the residents surveyed had experienced these expenditures and were aware of the costs imposed by the smoke event. However, more than a single survey is needed to assess the value of an individual day of smoke avoidance.

7.3.4 Displaced Recreation and Tourism

In economic studies, it is assumed that a specific recreation choice provides the greatest benefit to the consumer and that being required, or choosing, to recreate in a different place or at a different time period to avoid smoke will result in a loss to the consumer and possibly result in increased costs (e.g., if the distance traveled must be increased). Although there is evidence that tourism and travel expenditures are affected by wildfires, these effects have not been isolated specifically for smoke (Butry et al. 2001; Rahn 2009; Kootenay Rockies Tourism Association 2018), although we can assume that at least some of the impacts are due to smoke. Other summaries have noted the costs of event cancellations from smoke, particularly when events are scheduled outdoors (e.g., outdoor theaters and raceways) (Lehner 2018).

7.4 Social Acceptance and Risk Communication

7.4.1 Social Acceptability

Research specific to public perspectives and response to wildland fire smoke is limited, with most relevant findings emerging from specific questions in studies that were focused broadly on social acceptance of prescribed fire. This research has consistently found that approximately 30% of survey respondents indicate someone in their household has a health issue affected by smoke. For these individuals and households, smoke tends to be a particularly salient topic. However, for most individuals, smoke is less critical in shaping acceptance of prescribed burning. Instead, understanding the beneficial ecological effects of fire and trust in those who are implementing the prescribed burn are more critical. Research also has shown that concerns about prescribed fire smoke tend to decrease with greater understanding of the ecological benefits of fire (McCaffrey and Olsen 2012).

Two recent studies (Blades et al. 2014; Olsen et al. 2014) did focus on public perceptions and preferences around smoke. This work included populations in both rural and urban areas of different regions in the USA and found that the majority of respondents were tolerant of all types of wildland fire smoke, with the highest tolerance for smoke from wildfires that were being actively suppressed. Slightly lower tolerance levels were found for smoke from prescribed fires and managed naturally-ignited fires. However, tolerance for smoke from both types of fire was significantly higher when forest health objectives were mentioned. Overall, the two studies found that human health status was the strongest predictor of smoke tolerance, with those who had experienced personal health effects from wildfire smoke being less tolerant. No significant differences between rural and urban residents were found in the two studies, and the few differences between regions were attributed to specific local context (Engebretson et al. 2016). One of the studies conducted a conjoint analysis in which the relative importance of advance warning was consistently found to be (slightly) higher than negative health effects or smoke duration as an influence on smoke tolerance levels. This highlights the importance of the communication process in allowing people to mitigate or avoid potential smoke impacts (Blades et al. 2014).

7.4.2 Risk Communication

Risk communication focuses on identifying the most effective means of providing information about a potential risk or threat to the health and safety of people or the environment and in a way that encourages appropriate protective actions (Reynolds and Seeger 2005; Steelman and McCaffrey 2013). Effective risk communication ensures that information is timely, accurate, and credible. To date, most research at the intersection of risk communication and wildland fire has focused on communication about a specific fire event (e.g., Steelman and McCaffrey 2013). However, interest in research on risk communication about the public health impacts of smoke events is growing (Olsen et al. 2014).

Deploying risk communication messaging during a wildfire smoke event is an essential strategy for protecting public health (Hano et al. 2020a). The messaging is designed to increase awareness of the issue among at-risk populations and the general public in an effort to support individual decisions that can mitigate or minimize exposure to smoke (Elliott 2014; Fish et al. 2017). During a smoke event, the type of information provided to affected populations may include status of the wildfire, wildfire smoke forecasts, and what actions people can take to protect themselves. Depending on factors such as air quality levels and the populations most at risk, specific public health recommendations regarding actions to reduce smoke exposure may differ but generally include: stay indoors, reduce outdoor physical activity, wear a properly fitted particulate-filtering respirator, develop and activate an asthma/COPD action plan, and create an at-home clean air room (Elliott 2014; USEPA 2019c; Joseph et al 2020). Recommendations and alerts may be deployed using television, radio, community meetings, newspapers, social media, and mobile applications. The source of information can include trusted news outlets, public health professionals, governmental agencies, medical providers, and fire response teams.

Investigating the effectiveness of risk communication for encouraging the adoption of specific recommended behaviors in the context of wildland fire is a growing area of research. Models of health behavior provide some guidance for structuring and deploying risk communication messaging. Recent research suggests that, similar to other public health domains, an individual's attitudes and risk perception may influence individual decision-making regarding exposure-reducing behaviors (Hano et al. 2020b). Findings and recommendations from the few studies that have examined risk communication during smoke events are similar to those found in the broader literature on health-risk communication.

7.4.2.1 Protective Actions

Minimal research has examined the effectiveness of wildfire smoke-related public health messages in leading to appropriate protective action. Most studies are retrospective and observational, limiting the ability to infer causation (Fish et al. 2017). Initial evidence indicates that messages that use plain language and are non-technical—such as "stay indoors" and "reduce outdoor physical activity"—are more commonly recalled and understood, and have a higher compliance rate (Sugerman et al. 2012; Dix-Cooper 2014; Fish et al. 2017). A study conducted in New South Wales, Australia, found that 42.5% of respondents reported a behavior change due to hearing a public health message, with staying indoors and reducing outdoor activity being the most common actions (Burns et al. 2010).

There is some evidence that certain groups are more likely to follow recommended public health actions including: those with chronic respiratory disease or high socioeconomic status, young or middle-aged adults, parents with young children, and native or primary English speakers (Dix-Cooper 2014; Fish et al. 2017). Although initial evidence shows the most common action taken was to stay indoors, it is difficult to infer from current studies if this action would have been taken by the population regardless of advisories (Fish et al. 2017). Limited information is available on adoption of other recommended protective actions, such as use of high-efficiency particulate air (HEPA) filters, and no information is available on compliance with public health messages or recommendations related to the duration of a smoke event, frequency of messages released, or length of time a message has been issued (Fish et al. 2017).

Few studies have evaluated how effective different recommended protective actions are in decreasing health impacts from smoke. However, there is growing evidence that demonstrates the effectiveness of filtration devices and behavioral modifications for reducing exposure to $PM_{2.5}$ (Rajagopalan et al. 2020). There is some evidence that air cleaners with HEPA filters can decrease exposures to smoke (Barn et al. 2016) and reduce the odds of experiencing adverse health effects associated with exposure (Mott et al. 2002).

Occupational studies have also shown that properly fitted N95 masks can substantially reduce particle exposure, although limited data are available on the benefits of using N95 masks by individuals who do not have access to fit testing for masks. There is growing evidence that demonstrates individual behaviors, such as avoiding air pollution by staying indoors, closing windows, and modifying exercise and outdoor activities, are effective strategies to reduce exposure to $PM_{2.5}$ (Rajagopalan et al. 2020).

Finally, although individuals are generally advised to stay indoors, evidence is limited on the degree, and duration, that air quality is better indoors as compared to outdoors (Chap. 6), making it unclear when it is appropriate to provide guidance on sheltering people in locations without good ventilation systems (Shrestha et al. 2019). As wildfire seasons increase in length and more fires cause higher air pollution levels, it also will be important to know if recommending sheltering in place for extended periods offsets other public health concerns (e.g., mental health considerations).

7.4.2.2 Communication Channels and Timing

Risk communication literature suggests conventional mass communication channels tend to be the primary mechanisms for conveying health-risk communication during emergency response situations. Wolkin et al. (2019) recently analyzed a set of Community Assessments for Public Health Emergency Response conducted from 2014 to 2017 for chemical spill, harmful algal bloom, hurricane, and flood emergencies. Their study found that television was the main information source for most individuals, followed by social media, and word of mouth. The assessment concluded that using multiple communication methods increases the likelihood of reaching a large portion of the population, and that the use of social media and word of mouth can increase the timeliness of messaging and provide message confirmation from sources that communities trust.

A review of the limited literature examining smoke communication (Fish et al. 2017) found a wide range of communication channels were used for smoke health messages, including mass media, road signage, face-to-face communication, hotlines, and the Internet. The few studies focused on risk communication during wildfire smoke events support the wider literature with respect to public reliance on mass communication channels. Two studies (Kolbe and Gilchrist 2009; Sugerman et al.2012) found that television was the primary source for smoke information, whereas another found a preference for smoke information from local papers as well as from television (for those under 40) and government-funded radio (for those over 40) (Burns et al. 2010). One US study did find a preference for receiving smoke warnings via a phone call rather than mass media (Blades et al. 2014).

The limited evidence about which communication channels are most effective in reaching at-risk populations indicates that certain groups are less likely to receive smoke warnings, including those with lower income or education levels, individuals for whom English is a second language, and the elderly (Dix-Cooper 2014; Fish et al. 2017). Finally, there is a lack of studies investigating effective communication for reaching culturally and linguistically diverse communities (Fish et al. 2017).

Evidence on the use of social media is limited, although there are some indications that mobile applications (apps) and Twitter data may be useful means of monitoring real-time air quality (Fish et al. 2017; Sachdeva et al. 2018). A number of apps have been developed specific to national air quality and health including: AirRater (Australia; https://airrater.org), AQHI Canada (Canada; https://open.alberta.ca/interact/aqhi-canada), AirNow (USEPA; www.airnow.gov/airnow-mobile-app), and Smoke Sense (USEPA; www.epa.gov/air-research/smoke-sense-study-citizen-science-project-using-mobile-app). Early research related to these apps has focused on understanding users' exposures to adverse air quality conditions and concurrent health symptoms (Johnston et al. 2018; Rappold et al. 2019), and preliminary evaluations of the effectiveness of mobile apps for mitigating adverse health impacts related to air quality (Rappold et al 2019; Campbell et al. 2020; Hatch 2020).

Trust in both the source and channel is a critical component of risk communication, influencing the saliency of messaging for the intended audience (Reynolds and Seeger 2005; Fish et al. 2017). General wildfire studies have found that trust in those providing information or implementing a practice is associated with a more positive public response to a practice (McCaffrey and Olsen 2012). No studies to date have specifically assessed trustworthy sources specific to smoke messages, but research about trusted information sources during a fire event found that information from fire officials and, to a lesser degree, friends and family were found to be trustworthy by a high portion of respondents; few found information from mass or social media to be trustworthy information sources (Steelman et al. 2015). Timing may also influence the saliency of health-risk communication. Mott et al. (2002) found that timely messaging about reducing wildfire smoke exposure using multiple channels (e.g., health care providers, radio broadcasts, and telephone messages) was associated with fewer reported respiratory symptoms among the general population.

7.5 Key Findings

Specific to social and societal implications of wildland fire smoke, the broadest body of research to date has focused on health effects of short-term $PM_{2.5}$ exposures. Much of this understanding is derived from broader air pollution studies, with a growing body of research specific to smoke from wildland fires that generally has congruent findings.

- Current understanding of the health effects of wildland fire smoke is rooted in decades of research examining the health effects of ambient air pollution, specifically $PM_{2.5}$, with the strongest evidence for effects on the respiratory and cardiovascular systems, as well as mortality.
  - Studies of wildland fire smoke report consistent evidence of a positive association with respiratory effects such as asthma and COPD exacerbations, and related increases in medication usage and physician visits.
  - A growing body of research is providing consistent evidence of links between wildfire smoke, specifically wildfire-specific $PM_{2.5}$ exposure and adverse cardiovascular impacts, including ischemic heart disease, cardiac arrhythmia, and heart failure.

- Although wildland fire smoke exposure has the potential to elicit adverse health effects among the general population, some populations and life stages are potentially at greater risk. Health effects from smoke are more common for individuals with preexisting cardiovascular or respiratory disease, older adults, children, pregnant people, outdoor workers, and possibly for those of lower SEP.
- Although a complete characterization of economic costs of wildland fire smoke is unavailable, limited economic research suggests that health costs and losses could be considerable. Economic losses associated with the health impacts of wildfire smoke are estimated to range from $88 billion to $142 billion per year, with the majority of those losses arising from premature human mortality.
- There is initial evidence that smoke from wildland fires can have a number of significant economic impacts, including those occurring due to temporary disruptions to lives and businesses, visibility problems resulting in transportation issues, alteration of recreation and tourism destinations or timing, and business losses through inventory damages and supply chain disruption.
- There is good evidence that, for most individuals, smoke is not a critical concern in determining acceptance of prescribed fire, instead this acceptance is based primarily on understanding of associated ecological benefits and trust in those who are implementing a prescribed burn. However, for households with health issues, smoke is a significant influence on prescribed fire acceptance. In addition, those who have recently experienced personal health effects from smoke are less tolerant of smoke in general. There is limited evidence that most individuals are receiving smoke messages and following advice to stay indoors.

7.6 Key Information Needs

Broadly speaking, a number of data gaps and challenges affect our ability to routinely conduct health-risk assessments and risk communication activities, and also limit assessments of economic effects of smoke, in relation to both short-term and long-term exposure. There are four general areas for which more information, both observational and experimental, is needed:

7.6.1 Understudied Health Effects

- More research is needed to better understand the health effects attributed to high smoke exposures (i.e., high $PM_{2.5}$ concentrations) during wildland fire events, specifically the effect of smoke exposure on cardiovascular and reproductive outcomes, as well as fetal development (which could potentially affect future chronic disease development and have intergenerational impacts), and the identification of populations at increased risk of smoke-related health effects.

- Improved understanding of the impact of repeated exposures to high concentrations of wildfire smoke and long-term chronic exposures over multiple fire seasons is critical, as this can inform studies of health impacts due to both short-term and long-term exposures. Currently, there are significant challenges in our ability to develop robust study designs and access historical health and smoke concentration data at appropriate spatiotemporal scales. The development of an occupational health surveillance system for wildland firefighter may provide information on the long-term effects of exposure to smoke for both workers and the general public.
- To fully understand the health burden of wildland fire smoke exposure, it will be important to consider both physical and mental health effects resulting from fire events. Therefore, we need to improve our knowledge of mental health impacts of smoke, particularly in relation to long-duration events.
- An emerging issue is the need to understand how wildfires that burn human infrastructure may adversely affect the mixture of air pollutants (Chaps. 5 and 6) and potential health effects.

7.6.2 Health Benefits and Trade-Offs of Public Health Interventions

- A better understanding is needed of which protective actions individuals are taking in response to smoke, whether the actions taken are appropriate given an individual's health status (e.g., being classified as "at-risk"), and barriers to taking protective actions to mitigate adverse health impacts.
- Experimental research is needed to assess the effectiveness of various risk communication messages and strategies (including best management practices) around smoke, particularly for at-risk groups.
- A better understanding is needed of effective risk communication strategies when high smoke concentrations coincide with other environmental hazards (e.g., high smoke PM levels during a heat wave). An assessment is needed of the effects of public-safety power-shutoff procedures on public health, especially should that coincide with an extreme heat event.

7.6.3 Economic Impacts

Additional research is needed to validate estimates of the values of human mortality, the economic impacts of evacuations specifically attributable to smoke on local economies, and the economic impacts of smoke on transportation networks and local businesses.

7.6.4 Central Repository of Standards and Actions

A central repository with data collection standards for smoke predictions would provide much needed information to accurately and consistently characterize population-level exposures. The public health community typically relies on exposure models, using data derived from various environmental data sources, including air quality measurements from monitors, chemical transport models, and remotely sensed information. Although such smoke exposure modeling efforts are feasible, they are resource intensive. The paucity of literature on effective solutions suggests that more research is needed to gain insight on health consequences of smoke exposure.

7.7 Conclusions

More empirical information on the acute and chronic health effects and economic impacts of wildland fire smoke exposures, as well as the protective actions that are taken in response to smoke risk communication, will be critical to improving future outcomes and effectively building on current biophysical research efforts. However, despite the large and potentially growing societal impacts of wildland fire smoke on a range of values, only a limited amount of scientific data are available that are useful for assessing these impacts. In particular, there are two hurdles specific to social research that present challenges to an improved understanding.

First, it is difficult to attribute a given health or economic outcome to one or more high pollution-exposure events or to measure the impacts of exposure on specific human or societal outcomes. Scientific studies of the effects of smoke on humans and society are observational; identifying the effects of specific interventions requires accounting for the influences of correlated and potentially confounding environmental and human factors. Furthermore, such science is complicated by the need to measure smoke transport across large distances, quantify human exposures to smoke, and model how humans respond to smoke with mitigating and averting behaviors.

Second, micro-level studies on how humans are affected by smoke are constrained by controls on personal health information, including health histories, which may be critical to obtaining a comprehensive and accurate assessment of the impacts of smoke on people. Economic studies of the effects of smoke are similarly inhibited by the limited availability of high-frequency economic time-series data, which could accurately reveal the timing and magnitudes of responses to evacuation orders or mitigating actions by individuals and businesses in affected locations.

Although these hurdles are significant, many can be addressed but would require research efforts similar to those that have contributed to the current understanding of the biophysical and chemical aspects of smoke.

References

Abdo M, Ward I, O'Dell K et al (2019) Impact of wildfire smoke on adverse pregnancy outcomes in Colorado, 2007–2015. Int J Environ Res Public Health 16:3720

Achtemeier GL (2008) Effects of moisture released during forest burning on fog formation and implications for visibility. J Appl Meteorol Climatol 47:1287–1296

Adetona AM, Adetona O, Gogal RM et al (2017) Impact of work task-related acute occupational smoke exposures on select proinflammatory immune parameters in wildland firefighters. J Occup Environ Med 59:679–690

Adetona O, Dunn K, Hall DB et al (2011a) Personal PM(2.5) exposure among wildland firefighters working at prescribed forest burns in Southeastern United States. J Occup Environ Hyg 8:503–511

Adetona O, Hall DB, Naeher LP (2011b) Lung function changes in wildland firefighters working at prescribed burns. Inhalation Toxicol 23:835–841

Adetona AM, Martin WK, Warren SH et al (2019) Urinary mutagenicity and other biomarkers of occupational smoke exposure of wildland firefighters and oxidative stress. Inhalation Toxicol 31:73–87

Adetona O, Reinhardt TE, Domitrovich J et al (2016) Review of the health effects of wildland fire smoke on wildland firefighters and the public. Inhalation Toxicol 28:95–139

Adetona O, Simpson CD, Onstad G, Naeher LP (2013) Exposure of wildland firefighters to carbon monoxide, fine particles, and levoglucosan. Ann Occup Hyg 57:979–991

Alman BL, Pfister G, Hao H et al (2016) The association of wildfire smoke with respiratory and cardiovascular emergency department visits in Colorado in 2012: a case crossover study. Environ Health 15:64

Amegah AK, Quansah R, Jaakkola JJ (2014) Household air pollution from solid fuel use and risk of adverse pregnancy outcomes: a systematic review and meta-analysis of the empirical evidence. PloS One 9:e113920

Analitis A, Georgiadis I, Katsouyanni K (2012) Forest fires are associated with elevated mortality in a dense urban setting. Occup Environ Med 69:158–162

Barn PK, Elliott CT, Allen RW et al (2016) Portable air cleaners should be at the forefront of the public health response to landscape fire smoke. Environ Health 15:116

Black C, Tesfaigzi Y, Bassein JA, Miller LA (2017) Wildfire smoke exposure and human health: significant gaps in research for a growing public health issue. Environ Toxicol Pharmacol 55:186–195

Blades JJ, Shook SR, Hall TE (2014) Smoke management of wildland and prescribed fire: understanding public preferences and trade-offs. Can J for Res 44:1344–1355

Borchers AN, Horsley JA, Palmer AJ et al (2019) Association between fire smoke fine particulate matter and asthma-related outcomes: systematic review and meta-analysis. Environ Res 179:108777

Borgschulte M, Molitor D, Zou EY (2019) Air pollution and the labor market: evidence from wildfire smoke. Online working paper. https://www.aeaweb.org/conference/2020/preliminary/paper/YYnFtk9N#:~:text=We%20calculate%20that%2017.7%20days,%24147%20billion%20in%202018%20dollars. Accessed 6 Jan 2021

Brey SJ, Fischer EV (2016) Smoke in the city: how often and where does smoke impact summertime ozone in the United States? Environ Sci Technol 50:1288–1294

Brook RD, Rajagopalan S, Pope CA III et al (2010) Particulate matter air pollution and cardiovascular disease: an update to the scientific statement from the American Heart Association. Circulation 121:2331–2378

Burke M, Driscoll A, Heft-Neal S et al (2021) The changing risk and burden of wildfire in the United States. In: Proceedings of the national academies of sciences, USA, vol 118, p e2011048118

Burns R, Robinson P, Smith P (2010) From hypothetical scenario to tragic reality: a salutary lesson in risk communication and the Victorian 2009 bushfires. Aust N Z J Public Health 34:24–31

Butry DT, Mercer DE, Prestemon JP et al (2001) What is the price of catastrophic wildfire? J Forest 99:9–17

Buysse CE, Kaulfus A, Nair U, Jaffe DA (2019) Relationships between particulate matter, ozone, and nitrogen oxides during urban smoke events in the Western US. Environ Sci Technol 53:12519–12528

California Air Response Planning Alliance (CARPA) (2014) Wildfire smoke response coordination—best practices being implemented by agencies in California, Working draft. https://ww3.arb.ca.gov/smp/progdev/iasc/wildfireresponse.pdf. Accessed 18 June 2020

Campbell SL, Jones PJ, Williamson GJ et al (2020) Using digital technology to protect health in prolonged poor air quality episodes: a case study of the AirRater App during the Australian 2019–20 Fires. Fire 2020:40

Cançado JE, Saldiva PH, Pereira LA et al (2006) The impact of sugar cane–burning emissions on the respiratory system of children and the elderly. Environ Health Perspect 114:725–729

Candido da Silva AM, Moi GP, Mattos IE, de Souza Hacon S (2014) Low birth weight at term and the presence of fine particulate matter and carbon monoxide in the Brazilian Amazon: a population-based retrospective cohort study. BMC Pregnancy Childbirth 14:309

Cascio WE (2018) Wildland fire smoke and human health. Sci Total Environ 624:586–595

Centers for Disease Control and Prevention (CDC) (2017) Fast Stats: Asthma. https://www.cdc.gov/nchs/fastats/asthma.htm. Accessed 18 June 2020

Centers for Disease Control and Prevention (CDC) (2019) Basics about COPD. https://www.cdc.gov/copd/basics-about.html. Accessed 18 June 2020

Chen L, Verrall K, Tong S (2006) Air particulate pollution due to bushfires and respiratory hospital admissions in Brisbane, Australia. Int J Environ Health Res 16:181–191

DeCicca P, Malak N (2020) When good fences aren't enough: the impact of neighboring air pollution on infant health. J Environ Econ Manage 102:102324

DeFlorio-Barker S, Crooks J, Reyes J, Rappold AG (2019) Cardiopulmonary effects of fine particulate matter exposure among older adults, during wildfire and non-wildfire periods, in the United States 2008–2010. Environ Health Perspect 127:037006

Delfino RJ, Brummel S, Wu J et al (2009) The relationship of respiratory and cardiovascular hospital admissions to the southern California wildfires of 2003. Occup Environ Med 66:189–197

Dennekamp M, Straney LD, Erbas B et al (2015) Forest fire smoke exposures and out-of-hospital cardiac arrests in Melbourne, Australia: a case-crossover study. Environ Health Perspect 123:959–964

Dix-Cooper L (2014) Evidence review: reducing time outdoors during wildfire smoke events. BC Centre for Disease Control, Vancouver, British Columbia. http://www.bccdc.ca/resourcegallery/Documents/Guidelines%20and%20Forms/Guidelines%20and%20Manuals/HealthEnvironment/WFSG_EvidenceReview_ReducingTimeOutdoors_FINAL_v6trs.pdf. Accessed 1 July 2020

Dodd W, Scott P, Howard C et al (2018) Lived experience of a record wildfire season in the Northwest Territories, Canada. Can J Public Health 109:327–337

Domitrovich J, Broyles G, Ottmar RD et al (2017) Wildland fire smoke health effects on wildland firefighters and the public. U.S. Forest Service, Pacific Northwest Research Station, Final Report, JFSP13-1-02-14. https://www.firescience.gov/projects/13-1-02-14/project/13-1-02-14_final_report.pdf. Accessed 18 June 2020

Doubleday A, Schulte J, Sheppard L et al (2020) Mortality associated with wildfire smoke exposure in Washington state, 2006–2017: a case-crossover study. Environ Health 19:1–10

Duncan GJ, Daly MC, McDonough P, Williams DR (2002) Optimal indicators of socioeconomic status for health research. Am J Public Health 92:1151–1157

Elliott CT (2014) Guidance for BC public health decisions makers during wildfire smoke events. BC Centre for Disease Control, Vancouver, British Columbia. http://www.bccdc.ca/resourcegallery/Documents/Guidelines%20and%20Forms/Guidelines%20and%20Manuals/Health-Environment/WFSG_BC_guidance_2014_09_03trs.pdf. Accessed 1 July 2020.

Engebretson JM, Hall TE, Blades JJ et al (2016) Characterizing public tolerance of smoke from wildland fires in communities across the United States. J Forest 114:601–609

Fann N, Alman B, Broome RA et al (2018) The health impacts and economic value of wildland fire episodes in the US: 2008–2012. Sci Total Environ 610:802–809

Faustini A, Alessandrini ER, Pey J et al (2015) Short-term effects of particulate matter on mortality during forest fires in Southern Europe: results of the MED-PARTICLES Project. Occup Environ Med 72:323–329

Fish J, Peters M, Ramsey I et al (2017) Effectiveness of public health messaging and communication channels during smoke events: a rapid systematic review. J Environ Manage 193:247–256

Galobardes B, Shaw M, Lawlor DA et al (2006) Indicators of socioeconomic position (part 1). J Epidemiol Community Health 60:7–12

Gan RW, Ford B, Lassman W et al (2017) Comparison of wildfire smoke estimation methods and associations with cardiopulmonary-related hospital admissions. GeoHealth 1:122–136

Gan RW, Liu JC, Ford B et al (2020) The association between wildfire smoke exposure and asthma-specific medical care utilization in Oregon during the 2013 wildfire season. J Eposure Sci Environ Epidemiol 1:11

Gaughan DM, Piacitelli CA, Chen BT et al (2014) Exposures and cross-shift lung function declines in wildland firefighters. J Occup Environ Hyg 11:591–603

Gehring U, Tamburic L, Sbihi H et al (2014) Impact of noise and air pollution on pregnancy outcomes. Epidemiology 25:351–358

General S (2014) The health consequences of smoking—50 years of progress: a report of the surgeon general. US Department of Health and Human Services

Groß S, Esselborn M, Weinzierl B et al (2013) Aerosol classification by airborne high spectral resolution lidar observations. Atmos Chem Phys 13:2487–2505

Hano MC, Prince SE, Wei L et al (2020) Knowing your audience: a typology of smoke sense participants to inform wildland fire smoke health risk communication. Front Public Health 8:143

Hano MC, Wei L, Hubbell B, Rappold AG (2020) Scaling up: citizen science engagement and impacts beyond the individual. Citizen Sci Theory Pract 5:1

Hatch A (2020) WSU study will use phone app to help people with asthma during wildfire season. WSU Insider, 12 August. Washington State University, Pullman. https://news.wsu.edu/2020/08/12/wsu-study-will-use-phone-app-help-people-asthma-wildfire-season. Accessed 28 Jan 2021

Hejl AM, Adetona O, Diaz-Sanchez D et al (2013) Inflammatory effects of woodsmoke exposure among wildland firefighters working at prescribed burns at the Savannah River Site, SC. J Occup Environ Hyg 10:173–180

Henderson SB, Brauer M, MacNab YC, Kennedy SM (2011) Three measures of forest fire smoke exposure and their associations with respiratory and cardiovascular health outcomes in a population-based cohort. Environ Health Perspect 119:1266–1271

Henn SA, Butler C, Li J et al (2019) Carbon monoxide exposures among U.S. wildland firefighters by work, fire, and environmental characteristic and conditions. J Occup Environ Hyg 16:793–803

Holstius DM, Reid CE, Jesdale BM, Morello-Frosch R (2012) Birth weight following pregnancy during the 2003 Southern California wildfires. Environ Health Perspect 120:1340–1345

Hu CY, Huang K, Fang Y et al (2020) Maternal air pollution exposure and congenital heart defects in offspring: a systematic review and meta-analysis. Chemosphere 253:126668

Hutchinson JA, Vargo J, Milet M et al (2018) The San Diego 2007 wildfires and medi-cal emergency department presentations, inpatient hospitalizations, and outpatient visits: an observational study of smoke exposure periods and a bidirectional case-crossover analysis. PLoS Med 15:e1002601

Inouea T, Nunokawab N, Kurisub D, Ogasawara K (2020) Particulate air pollution, birth outcomes, and infant mortality: evidence from Japan's automobile emission control law of 1992. SSM Popul Health 11:100590

Ignotti E, Hacon SdeS, Junger WL et al (2010) Air pollution and hospital admissions for respiratory diseases in the subequatorial Amazon: a time series approach. Cadernos de Saude Publica 26:747–761

Jacobson LdaS, Hacon SdeS, Castro HA et al (2012) Association between fine particulate matter and the peak expiratory flow of schoolchildren in the Brazilian subequatorial Amazon: a panel study. Environ Res 117:27–35

Jacobson LdaS, Hacon SdeS, Castro HA et al (2014) Acute effects of particulate matter and black carbon from seasonal fires on peak expiratory flow of schoolchildren in the Brazilian Amazon. PloS One 9:e104177

Jalaludin B, Smith M, O'Toole B, Leeder S (2000) Acute effects of bushfires on peak expiratory flow rates in children with wheeze: a time series analysis. Aust N Z J Public Health 24:174–177

Johnston F, Hanigan I, Henderson S et al (2011) Extreme air pollution events from bushfires and dust storms and their association with mortality in Sydney, Australia 1994–2007. Environ Res 111:811–816

Johnston F, Purdie S, Jalaludin B et al (2014) Air pollution events from forest fires and emergency department attendances in Sydney, Australia 1996–2007: a case-crossover analysis. Environ Health 13:105

Johnston FH, Wheeler AJ, Williamson GJ et al (2018) Airrater Tasmania: using smartphone technology to understand local environmental drivers of symptoms in people with asthma and allergic rhinitis. J Allergy Clin Immunol 141:AB84

Jolly WM, Cochrane MA, Freeborn PH et al (2015) Climate-induced variations in global wildfire danger from 1979 to 2013. Nat Commun 6:7537

Jones BA (2018) Willingness to pay estimates for wildfire smoke health impacts in the US using the life satisfaction approach. J Environ Econ Policy 7:403–419

Jones CG, Rappold AG, Vargo J et al (2020) Out-of-hospital cardiac arrests and wildfire-related particulate matter during 2015–2017 California wildfires. J Am Heart Assoc 9:e014125

Joseph G, Schramm PJ, Vaidyanathan A, Breysse P, Goodwin B (2020) Evidence on the use of indoor air filtration as an intervention for wildfire smoke pollutant exposure. United States Centers for Disease Control and Prevention, Atlanta, GA. https://www.cdc.gov/air/wildfire-smoke/social media/Wildfire-Air-Filtration-508.pdf. Accessed July 2020

Kiser D, Metcalf WJ, Elhanan G et al (2020) Particulate matter and emergency visits for asthma: a time-series study of their association in the presence and absence of wildfire smoke in Reno, Nevada, 2013–2018. Environ Health 19:92

Kochi I, Donovan GH, Champ PA, Loomis JB (2010) The economic cost of adverse health effects from wildfire-smoke exposure: a review. Int J Wildland Fire 19:803–817

Kolbe A, Gilchrist KL (2009) An extreme bushfire smoke pollution event: health impacts and public health challenges. NWS Public Health Bulletin 20:19–23

Kollanus V, Prank M, Gens A et al (2016) Mortality due to vegetation fire-originated PM2. 5 exposure in Europe-assessment for the years 2005 and 2008. Environ Health Perspect 125:30–37

Kondo MC, De Roos AJ, White LS et al (2019) Meta-analysis of heterogeneity in the effects of wildfire smoke exposure on respiratory health in North America. Int J Environ Res Public Health 16:960

Kootenay Rockies Tourism Association. (2018) Economic impact of 2017 wildfires on tourism in the Kootenay Rockies tourism region (final). Kootenay Rockies Tourism Association, Kimberley, British Columbia. https://www.krtourism.ca/wp-content/uploads/2018/10/Wildfire-Recovery-Survey-2018.pdf. Accessed 22 June 2020

Lakshmi P, Virdi NK, Sharma A et al (2013) Household air pollution and stillbirths in India: analysis of the DLHS-II National Survey. Environ Res 121:17–22

Larsen AE, Reich BJ, Ruminski M, Rappold AG (2018) Impacts of fire smoke plumes on regional air quality, 2006–2013. J Eposure Sci Environ Epidemiol 28:319

Lee T-S, Falter K, Meyer P et al (2009) Risk factors associated with clinic visits during the 1999 forest fires near the Hoopa Valley Indian Reservation, California, USA. Int J Environ Health Res 19:315–327

Lehner J (2018) Wildfires: a preliminary economic assessment. Oregon Office of Economic Analysis, August 8. https://oregoneconomicanalysis.com/2018/08/08/wildfires-a-preliminary-economic-assessment. Accessed 2 Oct 2019

Li G, Li L, Liu D et al (2021) Effect of PM2.5 pollution on perinatal mortality in China. Sci Rep 11, 7596

Linares C, Carmona R, Salvador P, Díaz J (2018) Impact on mortality of biomass combustion from wildfires in Spain: a regional analysis. Sci Total Environ 622:547–555

Linares C, Carmona R, Tobías A et al (2015) Influence of advections of particulate matter from biomass combustion on specific-cause mortality in Madrid in the period 2004–2009. Environ Sci Pollut Res 22:7012–7019

Liu D, Tager IB, Balmes JR, Harrison RJ (1992) The effect of smoke inhalation on lung function and airway responsiveness in wildland fire fighters. Am Rev Respir Dis 146:1469–1473

Liu JC, Pereira G, Uhl SA et al (2015) A systematic review of the physical health impacts from non-occupational exposure to wildfire smoke. Environ Res 136:120–132

Liu JC, Wilson A, Mickley LJ et al (2017) Who among the elderly is most vulnerable to exposure to and health risks of fine particulate matter from wildfire smoke? Am Epidemiol 186:730–735

Martin KL, Hanigan IC, Morgan GG et al (2013) Air pollution from bushfires and their association with hospital admissions in Sydney, Newcastle and Wollongong, Australia 1994–2007. Aust N Z J Public Health 37:238–243

Martin TR, Bracken MB (1986) Association of low birth weight with passive smoke exposure in pregnancy. Am J Epidemiol 124:633–642

McCaffrey SM, Olsen CS (2012) Research perspectives on the public and fire management: a synthesis of current social science on eight essential questions. (General Technical Report NRS-GTR-104). U.S. Forest Service, Northern Research Station, Newtown Square

McClure CD, Jaffe DA (2018) US particulate matter air quality improves except in wildfire-prone areas. Proc Nat Acad Sci USA 115:7901–7906

Miller LA (2019) Are adverse health effects from air pollution exposure passed on from mother to child? (Final report CN 15-303, California Air Resources Board). Regents of the University of California, Davis. https://ww2.arb.ca.gov/sites/default/files/classic/research/apr/past/15-303.pdf. Accessed 5 Feb 2021

Miranda AI, Martins V, Cascão P et al (2010) Monitoring of firefighters exposure to smoke during fire experiments in Portugal. Environ Int 36:736–745

Moore D, Copes R, Fisk R et al (2006) Population health effects of air quality changes due to forest fires in British Columbia in 2003. Can J Public Health 97:105–108

Morgan AD, Zakeri R, Quint JK (2018) Defining the relationship between COPD and CVD: what are the implications for clinical practice? Ther Adv Respir Dis 12:1753465817750524

Morgan G, Sheppeard V, Khalaj B et al (2010) Effects of bushfire smoke on daily mortality and hospital admissions in Sydney, Australia. Epidemiology 47–55

Mott JA, Mannino DM, Alverson CJ et al (2005) Cardiorespiratory hospitalizations associated with smoke exposure during the 1997 Southeast Asian forest fires. Int J Hyg Environ Health 208:75–85

Mott JA, Meyer P, Mannino D et al (2002) Wildland forest fire smoke: health effects and intervention evaluation, Hoopa, California, 1999. West J Med 176:157–162

Naeher LP, Brauer M, Lipsett M et al (2007) Woodsmoke health effects: a review. Inhalation Toxicol 19:67–106

National Heart, Lung, and Blood Institute (NHBLI) (2012) NHLBI FY 2012 Fact Book. U.S. National Institutes of Health, Dept. of Health and Human Services, Chap. 4. https://www.nhlbi.nih.gov/files/docs/factbook/FactBook2012.pdf. Accessed 11 Feb 2021

National Interagency Fire Center (NIFC) (2020) Interagency standards for fire and fire aviation operations (NFES 2724). Interagency Standards for Fire and Fire Aviation Operations Group, Boise.https://www.nifc.gov/PUBLICATIONS/redbook/2020/RedBookAll.pdf. Accessed 5 Feb 2021

Navarro K (2020) Working in smoke: Wildfire impacts on the health of firefighters and outdoor workers and mitigation strategies. Clin Chest Med 41:763–769

Navarro KM, Kleinman MT, Mackay CE et al (2019) Wildland firefighter smoke exposure and risk of lung cancer and cardiovascular disease mortality. Environ Res 173:462–468

Neitzel R, Naeher L, Paulsen M et al (2009) Biological monitoring of smoke exposure among wildland firefighters: a pilot study comparing urinary methoxyphenols with personal exposures to carbon monoxide, particular matter, and levoglucosan. J Eposure Sci Environ Epidemiol 19:349–358

O'Dell K, Ford B, Fischer EV, Pierce JR (2019) Contribution of wildland-fire smoke to us PM2.5 and its influence on recent trends. Environ Sci Technol 53:1797–1804

Occupational Safety and Health Administration (OSHA) (2017) Permissible Exposure Limits, 2017; Vol. 29 CFR 1910.1000.

Olsen C, Mazzotta D, Toman E, Fischer A (2014) Communicating about smoke from wildland fire: challenges and opportunities for managers. Environ Manage 54:571–582

Organisation for Economic Co-operation and Development (OECD) (2016) The economic consequences of outdoor air pollution: Policy highlights. OECD Policy Highlights, June. www.oecd.org.environment/the-economic-consequences-of-outdoor-air-pollution-9789264257474-en.htm. Accessed 22 June 2020

Ortman JM, Velkoff VA, Hogan H (2014) An aging nation: the older population in the United States (Report P25-1140 May). United States Census Bureau, Economics and Statistics Administration, Washington, DC. https://www.census.gov/prod/2014pubs/p25-1140.pdf. Accessed 22 June 2020

Permissible Exposure Limits, 29 CFR 1910.1000 C.F.R. § Table Z-1-A (2017)

Prunicki M, Kelsey R, Lee J et al (2019) The impact of prescribed fire versus wildfire on the immune and cardiovascular systems of children. Allergy 74:1989–1991

Rahn M (2009) Wildfire impact analysis. San Diego State University, San Diego, CA. http://universe.sdsu.edu/sdsu_newscenter/images/rahn2009fireanalysis.pdf. Accessed 22 June 2020

Rajagopalan S, Brauer M, Bhatnagar A et al (2020) Personal-level protective actions against particulate matter air pollution exposure: a scientific statement from the American Heart Association. Circulation 142:e411–e431, e448

Rappold AG, Cascio WE, Kilaru VJ et al (2012) Cardio-respiratory outcomes associated with exposure to wildfire smoke are modified by measures of community health. Environ Health 11:71

Rappold AG, Hano MC, Prince S et al (2019) Smoke Sense initiative leverages citizen science to address the growing wildfire-related public health problem. GeoHealth 3:443–457

Rappold AG, Stone SL, Cascio WE et al (2011) Peat bog wildfire smoke exposure in rural North Carolina is associated with cardiopulmonary emergency department visits assessed through syndromic surveillance. Environ Health Perspect 119:1415–1420

Reid CE, Brauer M, Johnston FH et al (2016) Critical review of health impacts of wildfire smoke exposure. Environ Health Perspect 124:1334–1343

Reid CE, Considine EM, Watson GL et al (2019) Associations between respiratory health and ozone and fine particulate matter during a wildfire event. Environ Int 129:291–298

Reid CE, Jerrett M, Tager IB et al (2016) Differential respiratory health effects from the 2008 northern California wildfires: a spatiotemporal approach. Environ Res 150:227–235

Reinhardt TE, Broyles G (2019) Factors affecting smoke and crystalline silica exposure among wildland firefighters. J Occup Environ Hyg 16:151–164

Reinhardt TE, Ottmar RD (2000) Smoke exposure at western wildfires. (Research Paper PNW-RP-525). U.S. Forest Service, Pacific Northwest Research Station, Portland

Reinhardt TE, Ottmar RD (2004) Baseline measurements of smoke exposure among wildland firefighters. J Occup Environ Hyg 1:593–606

Reisen F, Brown SK (2009) Australian firefighters' exposure to air toxics during bushfire burns of autumn 2005 and 2006. Environ Int 35:342–352

Reisen F, Hansen D, Meyer CP (2011) Exposure to bushfire smoke during prescribed burns and wildfires: firefighters' exposure risks and options. Environ Int 37:314–321

Reynolds B, Seeger M (2005) Crisis and emergency risk communication as an integrative model. J Health Commun 10:43–55

Rittmaster R, Adamowicz WL, Amiro B, Pelletier RT (2006) Economic analysis of health effects from forest fires. Can J for Res 36:868–877

Sachdeva S, McCaffrey S (2018) Using social media to predict air pollution during California wildfires. In: Proceedings of the international conference on social media and society, July 18–20; SM Society, Copenhagen

Sacks JD, Stanek LW, Luben TJ et al (2011) Particulate matter–induced health effects: who is susceptible? Environ Health Perspect 119:446–454

Semmens EO, Domitrovich J, Conway K, Noonan CW (2016) A cross-sectional survey of occupational history as a wildland firefighter and health. Am J Ind Med 59:330–335

Shrestha PM, Humphrey JL, Carlton EJ et al (2019) Impact of outdoor air pollution on indoor air quality in low-income homes during wildfire seasons. Int J Environ Res Public Health 16:3535

Steelman TA, McCaffrey S (2013) Best practices in risk and crisis communication: Implications for natural hazards management. Nat Hazards 65:683–705

Steelman TA, McCaffrey SK, Velez A-L, Briefel JA (2015) What information do people use, trust, and find useful during a disaster? Evidence from five large wildfires. Nat Hazards 76:615–634

Sugerman D, Keir J, Dee D et al (2012) Emergency health risk communication during the 2007 San Diego Wildfires: comprehension, compliance, and recall. J Health Commun 17:678–712

Swiston JR, Davidson W, Attridge S et al (2008) Wood smoke exposure induces a pulmonary and systemic inflammatory response in firefighters. Eur Respir J 32:129–138

Tham R, Erbas B, Akram M et al (2009) The impact of smoke on respiratory hospital outcomes during the 2002–2003 bushfire season, Victoria, Australia. Respirology 14:69–75

Thelen B, French NH, Koziol BW et al (2013) Modeling acute respiratory illness during the 2007 San Diego wildland fires using a coupled emissions-transport system and generalized additive modeling. Environ Health 12:94

U.S. Environment Protection Agency (USEPA) (2010) Integrated science assessment (ISA) for carbon monoxide (Final Report, Jan 2010). U.S. Environmental Protection Agency (EPA/600/R-09/019F). Washington, DC 2010. https://cfpub.epa.gov/ncea/isa/recordisplay.cfm?deid=218686

U.S. Environmental Protection Agency (USEPA) (2013) Integrated science assessment (ISA) for ozone and related photochemical oxidants (EPA 600/R-10/076F). Washington, DC

U.S. Environmental Protection Agency (USEPA) (2014) Environmental benefits mapping and analysis program—community edition (BenMAP-CE). https://www.epa.gov/benmap. Accessed 24 Sept 2019

U.S. Environmental Protection Agency (USEPA) (2019a) Integrated science assessment (ISA) for particulate matter (EPA/600/R-19/188). Washington, DC

U.S. Environmental Protection Agency (USEPA) (2019b) Wildfire smoke and your patients' health. Washington, DC. https://www.epa.gov/wildfire-smoke-course. Accessed 14 July 2020

U.S. Environmental Protection Agency (USEPA) (2019c) Wildfire smoke: a guide for public health officials (EPA-452/R-19-901). Washington, DC. https://www3.epa.gov/airnow/wildfire-smoke/wildfire-smoke-guide-revised-2019.pdf. Accessed 14 July 2020

U.S. Environmental Protection Agency (USEPA) (2020) Integrated science assessment (ISA) for ozone and related photochemical oxidants (EPA/600/R-20/012). Washington, DC. https://cfpub.epa.gov/ncea/isa/recordisplay.cfm?deid=348522. Accessed 5 Feb 2021

Wang X, Zuckerman B, Pearson C et al (2002) Maternal cigarette smoking, metabolic gene polymorphism, and infant birth weight. JAMA 287:195–202

Wang G, Zhang R, Gomez ME et al (2016) Persistent sulfate formation from London Fog to Chinese haze. Proc Nat Acad Sci USA 113:13630–13635

Weber E, Adu-Bonsaffoh K, Vermeulen R et al (2020) Household fuel use and adverse pregnancy outcomes in a Ghanaian cohort study. Reprod Health 17:29

Wettstein ZS, Hoshiko S, Fahimi J et al (2018) Cardiovascular and cerebrovascular emergency department visits associated with wildfire smoke exposure in California in 2015. J Am Heart Assoc 7:e007492

Windham GC, Hopkins B, Fenster L, Swan SH (2000) Prenatal active or passive tobacco smoke exposure and the risk of preterm delivery or low birth weight. Epidemiology 11:427–433

Wolkin A, Schnall A, Nakata N, Ellis E (2019) Getting the message out: Social media and word-of-mouth as effective communication methods during emergencies. Prehosp Disaster Med 34:89–94

World Bank (2016) The cost of air pollution: strengthening the economic case for action. International Bank for Reconstruction and Development, The World Bank, Washington, DC. http://documents.worldbank.org/curated/en/781521473177013155/pdf/108141-REVISED-Cost-of-PollutionWebCORRECTEDfile.pdf. Accessed 1 July 2020

Wouter Botzen WJ, Deschenes O, Sanders M (2019) The economic impacts of natural disasters: a review of models and empirical studies. Rev Environ Econ Pol 13:167–188

Yao J, Eyamie J, Henderson SB (2016) Evaluation of a spatially resolved forest fire smoke model for population-based epidemiologic exposure assessment. J Eposure Sci Environ Epidemiol 26:233

Yao J, Stieb DM, Taylor E, Henderson SB (2019) Assessment of the air quality health index (AQHI) and four alternate AQHI-Plus amendments for wildfire seasons in British Columbia. Can J Public Health 111:96–106

Yao J, Brauer M, Wei J et al(2020). Sub-daily exposure to fine particulate matter and ambulance dispatches during wildfire seasons: a case-crossover study in British Columbia, Canada. Environ Health Perspect 128:067006

Youssouf H, Liousse C, Roblou L et al (2014) Non-accidental health impacts of wildfire smoke. Int J Environ Res Public Health 11:11772–11804

Open Access This chapter is licensed under the terms of the Creative Commons Attribution 4.0 International License (http://creativecommons.org/licenses/by/4.0/), which permits use, sharing, adaptation, distribution and reproduction in any medium or format, as long as you give appropriate credit to the original author(s) and the source, provide a link to the Creative Commons license and indicate if changes were made.

The images or other third party material in this chapter are included in the chapter's Creative Commons license, unless indicated otherwise in a credit line to the material. If material is not included in the chapter's Creative Commons license and your intended use is not permitted by statutory regulation or exceeds the permitted use, you will need to obtain permission directly from the copyright holder.

Chapter 8
Resource Manager Perspectives on the Need for Smoke Science

Janice L. Peterson, Melanie C. Pitrolo, Donald W. Schweizer,
Randy L. Striplin, Linda H. Geiser, Stephanie M. Holm, Julie D. Hunter,
Jen M. Croft, Linda M. Chappell, Peter W. Lahm, Guadalupe E. Amezquita,
Timothy J. Brown, Ricardo G. Cisneros, Stephanie J. Connolly,
Jessica E. Halofsky, E. Louise Loudermilk, Kathleen M. Navarro,
Andrea L. Nick, C. Trent Procter, Heather C. Provencio, Taro Pusina,
Susan Lyon Stone, Leland W. Tarnay, and Cynthia D. West

Abstract Smoke from wildland fire is a significant concern to resource managers who need tools, knowledge, and training to analyze, address, and minimize potential impacts; follow relevant rules and regulations; and inform the public of possible effects. Successful navigation of competing pressures to appropriately use fire on the

---

J. L. Peterson (✉)
US Forest Service, Pacific Northwest Region, Seattle, WA, USA
e-mail: janice.peterson@usda.gov

M. C. Pitrolo
US Forest Service, Southern Region, Asheville, NC, USA
e-mail: melanie.pitrolo@usda.gov

D. W. Schweizer
US Forest Service, Pacific Southwest Region, Bishop, CA, USA
e-mail: donald.schweizer@usda.gov

R. L. Striplin
US Forest Service, Pacific Southwest Region, Riverside, CA, USA
e-mail: randy.striplin@usda.gov

L. H. Geiser · J. M. Croft · P. W. Lahm
US Forest Service, National Headquarters, Washington, DC, USA
e-mail: linda.geiser@usda.gov

J. M. Croft
e-mail: jennifer.croft@usda.gov

P. W. Lahm
e-mail: peter.lahm@usda.gov

S. M. Holm
University of California, Pediatric Environmental Health Specialty Unit (Western States), San Francisco, CA, USA
e-mail: stephanie.holm@ucsf.edu

J. D. Hunter
Northern Sierra Air Quality Management District, Portola, CA, USA

This is a U.S. government work and not under copyright protection in the U.S.; foreign copyright protection may apply 2022
D. L. Peterson et al. (eds.), *Wildland Fire Smoke in the United States*,
https://doi.org/10.1007/978-3-030-87045-4_8

landscape to manage fire-adapted and fire-dependent ecosystems, while protecting public health and other air quality values, depends on credible science and tools conceived of and developed in partnership between managers and the research community. Fire and smoke management are made even more complex by the current

e-mail: juliedhunter1@gmail.com

L. M. Chappell
US Forest Service, Intermountain Region, Logan, UT, USA
e-mail: linda.chappell@usda.gov

G. E. Amezquita · R. G. Cisneros
US Forest Service, Pacific Southwest Region, Clovis, CA, USA
e-mail: guadalupe.amezquita@usda.gov

R. G. Cisneros
e-mail: ricardo.cisneros@usda.gov

T. J. Brown
Desert Research Institute, Reno, NV, USA
e-mail: tim.brown@dri.edu

S. J. Connolly
US Forest Service, Office of Sustainability and Climate, Parsons, WV, USA
e-mail: stephanie.connolly@usda.gov

J. E. Halofsky
US Forest Service, Pacific Northwest Research Station, Olympia, WA, USA
e-mail: jessica.halofsky@usda.gov

E. L. Loudermilk
US Forest Service, Southern Research Station, Athens, GA, USA
e-mail: eva.l.loudermilk@usda.gov

K. M. Navarro
Centers for Disease Control and Prevention, National Institute for Occupational Safety and Health, Denver, CO, USA
e-mail: knavarro@cdc.gov

A. L. Nick
US Forest Service, Pacific Southwest Region, Fawnskin, CA, USA
e-mail: andrea.nick@usda.gov

C. T. Procter
Pacific Southwest Region, US Forest Servhice, Porterville, CA, USA
e-mail: tprocter14@gmail.com

H. C. Provencio
US Forest Service, Southwestern Region, Williams, AZ, USA
e-mail: heather.provencio@usda.gov

T. Pusina
US Forest Service, Pacific Southwest Region, Redding, CA, USA
e-mail: taro.pusina@usda.gov

S. L. Stone
US Environmental Protection Agency, Research Triangle Park, NC, USA
e-mail: stone.susan@epa.gov

condition of ecosystems as a result of fire exclusion and the future implications of a changing climate. This chapter describes the scope of smoke management, social and regulatory contexts, and pathways through which scientific information and tools can improve the accuracy and timeliness of management and communication with the public.

Keywords Air quality · Emission reduction techniques · Modelling · Prescribed fire · Public health · Smoke · Wildfire

8.1 Introduction

In many parts of the USA, fire-prone ecosystems have been altered by a combination of past management and changing environmental conditions. Accumulation of wildland fuels from decades of wildfire exclusion (including suppression) is contributing to increased wildfire size and severity across much of western North America (Schoennagel et al. 2017). As a result, smoke impacts have been severe and widespread in recent decades, with serious consequences for human health, economic impacts from lost tourism, and visibility impairment on roads and at airports (Chaps. 4 and 7). In addition, climate change is contributing to greater amplitude of seasonal weather extremes and altered ecosystem functions (Box 8.1). Wildfire season is starting earlier in the year and ending later. If a warmer climate does indeed cause more area burned by wildfires (Abatzoglou and Kolden 2013; Jolly et al. 2015), an increased frequency of degraded air quality, or "smoke waves" (Larkin et al. 2015; Liu et al. 2016), is likely. Communities have become more aware of how to prepare for wildfire, but the concept of being prepared for smoke is new and not yet widespread. Smoke exposure cannot be eliminated, but with preparation and planning, people can take actions to mitigate adverse effects of wildland fire smoke.

> **Box 8.1 Managing Smoke in a Changing Climate**
> **Jessica E. Halofsky, Stephanie J. Connolly, Cynthia D. West**
>
> Climate change is expected to cause more frequent wildfires across larger areas of the USA, altering vegetation patterns and ecosystem function (Reidmiller et al. 2018; Vose et al. 2018; Halofsky et al. 2020). Extreme fire weather is expected to become more frequent, affecting primarily the arid and semiarid west, but also affecting the upper Midwest and portions of the northeastern

---

L. W. Tarnay
US Forest Service, Pacific Southwest Region, Truckee, CA, USA
e-mail: leland.tarnay@usda.gov

C. D. West
US Forest Service, Northern Research Station, Madison, WI, USA
e-mail: cynthia.west@usda.gov

and southern USA (Barbero et al. 2015). Annual area burned in the western USA may increase by 2–6 times by the mid-twenty-first century compared to historical area burned in the twentieth century, depending on ecosystem and local climate (Kitzberger et al. 2017; Litschert et al. 2012; McKenzie et al. 2004). If a warmer climate does indeed cause more area burned by wildfires, an increased frequency of degraded air quality, or "smoke waves" (Liu et al. 2016), is likely. Given these projections of increased emissions, a greater number and magnitude of human impacts from smoke could be expected in the future.

A range of environmental changes resulting from climate change may affect future smoke emissions. Longer fire-weather seasons will have commensurately longer periods of drier (dead and live) fuels (Jolly et al. 2015). As the frequency and extent of drought increases, lower soil moisture will accentuate the influence of soil organic matter (and mineral soil in very intense fires [Bormann et al. 2008]) on smoke chemistry and carbon dynamics (Martínez Zabala et al. 2014; Mayer et al. 2020), potentially limiting use of prescribed fire. A better understanding is needed of how climate change will alter the flammability of surface fuels, particularly on how pre-fire soil moisture and fine fuels influence the size and intensity of wildfires (Krueger et al. 2015).

Because more frequent fires will lead to more frequent smoke effects (Peterson et al. 2020), planning for and managing smoke emissions and dispersion will become increasingly difficult and require agencies and communities to consider new approaches to: (1) managing wildfires, (2) timing of management actions such as prescribed fire, and (3) mitigating health and economic risks from large wildfires through the effective use of fire. Air quality agencies will need to consider the potential increased emission load when approving a planned burn.

As extreme weather becomes more common, better meteorological data will be needed to respond to its effects. The Remote Automated Weather Station (RAWS) system deploys stations across forested areas, but real-time meteorological data are needed to guide fire management. Most national forests are above 200 m elevation in the eastern USA and above 1200 m in the western USA, but climate trends modeled by NOAA rely chiefly on meteorological data from low-elevation stations, contributing to potential errors.

Smoke management practices may need to be revised in some cases to address the potential effects of climate change but will generally not need to be greatly altered. However, a comprehensive approach to smoke would benefit fire and air quality management in this changing environment. Williamson et al. (2016) proposed the concept of a smoke regime, a transdisciplinary framework that would take into account the collective consequences of numerous smoke-related elements including the risk factors for each type of fire (wildfire and prescribed fire). These, combined with human population attributes such as exposure, size, vulnerability, and mitigation, could be used to describe overall population health impact.

Recognition is growing that prescribed fire—when carefully planned, timed, and managed—can reduce wildfire risk and lower potential environmental and public health impacts as compared to the unplanned nature of wildfires (WFLC 2020). Prescribed fire can help reduce fuels, enhance ecosystem resilience, and enhance community safety; however, smoke from prescribed fire can be a serious health threat for some individuals and is often a nuisance, even to those who are unlikely to have adverse health impacts. Prescribed fires offer the opportunity to adjust the timing of fire and some ability to manage the amount of smoke and its path, thereby potentially reducing the overall impacts of wildland fire emissions on public health and welfare (WFLC 2020).

In some places, federal land managers are tentatively using some naturally ignited wildfires to allow more fire on the landscape, although this approach can be controversial and difficult to implement. In California, a fee is assessed for wildfires used in this way. Air quality regulations focus on planned ignitions (prescribed fire) but do little to address unplanned ignitions (wildfire), although wildfire impacts are much worse in both magnitude and duration.

Managers need to take into account and minimize smoke impacts to remain within legal constraints for air quality protection and protect the public. In order to maintain societal support for use of prescribed fire, it can also be beneficial for managers to provide the public with information on the goals and effects of fire use (Olsen et al. 2014). Fire that mimics historic wildfire timing, size, and intensity can contribute to desired ecosystem functions. Alternatives to burning can be more socially desirable as immediate impacts of smoke can be avoided (Shindler and Toman 2003). Although smoke is reduced in the short term, land managers have little scientific evidence that using alternative techniques accomplishes the goals of reducing emissions while simultaneously sustaining desired ecosystem process and functions (Thompson et al. 2018). In addition, most alternatives to fire are more costly than prescribed fire and may have negative environmental effects of their own, such as soil compaction from the use of heavy equipment. Managers are challenged to understand how use of different emission reduction techniques affect forest structure following prescribed fire and how the techniques may work individually and in combination.

Managers have a responsibility to analyze the effects of fire management decisions, inform the public about planned actions that affect air quality, and alert the public during active smoke events that may affect human health. Ultimately, public support for actions taken to manage wildlands depends on sharing credible information about potential smoke effects from active fire management, including possible air quality implications of taking no action. Efforts to develop state-of-science smoke analysis tools, and to provide smoke science information to managers, play a vital role in managing ecosystems, while protecting people and communities from the impacts of smoke.

8.2 Managing Wildland Fire to Improve Ecosystem Conditions While Minimizing Smoke Impacts

Mandated in the Federal Land Assistance, Management, and Enhancement Act of 2009 (FLAME Act; US Congress H.R. 1404), an intergovernmental planning group involving stakeholders and the public adopted a national vision for wildland fire management for this century: "To safely and effectively extinguish fire when needed; use fire where allowable; manage our natural resources; and as a nation, to live with wildland fire" (National Science and Analysis Team 2014). Prescribed fire helps reduce fuel accumulation and improve ecosystem conditions, and scientific documentation can support efforts to increase use of prescribed fire as envisioned by the National Cohesive Wildland Fire Management Strategy (Cohesive Strategy). Analysis of when and where prescribed fire is most effective and most necessary helps managers maximize the benefits of fire, while considering alternatives to put less smoke in the air. In addition to prescribed fires, managers may utilize natural ignitions but need information to weigh short- and long-term benefits and costs of these fires, especially potential smoke impacts.

8.2.1 Smoke Concerns and Barriers to Prescribed Fire

Wildland fire management options to improve ecological conditions, including prescribed fire, have been supported by research and federal fire management policy and guidance for decades. However, resource managers have not been able to utilize these options at the larger spatial scales needed to increase ecological resilience in the West (Stephens et al. 2016; Thompson et al. 2018; Kolden 2019; Schultz et al. 2019). Two likely reasons for this include risk aversion and a systemic set of incentives that lead to an emphasis on minimizing short-term impacts while accepting negative future tradeoffs, thus deferring fire and smoke risk into the future (Maguire and Albright 2005; Schultz et al. 2019).

Fundamental changes in how the fire management community thinks about, learns from, plans for, and responds to wildfires may be needed before change is seen in the amount of fire on the landscape (Thompson et al. 2018). A recent survey of prescribed fire practitioners found that air quality was one of the top three perceived impediments to implementing prescribed fire (Melvin 2018), suggesting that there also will be a need to develop more effective options for mitigating wildfire smoke and air quality impacts.

A comprehensive set of signatories to the Cohesive Strategy (National Science and Analysis Team 2014) including federal, state, tribal, county, and municipal governments has established goals to increase the use of prescribed fire. However, prescribed fire may not always be the right tool for some locations and sociopolitical environments. To prioritize the use of fire and address potential short-term smoke emissions, managers need to answer numerous questions, including: (1) when and where is fire an irreplaceable tool?, (2) how do seasonal shifts in the use of prescribed fire affect the function of fire in an ecosystem?, and (3) where and how will the use of prescribed fire affect wildfire risk?

Current fire and air quality analysis tools are not always well suited to these needs. A significant amount of research has been completed to help estimate fuel consumption (Chap. 2) and emissions (Chap. 5) from various types of fires. However, because of the site- and unit-specific nature of prescribed burning operations, additional research and tools are needed to assist in local planning and reporting.

8.2.2 Applying Prescribed Fire Across Large Landscapes

Large areas of the western USA are far out of range of the natural fire cycle (based on historical conditions) and are at risk of experiencing high-intensity wildfires that can spread into communities and degrade other values of wildlands (Morgan et al. 2001). The current pace and spatial scale of prescribed fire cannot keep up with the need for reducing fuels, fire intensity, and smoke (Vaillant and Reinhardt 2017). Applying prescribed fire across larger landscapes, thousands of hectares instead of hundreds of hectares at a time, would also provide ecological benefits and return the role of more frequent fire (Schweizer et al. 2017).

However, air quality laws and limited land management resources make it unlikely that prescribed fire and alternatives to burning can be utilized to the extent that would be desirable for ecosystems. Managers need to both understand and communicate tradeoffs among the options of fire use, alternatives to burning, and restoration of desired ecological functions (e.g., resilience to future fire and climate change) in fire-prone ecosystems. This information will help resource managers better assess how to burn larger areas while minimizing air quality impacts. Support for burning large areas will depend on evidence that current approaches are inadequate, that larger burn areas will provide greater ecosystem benefits, and that large burns can be done safely and with acceptable smoke impacts.

8.2.3 Utilizing Wildfires and Natural Ignitions

Federal wildfire policy allows for a range of responses to a wildfire, from monitoring to full suppression (with all options along this spectrum available depending on assessment of each individual fire) while prioritizing firefighter and public safety first and using sound, risk-based decisions as the foundation for all fire management activities (National Science and Analysis Team 2014).[1] Each fire management

---

[1] In reality, the range of response options available to federal managers on individual land management units may be constrained by options spelled out in existing planning documents, some of which are many years old and may not reflect current priorities for wildland fire management.

strategy considers associated risks, short- and long-term tradeoffs, levels of uncertainty of effects on resources and assets, probability of success, and potential duration. For example, in remote locations, selecting a monitoring strategy for a wildfire with resource objectives may be an appropriate decision, having little immediate risk and exposure to firefighters, lower costs, and low smoke impacts. But those impacts may be of longer duration, and uncertainty about future smoke emissions increases over time.

In areas where communities or other highly valued resources and assets are threatened, selection of a full suppression strategy is likely. When firefighters actively engage in suppressing fire, their risk and exposure are high, cost per unit of time is high, and smoke impacts are relatively low or of short duration if the fire is quickly controlled. However, in fire-prone ecosystems, this potentially defers impacts, including for air quality, to future years (Box 8.2). Quantifying ecological and air quality tradeoffs in a manner that is understandable for policy makers and the public is a challenging proposition, requiring the assistance of scientists, because smoke impacts from prescribed fires are no more welcome than from wildfires.

Box 8.2 Managing Fires for Resource Benefits in California
Under California law (Title 17), lightning ignitions managed for multiple objectives, including resource benefit, are classified and regulated as prescribed fires. This law does not reflect contemporary federal definitions for wildland fires. The California law requires close coordination, compliance, and mitigation measures with local and state air quality regulatory authorities. Resource managers are often scrutinized and criticized for not aggressively suppressing all unplanned ignitions and for allowing smoke to adversely impact communities. This perspective often fails to recognize that absence of fire defers smoke to a later date when emissions and risk of impacts may be higher.

One option for reintroducing the natural role of fire is to aggressively suppress fewer wildfires. However, this could lead to smoke impacts with considerably longer durations than for prescribed burning. If prescribed fire is applied at a much larger scale, then smoke impacts can more easily be spaced out over time. Some studies have suggested that both wildfires allowed to burn for resource objectives and prescribed fires have substantially lower smoke impacts than large, fast-spreading wildfires (North et al. 2012; Schweizer and Cisneros 2017). However, Navarro et al. (2018) compared emissions from a number of studies and determined there is little difference in emissions between wildfires and prescribed fires for a given area. In any case, wildfires managed for resource objectives and prescribed fires are more able to align emission pulses with times of best dispersion, providing greater control of smoke impacts (Long et al. 2017).

8.2.4 Implications of Wildfire Response Actions and Suppression for Air Quality

Wildfire suppression and management may include the tactical use of intentionally ignited burnout operations to deprive an advancing wildfire of fuels. Although this technique can stop or slow wildfire spread, there are air quality implications that can be difficult to assess, raising questions about whether (1) aggressive use of intentionally ignited fires for wildfire control results in a significant decrease in the duration and/or size of the wildfire and air quality impacts, and (2) the public will support more smoke in the short term if it means fewer air quality impacts over the long term.

Fire management and suppression decisions must always be made with safety in mind. Although it is easier to understand risks to wildland firefighters during suppression actions, public health and safety considerations from smoke are more difficult to understand and communicate; better integration in decision-making would improve wildfire management.

8.2.5 Alternatives to Burning—Evaluating Emissions Reduction

Alternatives to burning (e.g., mechanical thinning, grazing) are considered in nearly every decision to use prescribed fire, and there is often pressure to increase the use of alternatives to reduce smoke impacts. Alternatives to burning cannot always be scaled to the extent needed to foster a resilient ecosystem because of the buildup of fuels from fire exclusion, particularly with the additional stress of climate change (Bradstock et al. 2012; Flannigan et al. 2009; Vaillant and Reinhardt 2017). In addition, surrogates of fire may not provide the same ecological benefits of fire (Keeley et al. 2005; Klocke et al. 2011; Kobziar et al. 2018). For example, undergrowth brush mastication reduces fuel loading but does not provide the heat needed to release seeds from the cones of serotinous pine species (e.g., lodgepole pine [*Pinus contorta* var. *latifolia*]). In this case, mastication would meet the societal need of reducing fire risk and smoke but not the ecological need for tree regeneration.

Information on the cost and benefits of large-scale emission reduction techniques using alternatives to burning is limited. Alternatives to burning can be expensive and may create additional, if different, air quality impacts (e.g., dust from mastication, transportation emissions associated with thinning or milling). Research is needed to quantify the effect of different emission reduction techniques. For example, allowing fires in remote areas or wilderness to progress with minimal intervention may reduce smoke exposure to nearby communities (Schweizer et al. 2019b), but little is known about collateral air quality impacts.

The ecological role of smoke and the effects of altering the smoke cycle with suppression are poorly understood. For example, one study found that loss of ground-level cooling due to smoke potentially increased stream temperatures and reduced the benefits of cold water to fisheries (David et al. 2018). Further study is needed to understand the tradeoffs of different fire management techniques, particularly when the decision to ignite is being used in part for air quality considerations (e.g., good smoke dispersal) (Schweizer et al. 2020).

Managers need emission reduction techniques that have a high probability of minimizing air quality impacts, along with a description of possible unintended consequences to other ecosystem processes. A rigorous procedure to evaluate tradeoffs and costs when choosing emission reduction techniques or alternatives to burning will ensure that those techniques align with land management goals. Numerous prescribed fire emission reduction techniques have been identified and described (Peterson et al. 2020), although effectiveness and applicability of specific techniques to specific locations, fuel types, meteorological conditions, lighting patterns, and land management objectives are uncertain.

8.2.6 Effects of Fuel Moisture on Emissions and Dispersion

Moisture content of components of wildland fuelbeds can affect consumption and emissions in different ways (Chaps. 2 and 5). The ability to target fuelbed components for consumption during prescribed fire while leaving other components largely in place can result in reduced emissions. A prescribed fire in moist fuels may result in less fuel consumed and less smoke, although if insufficient heat is produced, the smoke plume may stay near the ground, resulting in greater smoke impacts. Successful consumption of targeted fuelbed components can help to meet ecosystem objectives and ensure good smoke dispersion.

8.2.7 Fuel Type, Fuel Loading, and Fuel Consumption

Although good information about the type and amount of fuel burning is fundamental to fire management and smoke estimations, most vegetation datasets are not available at a sufficiently fine spatial scale to inform project implementation. LANDFIRE, although good for national and regional discussions, does not meet the needs of fuel managers for project planning and accurate smoke forecasting. There is no national product that maps actual fuels (species, fuel loading, moisture, etc.) that can be used for prescribed fire project planning. In addition, most datasets are not kept current, meaning they may be out of date within just a few years due to disturbances across the landscape. Without a maintenance loop, one-time data become unusable, and often after a few years, another one-time dataset is created for another project, leading to inefficiency. Research could help develop project-level data collection methods and standards for national vegetation layers that are updated annually, depending on disturbance extent.

Although some stand-scale models calculate smoke components (e.g., particulate matter [PM] emissions) (Rebain 2010), they do not account for most ecosystem processes and disturbance interactions. Incorporating feedbacks among ecosystem processes, disturbance interactions, and changing fuel and fire conditions would likely improve the accuracy of smoke production and transport modeling (Chaps. 4, 5, and 6). Incorporating feedbacks among climate change, vegetation, fire, and smoke will improve models used in smoke management and inform management strategies to minimize smoke impacts (McKenzie et al. 2014). Several in-progress studies are linking models such as Quic-Fire to complex, three-dimensional fuel properties, fire effects, and ecosystem responses and recovery through time, thus improving predictions of how future fires will influence smoke production and dispersion.

Current wildfire data collection, typically via ICS 209,[2] does not address fire intensity, severity, or unburned islands within the fire perimeter, which can result in overprediction of smoke emissions and impacts. Emissions estimates compiled after a fire is over, using area blackened rather than area within a fire perimeter, would be more accurate.

Another challenge is that each agency and landowner has their own prescribed fire standards for reporting post-burn fuel consumption. For example, in Utah, area blackened is reported with no estimate of severity or percent consumption, thus potentially overestimating actual emissions. Many other states have the same challenge. Lack of consistent data standards also affects reported annual emissions data, which differ for wildfire and prescribed fire, and for different agencies and states.

It is relatively straightforward to model fuel consumption for prescribed fire because a planned fire typically has known environmental variables. Wildfire fuel consumption is more challenging due to the need for field measurements during wildfires for model development. Remote sensing methods that use other parameters, such as heat release or fire radiative power, to ascertain fuel consumption may be the best option for wildfires (Chap. 5).

8.2.8 Techniques for Minimizing Smoke Impacts

Prescribed fire planning typically considers ways to reduce smoke impacts, and managers use various techniques to influence fire behavior in ways that maximize smoke transport and dispersion. Identifying the probability of finding best days to burn for smoke dispersion based on meteorological averages and trends can help managers and burners anticipate and maximize good dispersion windows. Robust, simple meteorological analysis tools would help managers customize burn techniques to take advantage of favorable weather conditions (e.g., inversions, winds at various

---

[2] An incident status summary (ICS 209) is used for reporting specific information on incidents of significance. The ICS 209 is an important reporting tool giving daily snapshots of the wildland fire management situation and individual incident information including costs, critical resource needs, fire behavior, and size.

levels, precipitation). Clarity is needed on appropriate and effective use of weather parameters (mixing height, ventilation indices, air quality alerts, air stagnation, etc.) and the performance and validation of a point fire-weather matrix (e.g., Atmospheric Dispersion Index, Low Visibility Occurrence Risk Index) and enhanced availability of such parameters for all landowners and managers.

Further research is needed on the effects of ignition techniques and ignition speed on plume rise, dispersion, smoke quantity, smoke dispersion, and air quality impacts, including effectiveness in producing desired outcomes. Firing patterns can affect prescribed fire emissions. Similar-sized units with the same fuelbeds burned under the same atmospheric conditions may display disparate downwind impacts based on the timing and technique used for burning. To better understand the effects of fire patterns on emissions, the following questions need more scientific study:

- What are the effects of different firing methods on emissions and heat release?
- What effect does aerial ignition have on emissions compared to hand ignition?
- How do emission profiles during each stage of combustion compare across different firing techniques?
- How does firing technique affect plume rise and downwind dispersion?

8.2.9 Components of Wildland Fire Smoke

Land managers and air quality regulators are somewhat limited in understanding the full range of potential smoke impacts on air quality. Fine particles in smoke understandably get much of the research attention, but other components in smoke are important and need to be quantified (Bytnerowicz et al. 2016; Clinton et al. 2006). Ratios of coarse-to-fine PM are extremely volatile during the course of a day and throughout the duration of a fire (Schweizer et al. 2019a). Wildland fire can contribute to ozone formation (Chaps. 1, 6) but can also reduce radiative production over urban areas, thus suppressing ozone formation (Burley et al. 2016; Bytnerowicz et al. 2010; Preisler et al. 2005). Smoke also contains other components that are health hazards (Chap. 6), but research and methods for estimating them are limited. Pollutants in smoke may remain aloft for days and travel hundreds of kilometers from the fire before affecting distant communities. Further research into the mixture and aging of the emitted air pollutants is needed (Chaps. 5 and 6).

8.2.10 Soils and Emissions

A better understanding of how organic soil moisture and other soil properties interact to affect emissions and smoke chemistry is needed (Chaps. 5 and 6). Emissions from organic soil combustion are shaped by current and antecedent meteorological conditions (Weise and Wright 2014) and by the composition and relative amounts of combustible vegetation, soil organic matter, and mineral soil. Understanding soil moisture influences on fire behavior, particularly smoldering and glowing combustion, in combination with fuel moisture levels can inform fire suppression and prioritization of fuel reduction activities. In addition, using prescribed fire when there is high moisture content of the organic horizon can be an effective emissions reduction technique if prescribed fire managers have the ability to predict this condition.

A better understanding of soil characteristics could also inform potential differences in human health impact from wildfire versus prescribed fire smoke. A low-moisture organic horizon releases more heat, promoting faster, more complete combustion and higher emissions of volatile organic compounds and gases. Not accounting for combustion of the organic horizon can result in underprediction of smoke emissions (Zhao et al. 2019).

Soil moisture is monitored primarily across agricultural lands, and in situ stations measuring soil moisture are expensive to implement and maintain. Integrating soil characteristics, drought monitoring, and soil moisture measurements with fire and fuels models would allow fire managers to more effectively analyze emissions under drought conditions and prioritize actions to mitigate fire and emissions risks.

8.2.11 Remote Sensing and Data for Fuels, Fire, and Smoke

Remote sensing of vegetation and fuel characteristics needed for estimating fuel consumption and smoke has resulted in advances for fire detection and smoke modeling. The Smoke and Emissions Model Intercomparison Project (SEMIP) analysis found that the largest sources of uncertainty in estimating fire emissions come from uncertainties in overall fuel loading, fuel loading in specific fuel categories, overall area burned, and area burned by type of fire (Larkin et al. 2012). Dense clouds of smoke can inhibit the ability of remote sensing to detect fires, meaning that some fires may be missed entirely or their size is underestimated.

A new area of remote sensing that is ripe for study comes from the rapid expansion of unmanned aerial vehicles or drones for wildfire surveillance, prescribed fire ignition, and other uses. The interest in these tools offers the opportunity to explore this technology for gathering data on such things as fuel loading, fuel consumption, and area burned.

Accurately linking vegetation and fuels to smoke and emissions is critical (Chaps. 2 and 5). Pre- and post-fire parameters are often measured in research projects but rarely during prescribed fires or wildfires. Large, high-quality datasets like those produced for the Fire and Smoke Model Evaluation Experiment (FASMEE; Prichard et al. 2019) (Fig. 8.1), the Prescribed Fire Combustion and Atmospheric Dynamics Research Experiment (RxCADRE) (Peterson and Hardy 2016), and other integrated campaigns need to be prioritized. FASMEE is a multiagency, interdisciplinary effort to identify and collect measurements of fuels, fire behavior, fire energy, meteorology, smoke, and fire effects that will be used to evaluate and advance fire and smoke models. This knowledge will promote better predictions of the production and spread of smoke and effects on human health and safety, as well as inform allocation

Fig. 8.1 Crown fire behavior and effects were intentionally created in an experimental burn on the Fishlake National Forest (Utah), implemented in 2019 as part of the Fire and Smoke Model Evaluation Experiment (FASMEE) (Photo by Kreig Rasmussen)

for firefighting resources. FASMEE currently focuses on only a few fuel types, and similar work is needed across other fuel types to provide a scientific foundation for implementing fuel treatments and managing smoke at large spatial scales.

Mapping fuel structure, composition, and condition (moisture and decay) along with behavior of the fires that consume those fuels is a priority research area (Parsons et al. 2011; Rowell et al. 2020) (Chaps. 2, 5 and 6). This fuel characterization can then be linked to combustion phases (smoldering and flaming) and their respective links to emission factors (Prichard et al. 2020; Chap. 5).

Empirical and computational fluid dynamics (CFD) models offer promise in estimating spatial and temporal patterns of smoke dispersion (Chap. 4). The fuel inputs to these computationally intensive applications are amenable to remotely sensed data that is increasingly available to managers on the ground (e.g., Moran et al. 2020).

8.2.12 Prescribed Fire Tracking Data

As prescribed burning activity increases, the ability to conduct real-time airshed analysis will also rise. Regional tools will be needed that track fire activity and assess the potential for additive downwind impacts from multiple burn units. The Airshed Management System (AMS) in Montana and Idaho, formed over 40 years ago to manage and limit the impacts of smoke from prescribed fire, may be a model for other states (https://mi.airshedgroup.org).

Prescribed fire data collection differs by land management organization. Data collected by federal agencies differs from that collected by states. Lack of data and parameter definitions discourages collaboration. More states are developing local systems (e.g., California, Florida, North Carolina, South Carolina, Washington) to track smoke and prescribed fire permits, and identify adjacent fire locations (Box 8.3), although other states have minimal information available, especially for agricultural burning and its contribution to smoke impacts.

> **Box 8.3 Prescribed Fire Tracking in California**
>
> Smoke management is coordinated year-round in California and depends in part on a system called the Prescribed Fire Information Reporting System (PFIRS 2020). PFIRS provides land managers and air regulators an opportunity to assess burning statewide and manage airsheds accordingly. PFIRS is designed to simplify and expedite communications on planned burns and their approvals and provide a public platform for viewing statewide burning for participating air districts. Land managers can use PFIRS to submit Smoke Management Plans, requests for ignition, and post-fire area burned.

8.2.13 Fire Emissions and the National Emissions Inventory

The US Environmental Protection Agency (USEPA) conducts the National Emissions Inventory (NEI) every three years, collecting data from point, area, mobile, and biogenic sources to assess air emissions status and trends throughout the USA. Emission inventories form the basis for rules and regulations designed to improve ambient air quality. Accurate emissions information allows the USEPA and state air regulatory agencies to focus air quality improvement on anthropogenic sources. Although it is relatively easy to calculate emissions from a point source, such as a coal-fired power plant or a paper mill, determining emissions from all prescribed fire activities over a large landscape is more difficult. Many fires are not reported, and states and the USEPA often rely on satellite detections to identify fire locations.

8.3 Wildland Fire and Smoke Decision Tools

Accurate emissions information is critical to managers for burn planning, dispersion modeling (Chap. 4), and emissions inventory (Chap. 5). Fire managers assess emissions to compare alternatives and evaluate the scale of a proposed management activity. Knowing the quantity of emissions that will be produced is needed for making informed decisions on whether to go forward with a burn. Dispersion model outputs are used to inform "go/no-go" decisions and to communicate air quality information to the public. In some locations, emissions must be reported to local, state, and/or federal agencies as part of emission inventory efforts.

Predictive tools that can help managers identify air quality implications of decisions, are critical for wildland fire use and management. Although existing tools provide valuable information, more research, development, and testing are needed to reduce uncertainty in predictions of smoke dispersion and impacts (Chaps. 4, 5, and 6). Improved models would facilitate wildfire and prescribed fire management decisions and better enable managers to proactively communicate likely smoke impacts to communities.

Better smoke modeling and decision support tools help managers take maximum advantage of meteorological windows in which fire use goals can be accomplished while protecting air quality. Determining where prescribed fire may be most useful, assessing the potential for scaling burn units to larger parcels, and having accurate emissions information and modeling platforms that evaluate downwind impacts from multiple burn units can help managers identify opportunities to improve the condition of natural resources while protecting air quality and human health.

When considering potential emissions and smoke impacts from a proposed prescribed fire activity, managers use emissions analyses in several ways:

- During the planning process to assess emissions, compare alternatives, and evaluate the scale of a proposal.
- During burn implementation, to make informed decisions on whether to go forward with a burn event, and whether to curtail an activity already initiated.
- As a communication tool for conveying smoke impact information to the public.
- In some areas, emissions are reported to local, state, and/or federal agencies as part of emission inventory efforts.

8.3.1 Multiple Fires and Airshed Analysis

Many tools help managers assess the downwind smoke impacts from their own burn unit, but no tools are available to assess fire activity across a geographic area in real or near-real time. The most widely used tools to predict downwind smoke impacts generally model effects for only one burn at a time. Workarounds exist to add other nearby burn activity to a modeling output but are time consuming and cumbersome. In the future, user-friendly tools will be needed to assess impacts from all fire activity within an airshed on a given day. Regional tools are needed that track proposed fire activity in real time and assess the potential for additive downwind impacts from multiple burn units.

8.3.2 Fire Growth Models and Smoke Dispersion

Current smoke dispersion models have built-in assumptions regarding wildfire growth, so emissions estimates are unable to take into account that a fire may grow at different rates or shrink in response to weather conditions, available fuels, terrain, or suppression actions. Fire behavior models (contained within the Wildland Fire Decision Support System and Interagency Fuel Treatment Decision Support System) provide fire growth predictions that could be integrated into smoke models, facilitating more accurate predictions of upcoming smoke impacts (Chaps. 2 and 3).

8.3.3 Background Air Quality Conditions

As most smoke dispersion models assume a clean atmosphere at initiation, including background air quality will improve their accuracy (Chaps. 5 and 6). This is especially important with respect to ozone impacts from wildfires which, unlike PM impacts, can be highest at a significant distance from the fire (Jaffe and Wigder 2012, Chap. 6). Including background air quality in models would help support states during exceptional events. With proper analysis and documentation, air quality data showing high smoke impacts from wildfires and some prescribed fires may be excluded from the record and prevent exceedance of an air quality standard. Understanding and including background air quality conditions can improve smoke impact forecasting and decisions about when to use prescribed fire so it does not cause air quality exceedances.

8.3.4 Smoke Models for Fire Planning

Most smoke models rely on current or near-term meteorology for dispersion predictions, however, managers often need to plan for fire use months or years in advance. Managers are increasingly provided outputs from advanced atmospheric forecast models linked to fire spread and fuel consumption models. These "coupled" models may resolve some of the more complex interactions between fuel consumption and emission production and dispersion. However, ensuring these outputs are evaluated to sufficiently resemble real-world observations is difficult (Chap. 5).

8.3.5 Use of Air Quality Measurements

State air quality monitoring networks typically emphasize areas with significant populations, so small towns and rural areas may have little permanent, high-quality monitoring of air quality. Significant smoke episodes may go unquantified in these areas, so the true impact of smoke may be unknown. An assessment of the air quality monitor network could be undertaken using a similar approach to the recent Remote

Automated Weather Station (RAWS) network assessment (Brown et al. 2012) and would be useful in creating a more robust flow of monitoring data (Box 8.4).

> **Box 8.4 Why Conduct Long-Term Monitoring for Smoke?**
>
> Rural monitoring sites can provide a local assessment of the public health burden of smoke, directly testing if fire management tactics are reducing smoke impacts. The USDA Forest Service has operated a rural site monitoring $PM_{2.5}$ in the community of Kernville, California, since 2007. This site has experienced a significant amount of smoke several times since then, with the annual 98th percentile for $PM_{2.5}$ exceeding the 35 $\mu g\ m^{-3}$ National Ambient Air Quality Standard (NAAQS) (see figure below). After high smoke exposure in 2007 and 2008, the site had lower smoke exposure in 2009 and 2010. This helped resource managers decide to manage the Lion Fire in 2011 largely for ecological benefits. Low smoke exposure occurred from 2012 to 2015, followed by the second Lion Fire in 2017 which had the highest smoke exposure (98th percentile) ever observed at the Kernville monitoring station. Fire managers took this high exposure into account when planning for wildland fire in succeeding years.
>
>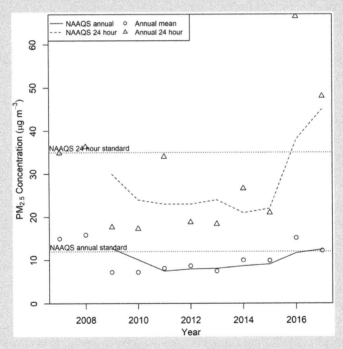
>
> Estimates of air quality conditions at Kernville compared to annual and 24-h NAAQS for $PM_{2.5}$ from 2007 to 2017.

8 Resource Manager Perspectives on the Need for Smoke Science

Information on rapidly changing smoke conditions can be used for early alerts to managers and the public to prepare for coming changes in air quality. Portable air quality monitoring instrumentation can help provide timely information during rapidly changing air quality conditions that often occur during smoke events. This information is important for developing health messages for schools, athletic directors, and other outdoor event planners when making decisions about outdoor activities. There are several types of inexpensive, portable monitoring instruments that can support managers. Testing and calibration of portable instruments based on comparison with official state monitors could improve accuracy.

8.3.6 Air Quality Impacts of Prescribed Fire Versus Wildfire

Assessing a quantitative difference between wildfire and prescribed fire emissions is challenging due to disparate $PM_{2.5}$ monitoring methods. Wildfire smoke tends to travel and disperse over much larger regions and for longer periods, whereas prescribed fire smoke tends to be more local and transient. Wildfire smoke measurements often are made a few to several hundred kilometers from the source, focusing on communities near the fire or more distant urban centers, whereas prescribed fire smoke tends to be measured much closer to the fire. Comparing emissions from a number of studies, Navarro et al. (2018) found little difference in emissions magnitude between wildfire and prescribed fire. This is counterintuitive to the concept that prescribed fire implies reduced fuel consumption and emissions. This highlights the importance of on-going research collaborations between wildland fire management and air quality agencies to further assess emission differences between fire types.

8.3.7 Smoke Model Performance and Accuracy

Fire and air quality managers need to know which smoke models work best and are most accurate in different wildland fire situations. Differences in terrain, seasonal weather patterns, local drainages, and other factors contribute to smoke dispersion, meaning that smoke models must be performance tested utilizing local topography and meteorology. Identifying major drainages and typical weather patterns in specific geographic locations would improve predictions of smoke plumes for local forecasting.

Smoke model testing using various fuel types, fuel loading, and topography would help in forecasting smoke impacts from firing operations and for locations where several prescribed burns are being conducted at the same time. These simulations would help to more accurately forecast air quality and inform decisions about the amount of burning that can be conducted in an area while avoiding an exceedance of an air quality standard.

All smoke model outputs have uncertainty due to inadequate meteorological or air quality inputs, physical or statistical inadequacies, or a lack of computational power that requires parameterizations (or "shortcuts") to produce an output. Transparency about the levels of smoke forecast uncertainty is critical for decision makers to effectively identify management actions most likely to lead to desired air quality outcomes. Managers will generally need assistance from scientists to quantify and communicate about uncertainty.

8.3.8 Long-Range Forecasts and Projections for Planning and Early Warning

Managers need more confidence in long-duration smoke forecasts to support use of fire in ecosystem restoration efforts, especially given anticipated effects of climate change. Current smoke prediction tools are typically short-term forecasts covering up to 72 h, which is useful for active wildfire incidents and daily tactical planning for prescribed fire. However, a number of environmental factors can be monitored or forecasted to provide an early warning of potential conditions beyond this short period, either from a wildfire or prescribed fire perspective. Several operational and experimental tools used by the scientific community could meet these management needs, but managers may not be aware of the practical applications of these tools or know how to access or use them. Focused efforts are needed to increase learning opportunities and develop more intuitive delivery systems for these tools.

8.3.9 Tools and Data Needs for the Future

Moving forward, managers may increase the use of prescribed fire across the landscape as well as increase the size of units and area burned. For example, instead of burning 100 ha in one 6-h period, they may want to burn 400 ha in that same time period. With larger burn areas, more pollutants are emitted, more heat is generated, and plume dynamics change. This may have implications for smoke impacts in nearby communities as well as communities farther downwind of the burn unit (Chap. 4) and additional planning tools, especially regarding smoke management and impacts on downwind communities, will be needed. In addition, how burns conducted by private landowners, which account for the majority of prescribed fire in the USA, are accounted for and integrated into such planning will need consideration as these systems develop.

Geographic screening or simple smoke dispersion models that are used during the burn planning process are not adequate to address hourly and cumulative airshed effects, although the conservative outputs are effective in predicting worst-case smoke management issues. More complex modeling tools that rely on forecast meteorology can provide robust outputs that help to facilitate go/no-go decisions for prescribed fire and help communicate impacts from wildfire. Ideally, managers would have access to complex modeling outputs earlier in the burn planning phase. This may include longer forecast intervals or the ability to generate realistic prescribed burn scenarios with more complex smoke models based on archived gridded meteorological data.

8.3.10 Identifying Areas at High Risk from Wildfire and Smoke

There have been multiple efforts to identify and map areas with high wildfire risk to help prioritize fuels management efforts. Depending on who is using the data and how the data are applied, there are multiple definitions of what constitutes "high wildfire risk," ranging from the "Wildfire Risk to Communities" website (https://wildfirerisk.org) to "Firesheds" and potential operational delineations (PODs). They address biophysical needs for treatment (Stratton 2020) but do not incorporate shared stewardship agreements, collaborative group involvement in projects, or risks caused by emissions and possible smoke impacts.

Wildfire risk analyses do not generally include public health vulnerability assessments related to smoke impacts. A public health vulnerability screen for smoke could include age, medical history, and similar information. This information could then be incorporated in project planning, early warnings, etc. Better tools that describe expected smoke outcomes and show the risk tradeoff between smoke from treatments versus potential smoke/wildfire impacts without treatments would provide a foundation for facilitating the restoration of fire. Identification of communities at risk from smoke could then be taken a step further to identify and map areas at greatest need of treatment based on reducing risk to public health vulnerabilities.

There is currently no standard process to identify and contact individuals whose health may be at most risk from smoke. Understandably, these are often the people most concerned about potential prescribed fire smoke impacts. However, with preparation and effective outreach, those who are most at risk can act to protect themselves and efforts to make communities "smoke ready" are just beginning (see Sect. 8.5).

Descriptions of smoke and air quality outcomes are needed long before implementing a prescribed burn. Communities need to be informed of expected smoke outcomes, the implications of first entry versus maintenance smoke levels, and the relevance of fire planning and smoke prediction. A smoke planning tool that has public-level graphics that are understandable for planning purposes and effectiveness outcomes is critical. The BlueSky Playground tool (Larkin et al. 2009) is a good start, but predicting smoke impacts is constrained by the capability of current predictive meteorological models. Some method is needed to predict future smoke impacts from prescribed fires in specific areas using estimates of historically determined probable meteorology. In addition, animated screens to show estimated smoke impacts throughout the life of the fire as video graphics could provide a valuable public communications tool.

8.4 Health, Safety, and Societal Impacts of Smoke

Smoke effects are a significant public health concern, particularly in relation to short-term $PM_{2.5}$ exposures which can lead to adverse respiratory and cardiovascular effects, including increased mortality (Chap. 7). Smoke can also have significant economic impacts on communities (e.g., business closures, displacement of tourism activities) but these are poorly quantified (Chap. 7) and a better understanding of both short- and long-term economic impacts due to smoke exposure is a critical need for scenario planning. Other social values that can be affected by smoke that managers need to take into account when making management decisions include safety impacts for transportation and visibility impairment in Class I areas.[3]

8.4.1 What is a Smoke-Affected Day?

"Smoke-affected day" is a term used in air quality management, fire management, and smoke health effects research. In air quality management, defining a day as smoke-affected allows the state or local agency to apply to have that day qualify as an "exceptional event." An exceptional event is a cause of poor air quality that is largely outside of human control such as a wildfire, volcano, or dust storm. By definition, an exceptional event is (1) something not reasonably controllable or preventable, (2) caused by human activity that is unlikely to recur at a particular location or a natural event, and (3) determined by the USEPA to qualify as an exceptional event (Environmental Protection Agency, 40 CFR Parts 50 and 51, 3 October 2016). The state air agency needs to prepare a demonstration showing that the event met the exceptional event criteria and prove a causal relationship between the event and the exceedances of air quality standards. Unless the event has regulatory significance (i.e., is likely to contribute to exceedance of an air quality standard), the local USEPA Region would not process the demonstration or remove the data from the official design value.

In Idaho and Montana, smoke-affected days have been used to determine wildland fire management strategies, and several national forests have used smoke-affected days in their press releases to explain the decision not to allow seasonal prescribed

---

[3] The Clean Air Act Amendments of 1977 gave special air quality and visibility protection to national parks larger than 6000 acres (2430 ha) and wilderness areas larger than 5000 acres (2020 ha) that were in existence at the time. These are called Class I areas.

burning. In health studies, the term "smoke-affected day" is used to define population exposure for the purpose of evaluating health outcomes. There is no standard approach for defining what constitutes a wildfire smoke-affected day in the health effects literature. Research to define wildfire smoke-affected time periods is needed to validate epidemiological methods (Doubleday et al. 2020).

8.4.2 Effects of Smoke Exposure on Human Health for Different Exposure Scenarios

Health effects from individual wildfire events are generally described by 24-h average exposures, or multiple events across a single location (Adetona et al. 2016; Reid et al. 2016). However, although existing research has identified a number of factors associated with smoke and increased susceptibility to poor health outcomes (e.g., pre-existing health conditions, specific life stages, and lower socioeconomic position) (Chap. 7), it is largely unknown how health effects of smoke exposure, for both firefighters and the public, may differ for different scenarios. These scenarios include (1) very-high concentration of smoke exposures across short durations (hours), (2) high concentration smoke exposures across limited durations (days to weeks), and (3) repeated high- to very-high exposures over the course of a fire season and multiple fire seasons (years). The duration and smoke concentration of an event are important, because they can lead to higher cumulative exposures to air contaminants (Navarro et al. 2018). This knowledge would be helpful for public health advisories and for making decisions about timing of prescribed fires and interpreting air quality sensor data.

8.4.3 Health Effects of Constituents of Smoke Beyond Particulate Matter

Wildland fire smoke is a complex mixture of individual air contaminants including many hazardous air pollutants (HAPs) from smoke and ash (Adetona et al. 2016; Chap. 6). Most epidemiologic studies have focused on exposure to particulate pollution, but some studies have found wildland fire smoke to be related to various health effects from ozone, carbon monoxide, and HAPs (Chap. 7). In addition, increased residential development increases the likelihood of large wildfires that burn structures where smoke is produced by sources other than vegetation, increasing the risk of adverse health impacts for the public and firefighters (Radeloff et al. 2018).

8.4.4 Smoke and Mental Health

Smoke from wildland fires may also have impacts on the mental health and emotional stress of communities. These impacts have been studied to some degree in the general context of fires, but the literature related to smoke is limited. To better understand the full health burden of smoke exposure, and to take steps to prevent or to reduce the mental health effects of a prolonged smoke event, it is important to understand (1) which aspects of a smoke event are associated with adverse mental health effects, (2) which strategies prevent or reduce those effects, and (3) the groups of people who are most likely to experience these effects, so that interventions can be directed to them. This information would assist in developing and deploying interventions to prevent or reduce smoke-related effects.

8.4.5 Smoke and Visibility Reduction on Roadways

Roadway visibility effects from heavy smoke can result in serious injuries or fatalities of the public and fire personnel. Better forecast tools can aid managers in their efforts to protect public safety by estimating the potential for smoke-caused impairment on roadways and at airports. The conditions that lead to dangerous levels of visibility impairment from smoke are relatively well understood. An especially dangerous occurrence is a mixture of smoke and high humidity known as "superfog" (Chaps. 4 and 7). Managers need methods that will help them anticipate and predict the occurrence of superfog and similar phenomena.

8.4.6 Visibility Conditions in Class I Areas

Visibility is a protected value in Class I areas as required by the federal Clean Air Act. States develop state implementation plans (SIPs) that contain regulatory measures used to protect visibility. Comparisons of current levels of visibility impairment due to wildland fire versus natural or historic visibility impairment from wildland fire inform this issue. As "natural" background visibility, including pre-suppression smoke levels, is poorly understood (Ford et al. 2018; Hamilton et al. 2018), research is needed to define historic visibility conditions, or what is "natural." If the frequency and extent of wildfires continue to increase, visibility likely will be even more reduced during large events. Although it is expected that greater use of prescribed fire could minimize duration and intensity of future visibility impairment, this needs to be empirically assessed.

8.5 Outreach and Messaging About Smoke

Science on wildland fire and smoke is an important foundation for effective outreach that can provide the public with information about smoke concentrations and episodes, potential health risks, and actions they can take to protect themselves (Chap. 7). Identifying potentially affected public sectors and engaging in targeted education and outreach also can assist in gaining the social license needed to support wildland fire management decisions. Effective communication requires trust, often based on existing relationships and procedures. Key stakeholders need to understand fire management decisions and their implications, preferably developed through collaborative efforts in which partners and the public have input into decisions. To facilitate discussions, research is needed to better identify the tradeoffs associated with periodically reduced air quality resulting from wildland fires.

Good outreach and transparent communications are critical to ensuring that appropriate information reaches the general public, especially to those most at-risk from smoke. Some populations are more adversely affected by smoke than others, however, research is needed to better understand how individuals who are most at-risk acquire, absorb, and act on information on projected smoke impacts. The Smoke Ready Community concept is an effort being developed by the federal government (USEPA, Centers for Disease Control and Prevention, US Forest Service) to reduce the public health burden of wildland fire smoke by integrating public health communication into wildland fire emergency management and preparedness.

8.5.1 Smoke Ready Interventions

Communities and individuals can undertake a number of actions to prepare for eventual smoke episodes and minimize health effects and community disruption. Managers can help prepare the public for smoke events by providing timely information about air quality, health effects, and exposure reduction measures. Common messaging about smoke and health precautions provides the public with information on how to protect themselves during smoke events (e.g., staying indoors, closing doors and windows, utilizing a "clean room," utilizing filtration devices if available) (Chap. 7).

Information provided to communities during smoke events include (1) data on pollutants of concern (especially fine PM and ozone), (2) expected timing and duration of smoke episodes, (3) predicted severity of smoke impacts using the AQI, (4) where and how to access real-time air quality advisories. Certain locations in communities (schools, hospitals, long-term care facilities, and athletic programs) are at a higher risk of adverse smoke outcomes and should be identified prior to wildfire season and provided with information on how to protect themselves during wildfire season. There is a need to understand barriers for different groups in taking appropriate protective actions to mitigate smoke impacts including how to best reach underserved communities.

8.5.1.1 Respirator Use by the Public

Various methods to help protect people from smoke are available, although effectiveness, expense, and public acceptance can be challenging. Respirators, such as an N95 mask, are in common use, inexpensive, and could be quite effective with proper training. To date, these have largely been used in occupational settings and with fit testing respirators can be quite effective. Few studies have assessed respiratory protection in the absence of fit testing, which is how they likely would be used by the general public.

N95 facepiece respirators are safe for healthy people and those with mild chronic illnesses. The few studies that have assessed adults wearing respirators during mild exercise or wearing them for multiple hours at a time suggest that respirators increase facial temperature and can be uncomfortable for this reason, but physiological parameters do not seem to change markedly (Harber et al. 2009; Rebmann et al. 2013; Roberge et al. 2010). However, if widespread use of filtering-facepiece respirators is to be recommended, larger studies are needed on their use under a variety of conditions (including different levels of exercise). There is also a paucity of data to clarify which people are too ill to use respirators and, in the absence of further study, they should be recommended only for people with the cognitive and physical ability to remove the respirator if they feel lightheaded or short of breath.

8.5.1.2 School-Based Information

Because children are a particularly at-risk population, scientific knowledge on how to provide healthy air at schools during smoke events will be a necessary component for communities to be smoke ready. Research-based guidance to schools is needed on:

- Use of low-cost air sensors to determine air quality in indoor and outdoor spaces.
- Air quality thresholds for determining appropriate indoor and outdoor activities, including recess and athletic practices.
- Ways to maximize indoor air cleaning using existing HVAC systems or supplemental air filtration for different building types.
- Use and effectiveness of different models of portable air cleaners (must not produce ozone).
- Best ways to predict smoke levels in different spaces during smoke events.
- Use of short-term air quality measurements (1-h average or less) for health protective behavioral messages.

8.5.1.3 Medical Professionals and Other Partners

Medical professionals are key partners and trusted communicators in preparing communities for smoke events. Outreach and education for medical professions to become proficient in smoke effects and interventions help facilitate protection of public health. Tools that can be used to reach the public during smoke events and increase the effectiveness of messaging include (1) "Wildfire Smoke: A Guide for Public Health Officials" and related factsheets (https://www.airnow.gov/wildfire-smoke-guide-publications), (2) "Particle Pollution and Your Patients' Health" web course (https://www.epa.gov/pmcourse), and (3) "Wildfire Smoke and Your Patients' Health" web course (https://www.epa.gov/wildfire-smoke-course).

Additional resources are needed for medical professionals who treat members of at-risk groups prior to or during smoke events. For instance, factsheets and online courses could be developed for older adults and people with heart or lung disease that include discussion of the risks of smoke exposure and how to prevent or reduce effects. Even simple tools that can be used in medical offices such as posters, care-plan forms, and magnets with exposure reduction measures would help to disseminate information.

8.5.2 Air Quality Conditions and Advisories

Tools that are easy for the public to understand (e.g., simple displays of current air quality conditions) motivate appropriate protective actions during smoke episodes. The Air Quality Index (AQI) is a nationally uniform index for reporting and forecasting daily air quality conditions across the USA. As smoke events become longer, in some cases lasting weeks or months, the concentrations at which effects occur during these longer-duration events are often uncertain. The AQI is designed to caution the public about daily pollution exposures. But when smoke lingers for days or weeks, should additional cautionary language, including additional exposure reduction measures, be recommended sooner?

Several national products provide air quality managers and the public with information to track and understand smoke impacts in most areas of the country. The multiagency AirNow website (www.airnow.gov) is administered by the USEPA and reports air quality using the AQI. The AirNow program accepts, stores, and displays monitoring data provided by state, local, and federal air quality agencies. Agencies submit continuous PM data to AirNow from over 1,200 $PM_{2.5}$ monitors and 500 PM_{10} monitors, plus temporary monitors, on an hourly basis. These data are available to the public via an interactive map on airnow.gov and through email notifications, software widgets, and smart-phone apps. The Fire and Smoke Map (Fig. 8.2), a joint USEPA and USFS effort, has data layers with information from ambient $PM_{2.5}$ monitors, satellite smoke plumes, and fire detections from the National Oceanic and

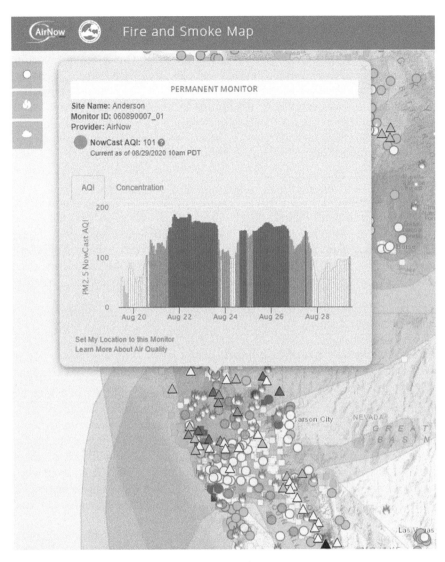

Fig. 8.2 Fire and Smoke Map—developed in partnership with the USEPA and US Forest Service with input from state, tribal, and local air regulatory agencies—provides air quality and wildland fire information for the public in a single location

Atmospheric Administration (NOAA) Hazard Mapping System, as well as smoke advisories and information about the location of the nearest monitors, smoke plumes, and fire detections.

Fig. 8.3 Social media posts by the National Weather Service office in Reno, Nevada, during the 2013 American Fire and Rim Fire (left), and the 2014 King Fire (right)

8.5.3 National Weather Service

Partnerships with the National Weather Service (NWS) and/or state air quality forecasters help inform the public of potential smoke impacts. The NWS National Air Quality Forecast Capability develops and implements operational air quality forecast guidance for ozone, smoke, dust, and $PM_{2.5}$. Local NWS offices deliver air quality predictions to the public and air quality management agencies to inform health warnings and individual actions to limit exposure to poor air quality, and the NWS can issue an air quality alert when warranted, although they do not use standard AQI categories (Fig. 8.3). Air quality agencies and health departments benefit from partnerships with the NWS and land managers that develop and share consistent messaging through various media platforms when smoke puts public health at risk.

8.5.4 Interagency Wildland Fire Air Quality Response Program and Air Resource Advisors

Recognition of the growing impact of wildland fire smoke on public health and safety has resulted in a proactive response led by the USDA Forest Service and partner agencies and authorized by the Dingell Act of 2019. The Interagency Wildland Fire Air Quality Response Program works to directly assess, communicate, and address risks posed by wildland fire smoke to the public and fire personnel. Informed by smoke science and tool development, the program depends on four primary components: (1) specially trained personnel, known as an Air Resource Advisor (ARA) (Box 8.5), (2) air quality monitoring, (3) smoke concentration and dispersion modeling, and (4) coordination and cooperation with agency partners.

Box 8.5 Air Resource Advisors
Air Resource Advisors (ARAs) are technical specialists trained to support wildland fire response efforts by addressing smoke and air quality concerns of both the public and fire personnel. ARAs are often assigned directly to wildfire incident management teams and work closely with incident personnel with expertise in meteorology, fire behavior, safety, planning, and public information. ARAs provide expertise in air quality rules and regulations, smoke dispersion modeling, air quality monitoring techniques, health effects of smoke, and public communications. The primary public outreach product produced by ARAs is the daily smoke outlook (Fig. 8.4), which provides (1) a text discussion of the status of nearby fires, (2) an hourly bar chart of Air Quality Index (AQI) (NowCast) from nearby monitors for the prior day, (3) the AQI for the prior day at each nearby location of interest, (4) AQI predictions for the next two days, and (5) any relevant comments by location.

Predicting the AQI for the coming day at a specific location requires a range of scientific concepts and tools. This is illustrated by the approach used by ARAs:

(1) Start with satellite-detected fire size and location.
(2) Crosswalk this information with a fuels map to assign fuel type and fuel loading.
(3) Link this to models that calculate fuel consumption and plume rise.
(4) Access meteorological predictions to disperse smoke downwind and estimate ground-level concentrations of $PM_{2.5}$.
(5) Communicate this information to the public in a form that is understandable and will motivate the public to take precautions to protect their health.

8.6 Transfer of Smoke and Air Quality Science and Tools to Managers

Effective technology transfer between researchers and practitioners is vital for continued improvement of best management practices and to ensure that science findings and new tools are used on the ground. Scientists working on wildland fire smoke are encouraged to emphasize outreach and technology transfer to landowners, managers, line officers, and agency administrators about smoke management techniques, smoke production and dispersion model use, air quality rules and regulations, air quality and smoke monitoring, and reporting of smoke impacts. Technology transfer can be prioritized by asking the following questions:

- How can burners and managers be effectively engaged in defining scientific objectives?
- Which communication and delivery methods are most effective for different audiences and products?

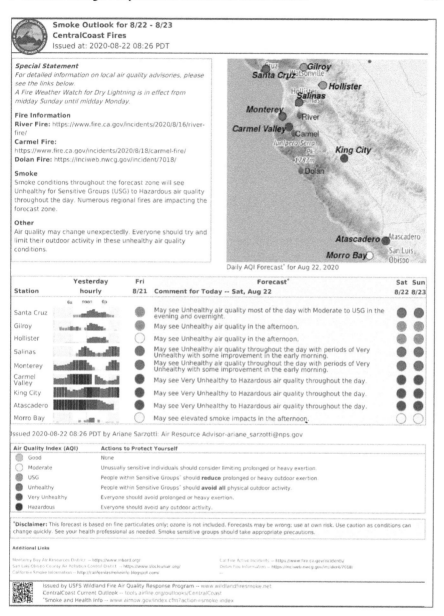

Fig. 8.4 "Daily smoke outlook" for a portion of the southwestern USA, produced by an Air Resource Advisor assigned to a wildland fire. A large amount of fire and smoke science and tool development underlies production of this one-page communications product

- Which outreach approaches (e.g., training courses, webinars, lay-audience newsletters, conferences) effectively communicate new science and products to managers in addition to publishing a research paper?

In order to maximize effectiveness, science-management connections need to be continuous and mainstreamed in the fields of fire and air quality. Greater involvement of field-level resources can help to identify (1) new research needed, (2) the best tools for making sound decisions, and (3) more effective systems for communicating new information.

8.6.1 Formal Fire and Smoke Training Opportunities

Many federal and state fire practitioners are required to attend several in-person training courses sponsored by the National Wildfire Coordinating Group (NWCG). Incorporation of new science and best management practices helps meet the needs of the Interagency Prescribed Fire Planning and Implementation Procedures Guide (NWCG 2017), which is used to ensure prescribed fires comply with the Clean Air Act. In addition to field training and documentation of critical competencies, the following coursework is required to become qualified as a prescribed fire practitioner: Prescribed Fire Plan Preparation (RX341), Smoke Management Techniques (RX410), and Fire Program Management (M-581). These courses integrate discussions on smoke management, air quality indicators, program direction and policy, and use of models to calculate emissions and dispersion. In addition, specialized national training courses, such as those for ARAs and prescribed fire line officers, are designed for those who provide smoke impact forecasts and ignition approval.

As in-person discussion among scientists, landowners, and managers is mutually beneficial, opportunities are needed for scientists and managers to come together to jointly identify priority needs and best management practices. Fire and smoke specialists need to be active participants in understanding climate change effects on wildland fire behavior, smoke emissions, and ecosystem management decisions. Expanding the traditional NWCG community to include climate scientists would better prepare fire managers to monitor climate change indicators and thresholds that trigger a change in action.

8.6.2 Informal Training and Collaboration Opportunities

Managers benefit from scientists bringing the latest research and analytical processes to smoke-related projects, such as updates to a state smoke management plan, analysis of a complex National Environmental Policy Act (NEPA) project, scientifically defensible comparisons of air quality impacts from prescribed fire versus wildfire, or implications of policy on wildland fire.

Agency administrators may be unaware of what they need to know about smoke management. Supplementing existing regional training with smoke management and air quality knowledge would ensure that up-to-date information is used. Research needs can be identified by encouraging more active dialogue among scientists, program managers, administrators, practitioners, and state air quality regulators.

8.6.3 Websites, Webinars, Etc.

Websites, webinars, newsletters, and lessons-learned documents are an increasingly popular way to reach a large and diverse audience and introduce private to federal practitioners and other interested parties to new research and tools. Recorded webinars can be viewed repeatedly and on a schedule that suits the individual.

A major source for fire-related research documents, summaries, newsletters, and recorded webinars is the Joint Fire Science Program (JFSP) and its associated Fire Science Exchange Network (https://www.firescience.gov/JFSP_exchanges.cfm), which regionalizes access to information for 15 areas of the USA (Fig. 8.5). These networks include practitioners and scientists from federal, state, private, and tribal entities. The ability to access the latest publication, workshop, science finding, or management need is important for all parties involved. Although the JFSP and

Fig. 8.5 Fire Science Exchange Network (https://www.firescience.gov/JFSP_exchanges.cfm) encompasses 15 geographic areas of the USA (see text for details)

exchanges address all wildland fire science topics, smoke science and management are emphasized only in specific research projects.

Other fire science organizations maintain websites and send research summaries via subscription newsletters, including common sources such as US Forest Service research stations, the Association for Fire Ecology, and the International Association of Wildland Fire. None of these is specific to smoke-related science, although all share smoke content.

A good resource for smoke-specific information, documents, websites, and training materials is the Emissions and Smoke Portal (https://www.frames.gov/smoke/home), sponsored by the NWCG Smoke Committee and hosted on the Fire Research and Management Exchange System (FRAMES) website which provides categorized hosting of all things fire and fuels.

8.6.4 Learning Pathways

Fire practitioners frequently seek additional information through a series of learning pathways within the fire community. By following the adult learning pathways of tactile interactions, virtual course delivery, and fire community-based sharing, there is a higher likelihood of connecting science and management.

Most modeling software is available on the FRAMES website. Links to the Wildland Fire Learning Portal (https://wildlandfirelearningportal.net/) and Northwest Knowledge Network (https://www.northwestknowledge.net/home) connect users to online coursework and information. Effective learning pathways for smoke management tools and science will always benefit both the scientific and management communities. Smoke management and modeling are available through the Wildland Fire Learning portal, a central location for sharing new science, new modeling tools, and best management practices. In-person training and interactive presentations at a regional or local level can be an effective way for scientists and managers to share new information, tools, and science.

8.6.5 Maintaining Contact

For both scientists and managers, attending a single conference or training course is not enough to ensure a scientific finding or tool is well integrated into a management process. Continual interactions, training, updates, and feedback are needed. Scientists can provide training when models are developed and updated, develop guidance documents and online hosting, and provide notices of model updates and enhancements.

8.7 Managing Smoke in a Changing Environment

Ecosystems and fire regimes will continue to change in response to a warmer climate (Halofsky et al. 2020, Box 8.1). Environmental risks from climate change are becoming more evident and happening quickly, requiring resource managers to incorporate climate considerations and adjust how they manage fire and natural resources. Because of expected increases in emissions, combined with increasing evidence of a multitude of potential health impacts associated with smoke exposure, effective smoke management and communication will be a priority.

Because more frequent fires will lead to more frequent smoke effects from wildfire (Peterson et al. 2020), planning for and managing smoke emissions and dispersion likely will become increasingly difficult. The management and mitigation of smoke impacts rely on continued advancements in the field of smoke science, including our understanding of smoke emissions and predictive modeling tools. Community preparedness for smoke from wildfire and prescribed fire will be critical, and managers will need to incorporate new knowledge and tools into decision-making and in their work with stakeholders, communities, and policy makers; appropriate protective actions and evaluation of the benefits of new management approaches will be needed. Active learning through testing of new approaches can accelerate the learning process, and an open dialogue with all stakeholders can help to identify effective local solutions that will improve smoke management and protect key social values.

References

Abatzoglou JT, Kolden CA (2013) Relationships between climate and macroscale area burned in the western United States. Int J Wildland Fire 22:1003–1020

Adetona O, Reinhardt TE, Domitrovich J et al (2016) Review of the health effects of wildland fire smoke on wildland firefighters and the public. Inhalation Toxicol 28:95–139

Barbero R, Abatzoglou JT, Larkin NK et al (2015) Climate change presents increased potential for very large fires in the contiguous United States. Int J Wildland Fire 24:892–899

Bormann BT, Homann PS, Darbyshire RL, Morrissette BA (2008) Intense forest wildfire sharply reduces mineral soil C and N: the first direct evidence. Can J for Res 38:2771–2783

Bradstock RA, Cary GJ, Davies I et al (2012) Wildfires, fuel treatment and risk mitigation in Australian eucalypt forests: insights from landscape-scale simulation. J Environ Manage 105:66–75

Brown TJ, Horel JD, McCurdy GD et al (2012) What is the appropriate RAWS network? PMS 1003, Report to the National Wildfire Coordinating Group. https://www.nwcg.gov/sites/default/files/publications/pms1003.pdf. 3 Sept 2020

Burley JD, Bytnerowicz A, Buhler M et al (2016) Air quality at Devils Postpile National Monument, Sierra Nevada Mountains, California, USA. Aerosol Air Qual Res 16:2315–2332

Bytnerowicz A, Cayan D, Riggan P et al (2010) Analysis of the effects of combustion emissions and Santa Ana winds on ambient ozone during the October 2007 southern California wildfires. Atmos Environ 44:678–687

Bytnerowicz A, Hsu Y-M, Percy K et al (2016) Ground-level air pollution changes during a boreal wildland mega-fire. Sci Total Environ 572:755–769

Clinton NE, Gong P, Scott K (2006) Quantification of pollutants emitted from very large wildland fires in Southern California, USA. Atmos Environ 40:3686–3695

David AT, Asarian JE, Lake FK (2018) Wildfire smoke cools summer river and stream water temperatures. Water Resour Res 54:7273–7290

Doubleday A, Schulte J, Sheppard L et al (2020) Mortality associated with wildfire smoke exposure in Washington state, 2006–2017: a case-crossover study. Environ Health 19:4

Flannigan MD, Krawchuk MA, de Groot WJ et al (2009) Implications of changing climate for global wildland fire. Int J Wildland Fire 18:483–507

Ford B, Val Martin M, Zelasky SE et al (2018) Future fire impacts on smoke concentrations, visibility, and health in the contiguous United States. GeoHealth 2:229–247

Halofsky JE, Peterson DL, Harvey BJ (2020) Changing wildfire, changing forests: the effects of climate change on fire regimes and vegetation in the Pacific Northwest, USA. Fire Ecol 16:4

Hamilton DS, Hantson S, Scott CE et al (2018) Reassessment of pre-industrial fire emissions strongly affects anthropogenic aerosol forcing. Nat Commun 9:3182

Harber P, Bansal S, Santiago S et al (2009) Multidomain subjective response to respirator use during simulated work. J Occup Environ Med 51:38–45

Jaffe DA, Wigder NL (2012) Ozone production from wildfires: a critical review. Atmos Environ 51:1–10

Jolly WM, Cochrane MA, Freeborn PH et al (2015) Climate-induced variations in global wildfire danger from 1979 to 2013. Nat Commun 6:1–11

Keeley JE, McGinnis TW, Bollens KA (2005) Seed germination of Sierra Nevada postfire chaparral species. Madroño 52:175–181

Kitzberger T, Falk DA, Westerling AL et al (2017) Direct and indirect climate controls predict heterogeneous early-mid 21st century wildfire burned area across western and boreal North America. PloS One 12:e0188486

Klocke D, Schmitz A, Schmitz H (2011) Fire-adaptation in *Hypocerides nearcticus* Borgmeier and *Anabarhynchus hyalipennis hyalipennis* Marquart and new notes about the Australian "Smoke Fly" *Microsania australis* Collart (Diptera: Phoridae, Therevidae and Platypezidae). Open Entomol J 5:10–14

Kobziar LN, Pingree MRA, Larson H et al (2018) Pyroaerobiology: The aerosolization and transport of viable microbial life by wildland fire. Ecospheres 9:e02507

Kolden CA (2019) We're not doing enough prescribed fire in the western United States to mitigate wildfire risk. Fire 2:30

Krueger ES, Ochsner TE, Engle DM et al (2015) Soil moisture affects growing-season wildfire size in the Southern Great Plains. Soil Sci Soc Amer J 79:1567–1576

Larkin NK, Abatzoglou JT, Barbero R et al (2015) Future megafires and smoke impacts (Final report JFSP-11-1-7-4). Boise: Joint Fire Science Program. https://www.firescience.gov/projects/11-1-7-4/project/11-1-7-4_final_report.pdf. 1 July 2020

Larkin NK, O'Neill SM, Solomon R et al (2009) The BlueSky smoke modeling framework. Int J Wildland Fire 18:906–920

Larkin NK, Strand TM, Drury SA et al (2012) Smoke and emissions model intercomparison project (SEMIP): Creation of SEMIP and evaluation of current models (Final report, JFSP 08-1-6-10). Boise: Joint Fire Science Program. https://www.firescience.gov/projects/08-1-6-10/project/08-1-6-10_final_report.pdf. 1 July 2020

Litschert SE, Brown TC, Theobald DM (2012) Historic and future extent of wildfires in the Southern Rockies ecoregion, USA. For Ecol Manage 269:124–133

Liu JC, Mickley LJ, Sulprizio MP et al (2016) Particulate air pollution from wildfires in the western US under climate change. Clim Change 138:655–666

Long JW, Tarnay LW, North MP (2017) Aligning smoke management with ecological and public health goals. J for 116:76–86

Maguire LA, Albright EA (2005) Can behavioral decision theory explain risk-averse fire management decisions? For Ecol Manage 211:47–58

Martínez Zabala LM, Celis García R, Jordán López A (2014) How wildfires affect soil properties: a brief review. Cuadernos De Investigación Geográfica 40:311–331

Mayer M, Prescott CE, Abaker WE et al (2020) Influence of forest management activities on soil organic carbon stocks: a knowledge synthesis. For Ecol Manage 466:118–127

McKenzie D, Gedalof Z, Peterson DL et al (2004) Climatic change, wildfire, and conservation. Conserv Biol 18:890–902

McKenzie D, Shankar U, Keane RE et al (2014) Smoke consequences of new wildfire regimes driven by climate change. Earth's Future 2:35–59

Melvin M (2018) National prescribed fire use survey report (Technical Report 03–18). Coalition of Prescribed Fire Councils. https://www.stateforesters.org/wp-content/uploads/2018/12/2018-Prescribed-Fire-Use-Survey-Report-1.pdf. 4 June 2020

Moran CJ, Kane VR, Seielstad CA (2020) Mapping forest canopy fuels in the Western United States with LiDAR–Landsat covariance. Remote Sens 12:1000

Morgan P, Hardy CC, Swetnam TW et al (2001) Mapping fire regimes across time and space: understanding coarse and fine-scale fire patterns. Int J Wildland Fire 10:329–342

National Science and Analysis Team (2014) The national strategy: The final phase in the development of the National Cohesive Wildland Fire Mangement Strategy. https://www.forestsandrangelands.gov/documents/strategy/strategy/CSPhaseIIINationalStrategyApr2014.pdf. 10 Mar 2021

National Wildfire Coordinating Group (NWCG) (2017) Interagency prescribed fire planning and implementation procedures guide (PMS 484). https://www.nwcg.gov/publications/484. 15 Sept 2020

Navarro KM, Schweizer D, Balmes JR, Cisneros R (2018) A review of community smoke exposure from wildfire compared to prescribed fire in the United States. Atmosphere 9:1–11

North M, Collins BM, Stephens S (2012) Using fire to increase the scale, benefits, and future maintenance of fuels treatments. J for 110:392–401

Olsen CS, Mazzotta DK, Toman E, Fischer AP (2014) Communicating about smoke from wildland fire: challenges and opportunities for managers. Environ Manage 54:571–582

Parsons RA, Mell WE, McCauley P (2011) Linking 3D spatial models of fuels and fire: effects of spatial heterogeneity on fire behavior. Ecol Model 222:679–691

Peterson J, Lahm P, Fitch M et al (2020) NWCG smoke management guide for prescribed fire (PMS 420–3/NFES 001279). National Wildfire Coordinating Group. https://www.nwcg.gov/sites/default/files/publications/pms420-3.pdf. 10 Mar 2021

Peterson DL, Hardy CC (2016) The RxCADRE study: a new approach to interdisciplinary fire research. Int J Wildland Fire 25:i–i

Prescribed Fire Information Reporting System (PFIRS) (2020) [WWW Document]. https://ssl.arb.ca.gov/pfirs. 1 July 2020

Preisler HK, Grulke NE, Bytnerowicz A, Esperanza A (2005) Analyzing effects of forest fires on diurnal patterns of ozone concentrations. Phyton 45:33–39

Prichard SJ, Larkin NS, Ottmar RD et al (2019) The fire and smoke model evaluation experiment: a plan for integrated, large fire-atmosphere field campaigns. Atmosphere 10:66

Prichard SJ, O'Neill SM, Eagle P et al (2020) Wildland fire emission factors in North America: synthesis of existing data, measurement needs and management applications. Int J Wildland Fire 29:132–147

Radeloff VC, Helmers DP, Kramer HA et al (2018) Rapid growth of the US wildland-urban interface raises wildfire risk. Proc Natl Acad Sci USA 115:3314–3319

Rebain SA (2010) The fire and fuels extension to the forest vegetation simulator: updated model documentation (Internal report, revised 23 March 2015). Fort Collins, U.S. Forest Service, Forest Management Service Center

Rebmann T, Carrico R, Wang J (2013) Physiologic and other effects and compliance with long-term respirator use among medical intensive care unit nurses. Am J Infect Control 41:1218–1223

Reid CE, Brauer M, Johnston FH et al (2016) Critical review of health impacts of wildfire smoke exposure. Environ Health Perspect 124:1334–1343

Reidmiller DR, Avery CW, Easterling DR et al (eds) (2018) Impacts, risks, and adaptation in the United States: fourth national climate assessment, vol II. U.S. Global Change Research Program, Washington, DC

Roberge RJ, Coca A, Williams WJ et al (2010) Physiological impact of the N95 filtering facepiece respirator on healthcare workers. Respir Care 55:569–577

Rowell E, Loudermilk EL, Hawley C et al (2020) Coupling terrestrial laser scanning with 3D fuel biomass sampling for advancing wildland fuels characterization. For Ecol Manage 462:117945

Schoennagel T, Balch JK, Brenkert-Smith H et al (2017) Adapt to more wildfire in western North American forests as climate changes. Proc Natl Acad Sci USA 114:4582–4590

Schultz CA, Thompson MP, McCaffrey SM (2019) Forest Service fire management and the elusiveness of change. Fire Ecol 15:13

Schweizer DW, Cisneros R (2017) Forest fire policy: Change conventional thinking of smoke management to prioritize long-term air quality and public health. Air Qual Atmos Health 10:33–36

Schweizer D, Cisneros R, Buhler M (2019a) Coarse and fine particulate matter components of wildland fire smoke at Devils Postpile National Monument, California, USA. Aerosol Air Qual Res 19:1463–1470

Schweizer D, Cisneros R, Navarro K (2020) The effectiveness of adding fire for air quality benefits challenged: a case study of increased fine particulate matter from wilderness fire smoke with more active fire management. For Ecol Manage 458:117761

Schweizer D, Cisneros R, Traina S, Ghezzehei TA, Shaw G (2017) Using National Ambient Air Quality Standards for fine particulate matter to assess regional wildland fire smoke and air quality management. J Environ Manage 201:345–356

Schweizer D, Preisler HK, Cisneros R (2019b) Assessing relative differences in smoke exposure from prescribed, managed, and full suppression wildland fire. Air Qual Atmos Health 12:87–95

Shindler B, Toman E (2003) Fuel reduction strategies in forest communities: a longitudinal analysis of public support. J for 101:8–15

Stephens SL, Collins BM, Biber E, Fulé PZ (2016) U.S. federal fire and forest policy: emphasizing resilience in dry forests. Ecosphere 7:e01584

Stratton RD (2020) The path to strategic Wildland fire management planning. Wildfire, Jan–Mar 24–31

Thompson MP, MacGregor DG, Dunn CJ et al (2018) Rethinking the wildland fire management system. J Forest 116:382–390

Vaillant NM, Reinhardt ED (2017) An evaluation of the Forest Service hazardous fuels treatment program—Are we treating enough to promote resiliency or reduce hazard? J Forest 115:300–308

Vose JM, Peterson DL, Domke GM et al (2018) Forests. In: Reidmiller DR, Avery CW, Easterling DR et al (eds) Impacts, risks, and adaptation in the United States: Fourth national climate assessment, vol II. U.S. Global Change Research Program, Washington, DC, pp 232–267

Weise DR, Wright CS (2014) Wildland fire emissions, carbon and climate: characterizing wildland fuels. For Ecol Manage 317:26–40

Wildland Fire Leadership Council (WFLC) (2020) Joint vision and key messsages on relative benefits of prescribed fire to wildfire. https://southernfireexchange.org/joint-vision-statement-on-relative-benefits-of-prescribed-fire-to-wildland-fire. 10 Mar 2021

Williamson GJ, Bowman DMS, Price OF et al (2016) A transdisciplinary approach to understanding the health effects of wildfire and prescribed fire smoke regimes. Environ Res Lett 11:125009

Zhao F, Liu Y, Goodrick S et al (2019) The contribution of duff consumption to fire emissions and air pollution of the Rough Ridge Fire. Int J Wildland Fire 28:993–1004

Open Access This chapter is licensed under the terms of the Creative Commons Attribution 4.0 International License (http://creativecommons.org/licenses/by/4.0/), which permits use, sharing, adaptation, distribution and reproduction in any medium or format, as long as you give appropriate credit to the original author(s) and the source, provide a link to the Creative Commons license and indicate if changes were made.

The images or other third party material in this chapter are included in the chapter's Creative Commons license, unless indicated otherwise in a credit line to the material. If material is not included in the chapter's Creative Commons license and your intended use is not permitted by statutory regulation or exceeds the permitted use, you will need to obtain permission directly from the copyright holder.

Appendix A
Regional Perspectives on Smoke Issues and Management

David L. Peterson and Linda Geiser

Current resource management, policy, and social issues provide an important context for assessing smoke science. This context must be considered when evaluating the rigor and adequacy of scientific data, tools, and other information for various applications. Wildland fire and smoke managers deal with different challenges throughout the USA. Here, we summarize (1) the ecological and social context, (2) prescribed fire issues, and (3) scientific activities and needs relative to wildland fire and smoke in nine different U.S. Forest Service Regions. We elicited this information from smoke, fuels, and fire ecology specialists in each U.S. Forest Service Region (Fig. A.1) with a standard set of questions, asking them to consider issues from the perspective of federal resource managers but with stakeholder issues in mind. Not every question was equally relevant for each Region, so different levels of detail are seen in the narratives.

The following summaries provide fire and smoke researchers with regional context that can inform development of future scientific priorities and activities. Ideally, researchers will collaborate directly with smoke and fire specialists on smoke research, model development, and real-world applications.

D. L. Peterson (✉)
School of Environmental and Forest Sciences, University of Washington, Seattle, WA, USA
e-mail: wild@uw.edu

L. Geiser
U.S. Forest Service, Air Program, National Headquarters, Washington, DC, USA
e-mail: linda.geiser@usda.gov

Fig. A.1 U.S. Forest Service Regions. Hawai'i (not shown) is in the Pacific Southwest Region

Alaska Region

Neil Stichert and Mark Cahur

Ecological and Social Context

What are the dominant forest/range/grassland ecosystems with which you work? What is the ecological role of fire in these systems? Which fire-related issues or widespread problems (forest health, invasive species, wildland–urban interface, etc.) are of greatest concern?

Tongass National Forest and Chugach National Forest encompass multiple ecoregions in Alaska, including hyper-maritime forests of the Gulf Coast of Alaska, the Alexander Archipelago, the Boundary Range, and the Chugach-St. Elias Range. These ecoregions are dominated by coniferous forests of Sitka spruce (*Picea sitchensis*), western hemlock (*Tsuga heterophylla*), mountain hemlock (*T. mertensiana*), western redcedar (*Thuja plicata*), Alaska cedar (*Callitropsis nootkatensis*), and red alder (*Alnus rubra*). Lowlands, outwash plains, and coastal forelands are dominated by a diverse mix of riparian plants, shrubs, and mosses.

In coastal forest regions, wind, insects, and pathogens are the dominant natural disturbance factors, with fire occurring in small, isolated events. In areas such as the Kenai Peninsula lowlands, fire occurs more frequently and can make episodic contact with dispersed urban areas and/or infrastructure (highways, power lines, etc.).

What have been the recent trends in wildfire occurrence? What are the projections for the future?

The average annual temperature in Alaska has increased by 1.7 °C over the past 60 years, more than twice as fast as the rest of the USA. By 2050, temperatures are projected to increase an additional 1.1–2.2 °C, with the Arctic region seeing the biggest increases. These rising temperatures are expected to increase wildfire occurrence in Alaska. At the current level of greenhouse gas emissions, the amount of area burned in Alaskan wildfires is projected to double by 2050 and triple by 2100. Analysis of 65 years of Alaska wildfire data found that the number of large wildfires (>400 ha) increased in the 1990s, while the 2000s saw nearly twice as many large wildfires as the 1950s and 1960s.

N. Stichert
U.S. Forest Service, Alaska Region, Juneau, AK, USA

M. Cahur
U.S. Forest Service, Alaska and Pacific Northwest Regions, Anchorage, AK, USA

During the past 20 years or so, which smoke events have been noteworthy (especially related to public health and safety)?

Smoke affecting public health and safety can occur year around in Fairbanks due to oil and wood heating emissions in winter and wildfire in summer. As a result, the greater Fairbanks area has been designated a $PM_{2.5}$ non-attainment area. Fairbanks Northstar Borough recorded "exceptional events" in 2009, 2010, 2013, and 2015 due to wildfire smoke, and the Alaska Department of Environmental Conservation (ADEC) requested $PM_{2.5}$ waivers from the U.S. Environmental Protection Agency (USEPA) due to exceedances. In 2019, the 68,000-ha Swan Lake Fire in Kenai National Wildlife Refuge and Chugach National Forest created significant smoke impacts for the communities of Cooper Landing, Sterling, and Soldotna, as well as dispersed effects in Anchorage.

What is the current regulatory environment? What are the major interactions with states and other governmental bodies?

The regulatory environment in Alaska consists of federal and state ambient air quality standards, which are monitored and enforced by the ADEC. Additional regulatory requirements for prescribed fire are incorporated in the Alaska Enhanced Smoke Management Plan (ESMP) prepared by the ADEC. The ESMP outlines processes and identifies issues that need to be addressed by the ADEC and land management agencies and private landowners/corporations to ensure that prescribed fire activities minimize smoke and air quality problems.

Both the U.S. Forest Service Alaska Region Air Program and Fuels Program work with the ADEC, Alaska Interagency Coordination Center, Alaska Division of Forestry, and Bureau of Land Management-Alaska Fire Service on environmental monitoring of air quality and wildland fire, smoke prediction, and outreach.

What are the current perspectives of stakeholders and the general public about smoke and wildfire?

Wildfire smoke is a serious issue in south-central and interior Alaska, with direct and/or dispersed effects to nearby villages and population centers in Fairbanks, Mat Su Valley, Anchorage, and the Kenai Peninsula. During prolonged high-pressure weather events, smoke from fires in the Yukon Territory and British Columbia can disperse to south-coastal Alaska communities.

Alaska tends to have episodic fire events, or "fire years," so smoke impacts tend to dominate the general public's mindset only for a period of time. Smoke impacts during Alaska's short summer and recreational tourism season can be particularly problematic.

The ADEC, Alaska Interagency Coordination Center, University of Alaska–Fairbanks, and the larger municipalities maintain online air quality advisories, real-time data, smoke models, and other products related to smoke, particulate matter, and human health. In the last decade, Alaskans also have tolerated ashfall from volcanic eruptions, as well as significant earthquake events.

Prescribed Fire and Smoke

What are the current uses of prescribed fire in different ecosystems? What are the goals for future use of prescribed fire and associated management actions (e.g., forest thinning)?

The coastal national forests in Alaska use prescribed fire in limited circumstances to burn activity-created slash that is piled after timber harvest or fuel reduction projects. Much of this type of burning occurs near the road system in Chugach National Forest. Similar prescribed fire activities occur within neighboring municipalities and boroughs.

The interior regions of Alaska use prescribed fire primarily to reduce hazardous fuels and restore fire-adapted ecosystems. Broadcast prescribed fire is mainly used to reduce fuels on military installations, thus limiting the potential for ordnance fires caused by military exercises. Burning of hand piles or machine piles for fuels reduction near communities is also utilized. Ecosystem restoration and habitat burning occurs in interior Alaska, although the mechanism is often not prescribed fire but natural fire occurrence that meets resource objectives.

To what degree is smoke a barrier to use of prescribed fire (regulatory process, air quality concerns, conflicts with other burners, etc.)?

Prescribed fire use by federal agencies is dictated by resource management objectives identified in National Environmental Policy Act (NEPA) documents and project-specific implementation plans. Additional regulatory requirements are incorporated in the Alaska ESMP prepared by the ADEC. The ESMP outlines processes and identifies issues that need to be addressed by the ADEC and land management agencies or private landowners/corporations to ensure that prescribed fire activities minimize smoke and air quality problems. Methods are further refined by factors such as fuel loading, topography, location, and unit size. Weather is a key barrier to prescribed fire in Alaska due to arrow seasonal and atmospheric windows appropriate for prescribed fire, and logistics in remote locations can be challenging. Identification of real and perceived barriers to social acceptance of prescribed fire and smoke is ongoing.

What are the current perspectives of stakeholders and the general public about smoke and prescribed fire?

The public, tourism, and aviation sectors pay attention to smoke issues but, given that most smoke originates from remote wildfires, the public largely considers smoke an uncontrollable nuisance. Smoke impacts during Alaska's short summer and recreational tourism season can be particularly problematic.

Smoke Research Needs and Scientific Efforts Applicable to the Region

Which new knowledge, research, modeling, and other tools are needed to predict and manage smoke from wildfire and prescribed fire in dominant ecosystems in the region?

Alaska vegetation modeling is lacking in most areas, with the most updated mapping recently completed on Kenai Peninsula. Significant effort is needed to provide a base vegetation layer for Alaska that could be used for many activities, including smoke and fuel modeling. Modeling and scaling of potential climate effects, forest pathogens, and disease in Tongass National Forest and boreal transition areas on the Kenai Peninsula would also be informative.

Which current scientific efforts (federal, state, tribal, universities, NGO's, etc.) will help address these needs?

The ADEC, Alaska Interagency Coordination Center, University of Alaska–Fairbanks, and larger municipalities maintain air quality advisories, real-time data, smoke models, and a variety of outreach products regarding smoke, particulate matter, and human health. A three-year study funded in 2019 by the NASA Arctic research program is examining the impact of smoke from Alaska wildfires on respiratory and cardiovascular health. The University of Maryland will examine wildfire smoke patterns and hospitalization records across years with different fire severities.

Eastern Region

Trent Wickman, Ralph Perron, and Jeremy Ash

Ecological and Social Context

What are the dominant forest/range/grassland ecosystems with which you work? What is the ecological role of fire in these systems? Which fire-related issues or widespread problems (forest health, invasive species, wildland–urban interface, etc.) are of greatest concern?

T. Wickman
U.S. Forest Service, Eastern Region, Superior National Forest, Duluth, MN, USA

R. Perron
U.S. Forest Service, Eastern Region, White Mountain National Forest, Campton, NH, USA

J. Ash
U.S. Forest Service, Eastern Region, Milwaukee, WI, USA

The Eastern Region contains diverse ecosystems in terms of general structure and species diversity. Eastern hardwoods dominate much of the Northeast where fire has not played a significant role in sustaining the ecosystem. Oak-hickory forests, found along the southern tier of the Region, are a fire-dependent ecosystem. The Lake States contain mixed-wood ecosystems with many fire-adapted species, and the Region also has boreal forest throughout. Maintaining the role of fire in ecosystems to benefit fire-adapted species is a desirable but challenging objective.

The Eastern Region contains over 40% of the US population. Although fire-related smoke has not created significant issues to date, future efforts to increase the prescribed fire program may create more discussion about fire-related issues in the wildland–urban interface (WUI).

What have been the recent trends in wildfire occurrence? What are the projections for the future?

Although temperatures are increasing, the increases primarily cause warmer winter lows that do not affect fire. Temperatures are expected to be moderated near the Great Lakes compared to other areas. It is unclear how precipitation will change in the future, but it may have higher and lower extremes. Wildfire occurrence in the Eastern Region has not created significant challenges for fire management or specifically affected smoke management. The WUI is expected to grow in some areas, which will likely increase human-caused ignitions. Large fire growth in the Eastern Region is not expected to increase significantly in the next few decades.

During the past 20 years or so, which smoke events have been noteworthy (especially related to public health and safety)?

In 2011, the Pagami Creek Fire (northern Minnesota) dispersed smoke southward to Milwaukee and Chicago. Smoke intrusions from Canada and the western USA have become a near-annual occurrence, contributing to high-particulate matter (PM) days in Minnesota. However, the number of high-smoke days is relatively small.

What is the current regulatory environment? What are the major interactions with states and other governmental bodies?

Smoke is viewed as irrelevant by state and USEPA air quality regulators because it is rare and viewed as uncontrollable. Industrial emissions are the primary focus.

What are the current perspectives of stakeholders and the general public about smoke and wildfire?

Large wildfires are generally rare, so people are not very interested in smoke and wildfire.

Prescribed Fire and Smoke

What are the current uses of prescribed fire in different ecosystems? What are the goals for future use of prescribed fire and associated management actions (e.g., forest thinning)?

The primary uses are maintaining the role of fire in ecosystems to benefit fire-adapted species and reducing fuel loads to protect values at risk.

To what degree is smoke a barrier to use of prescribed fire (regulatory process, air quality concerns, conflicts with other burners, etc.)?

Smoke is not a primary concern, although we have the most WUI in the USA. We have been proactive in smoke management, providing consistent smoke modeling across the Eastern Region and providing training for states and other partners. There are concerns about the effects of smoke on bat hibernacula (caves and other shelters used by bats), especially given that many bat species are currently stressed by white nose syndrome (caused by the fungus *Pseudogymnoascus destructans*).

Limited human and financial resources and lack of available windows for burning are significant barriers. The primary prescribed burning windows are in spring and fall. The spring windows (post snowmelt, before leaves emerge) have been too wet in recent years. The fall window (September–October) can be affected by transfer of resources to fight wildfires in the western USA, insufficient funds remaining toward the end of the federal fiscal year (September 30), and end-of-fire-season fatigue among fire field personnel. We are considering doing more burning in the summer.

What are the current perspectives of stakeholders and the general public about smoke and prescribed fire?

Prescribed fire is considered more of a curiosity than a concern at this point: Smoke impacts are rare.

Smoke Research Needs and Scientific Efforts Applicable to the Region

Which new knowledge, research, modeling, and other tools are needed to predict and manage smoke from wildfire and prescribed fire in dominant ecosystems in the region?

Consumption of soil organic matter is a problem in the Lake States. Although it usually dominates emissions, current models cannot produce accurate estimates, making smoke modeling in general ineffective. Resolving the effect of lake breezes is challenging, near the Great Lakes as well as for large inland lakes. Fine-scale meteorology is important but often ignored because most of the Region is perceived as not having complex terrain. Having the ability to measure mixing height in real time on burn sites would be useful. Finally, better scientific information on the impacts of smoke on bats would help conservation efforts in the Region.

Which current scientific efforts (federal, state, tribal, universities, NGO's, etc.) will help address these needs?

Forest Service research is aware of the soil organic matter consumption issue in the Eastern Region, but it is unclear if there are good solutions. An increase in computing power is facilitating finer-scale meteorology grids, which over time should help quantify lake breeze issues. On-site mixing instruments would also contribute to greater accuracy in quantifying the effect of lake breezes.

Intermountain Region

Linda Chappell

Ecological and Social Context

What are the dominant forest/range/grassland ecosystems with which you work? What is the ecological role of fire in these systems? Which fire-related issues or widespread problems (forest health, invasive species, wildland–urban interface, etc.) are of greatest concern?

The Great Basin encompasses a broad range of elevations, topography, vegetation types, climate, and fire history. Elevations range from 600 m in Nevada's deserts to over 4000 m in several locations. The basin-and-range topography drives a diverse network of ecosystems, including desert scrub/shrub; sagebrush–grass; pinyon–juniper (*Pinus edulis–Juniperus occidentalis*) woodland; chaparral; mountain shrubland; ponderosa pine (*Pinus ponderosa*), lodgepole pine (*P. contorta* var. *latifolia*), mixed conifer, and subalpine forest; subalpine meadows; and grasslands. In desert and low-precipitation areas, the amount of water available is the strongest driver of vegetation patterns. Coverage and density of pinyon/juniper woodland, a major component at mid elevations, have varied over time.

Fire-return interval fire patterns and patch size support functional fire-adapted ecosystems in this Region. Over the past century, fire exclusion, livestock grazing, reductions in native ungulate populations, uncharacteristic insect outbreaks, invasive species, and a growing human population have all altered how ecosystems and fire interact.

Ponderosa pine forest is a short, frequent-fire-return system, and its fire ecology in the Great Basin is well known. Many areas have not had regular fire in the last 100+ years, making restoration and fuel treatments challenging and expensive. Quaking

---

L. Chappell
U.S. Forest Service, Intermountain Region, Ogden, UT, USA

aspen (*Populus tremuloides*), mixed conifer, Douglas-fir (*Pseudotsuga menziesii*), and lodgepole pine forests are all also dependent on periodic fire for regeneration, although at longer fire-return intervals than ponderosa pine.

Range and grasslands are dominated by sagebrush-steppe systems. Salt desert shrub systems, a minor component of these areas, historically show little fire influence; however, invasive grasses are driving uncharacteristic wildfires. There are small pockets of montane grassland meadows, a minor but critical component in limiting fire patch size at upper elevations. Chaparral is limited to mixed-brush systems which are often intermixed among the other mid-elevation vegetation types. Little is known about the role of fire here.

What have been the recent trends in wildfire occurrence? What are the projections for the future?

In general, there has been an increase in both number of fires and area burned per year.

During the past 20 years or so, which smoke events have been noteworthy (especially related to public health and safety)?

Due to a relatively low overall population density, this Region has had few major smoke events. However, increasing populations, land-use changes, and increased wildfire occurrence are anticipated to contribute to more future smoke events. Smoke production from wood burning in the winter, as well as vehicle emissions, often lead to elevated PM values, which affect public health. In addition, the Great Basin is downwind of states with large amounts of emissions and large wildfires that, combined with local emissions, can cause elevated smoke. This is exacerbated by local topography that can trap pollutants for an extended period of time.

The 2003 Cascade II prescribed fire escape along the southern Wasatch Front in Utah led to unhealthy smoke exposure in two counties. Had the fire not escaped, it is unlikely that this exposure would have occurred. Perceived lowered public trust reduced the ability to burn in the area for several years, but that trust has since recovered.

The 2012 Trinity Ridge Fire (Boise National Forest) created dense smoke in Featherville, Idaho and in fire camp where carbon monoxide (CO) exposure directly inhibited decision making from the command and general staff. The $PM_{2.5}$ level exceeded 900 $\mu g\ m^{-3}$ at night, increasing to 1000–1800 $\mu g\ m^{-3}$ the next morning, then decreasing to <10 $\mu g\ m^{-3}$ after the inversion lifted.

The 2012 Mustang Complex Fire produced large amounts of smoke in Salmon, Idaho, causing public complaints due to long exposure to poor air quality.

Appendix A: Regional Perspectives on Smoke Issues and Management 289

Prescribed Fire and Smoke

What are the current uses of prescribed fire in different ecosystems? What are the goals for future use of prescribed fire and associated management actions (e.g., forest thinning)?

The Intermountain Region uses prescribed fire primarily to reduce hazardous fuels and restore fire-adapted ecosystems.

To what degree is smoke a barrier to use of prescribed fire (regulatory process, air quality concerns, conflicts with other burners, etc.)?

The regulatory process, air quality concerns, and conflicts with other burners all affect the implementation of prescribed fire, depending on the location. Each state has different regulatory requirements and interests, although Nevada and Wyoming have fewer regulatory hurdles. Utah and Idaho are currently rewriting their fire and smoke regulations, which are expected to have fewer limitations for prescribed fire. Barriers for any given project are often local social acceptance and support.

Smoke Research Needs and Scientific Efforts Applicable to the Region

Which new knowledge, research, modeling, and other tools are needed to predict and manage smoke from wildfire and prescribed fire in dominant ecosystems in the region?

Knowing the amount of smoke produced by each fuel type would be valuable, because most ecosystems, and therefore fuels, need fire to maintain functionality. Although we can model likely outputs, scientific information on smoke production by Great Basin fuels is rare.

Since fire is a regular recurrence, it would be very useful to have a database of geospatial layers that could identify sites likely to be smoke-free in order to best situate wildland fire camps for large projects (wildfire and prescribed fires) in locations more likely to assist firefighters in staying healthy. Better carbon monoxide monitoring is needed to understand exposure to firefighters and local populations and to reduce air quality impairment.

We need to know the best placement for monitors, for both prescribed fire and wildfire, in order to optimize data collection and understand smoke emissions.

More information is needed on outputs of smoldering fuels, especially in the Great Basin. Better knowledge about fire consumption of rotten fuels is critical. Currently, it is difficult to inform the public of potential smoke outcomes when stands with a high percentage of rotten fuels are burned, thus potentially reducing the social acceptability of burning.

We need more information on germination and other plant processes in order to better understand how fire and smoke interact with ecosystem function. Data are also needed on the effects of post-fire dust and ash on local air quality.

Which current scientific efforts (federal, state, tribal, universities, NGO's, etc.) will help address these needs?

The Fire and Smoke Model Evaluation Experiment (FASMEE) is analyzing new data on smoke in mixed conifer/aspen systems in the Intermountain Region. Similar efforts in other fuel types would be helpful in addressing fire and smoke concerns across the Great Basin.

Fire Influence on Regional to Global Environments and Air Quality (FIRE-AQ) has ongoing studies focused on emission factors and plume rise characteristics to enhance current smoke monitoring.

The Joint Fire Science Program has been heavily involved in working with resource managers to determine their needs in the social and scientific arenas around smoke management and air quality issues.

Northern Region

Jill Webster and Seth Morphis

Ecological and Social Context

During the past 20 years or so, which smoke events have been noteworthy (especially related to public health and safety)?

The summer of 2017 was noteworthy due to fire activity in the Northern Rocky Mountains as well as impacts to air quality in both Montana and Idaho.

What is the current regulatory environment? What are the major interactions with states and other governmental bodies?

The Montana/Idaho Airshed Management Group provides a framework for regular interaction between prescribed fire practitioners and the Montana Department of Environmental Quality (DEQ)/Idaho DEQ during the spring and fall prescribed fire seasons. In addition, communication is maintained with the DEQs during wildfire season regarding smoke impacts. The Idaho DEQ is in the process of developing a state smoke management program through a negotiated rulemaking process. The Montana DEQ regulates "major" prescribed fire practitioners by issuing an annual permit that contains work practice and reporting requirements.

J. Webster · S. Morphis
U.S. Forest Service, Northern Region, Missoula, MT, USA

What are the current perspectives by stakeholders and the general public about smoke and wildfire?

The public is generally tolerant about smoke from wildfire. Smoke is perceived as the result of a natural process and, as long as public opinion is that the incident is being adequately managed, the resulting smoke is generally patiently endured. This can change if smoke impacts from wildfires are particularly bad and the management strategy for fire suppression is perceived as insufficient. Health impacts from wildfire smoke is an ongoing topic of discussion.

Prescribed Fire and Smoke

To what degree is smoke a barrier to use of prescribed fire (regulatory process, air quality concerns, conflicts with other burners, etc.)?

In some areas (e.g., narrow mountain topography with populated valleys), air quality concerns and regulatory standards can limit the use of prescribed fire during a given burn period. However, this can vary based on weather conditions and wind direction. More commonly, the number of prescribed fire actions is limited by prescription windows and resource availability.

What are the current perspectives of stakeholders and the general public about smoke and prescribed fire?

Public perception about smoke from prescribed fire can be generalized as two positions: (1) those who accept prescribed fire as a needed tool and are willing to tolerate the associated smoke, and (2) those who are less tolerant. Those who oppose or are less tolerant to smoke from prescribed fire are typically focused on air quality and public health, but may also disagree with current land management practices.

Smoke Research Needs and Current Efforts Applicable to the Region

Which new knowledge, research, modeling, and other tools are needed to predict and manage smoke from wildfire and prescribed fire in dominant ecosystems in the region?

Improved dispersion models for wildfire and prescribed fire are needed to increase accuracy and reliability of air quality predictions for local communities. Priority needs include (1) increasing the availability of high-resolution meteorology as inputs for dispersion models, and (2) improving the accuracy of modeling of the effects of smoldering fuels at night.

Which current scientific efforts (federal, state, tribal, universities, NGO's, etc.) will help address these needs?

The Northern Region uses various products from U.S. Forest Service Research and Development, particularly those provided by the AirFire team (Pacific Northwest Research Station, Seattle, WA).

Pacific Northwest Region

Rick Graw

Ecological and Social Context

What are the dominant forest/range/grassland ecosystems that you work with? What is the ecological role of fire in these systems? Which fire-related issues or widespread problems (forest health, invasive species, wildland–urban interface, etc.) are of greatest concern?

The Region contains temperate desert, marine, Mediterranean, and temperate steppe dominate a, ranging from alpine systems to dry forest to temperate rainforest to sagebrush. Fire is the dominant disturbance in all but the marine systems. Smoke dispersion into communities during wildfires is a major concern.

What have been the recent trends in wildfire occurrence? What are the projections for the future?

In the past few decades, wildfires have become larger and more area has burned than in previous decades, although fire frequency has not increased. Several individual fires have burned over 50,000 ha, and one (Biscuit Fire) burned 200,000 ha. In 2020, four wildfires on the west side of the Cascade Range in Oregon burned 290,000 ha. The large wildfires contained significant areas with crown fires that burned with high intensity. The climate models are predicting warmer and drier summers across the Region, which will likely continue the general trend of higher area burned. Fires burning in areas with high fuel loadings will burn with high intensity and for long periods of time.

During the past 20 years or so, which smoke events have been noteworthy (especially related to public health and safety)?

There have been many smoke events, especially when a thermal ridge persists over the west side of the Region, creating an east wind that brings smoke from the central and eastern portion of the Region into the large population centers on the West Coast.

R. Graw
U.S. Forest Service, Pacific Northwest Region, Portland, OR, USA

Smoke was particularly bad in 2012, 2013, 2015, 2017, 2018, and 2020. In 2018, Medford, Oregon, experienced 33 days when the air quality index was unhealthy for sensitive groups or worse; Shady Cove, Oregon experienced 40 such days in the same year. In 2020, 3 million people in western Oregon were exposed to a week of air quality in the hazardous category while four large wildfires burned. A recent study in Washington state attributed 600 deaths to wildfire smoke in 2017.

What is the current regulatory environment? What are the major interactions with states and other governmental bodies?

The current regulatory environment consists of federal and state ambient air quality standards, which are enforced by state air quality regulatory programs. However, the Washington and Oregon air quality regulatory programs have delegated silvicultural burning programs to the Washington Department of Natural Resources and Oregon Department of Forestry, respectively. Each of these implements the state smoke management plan, which informs which prescirbed burns can be authorized.

In 2019, Oregon revised its smoke management plan to allow slightly more smoke in communities from prescribed burning, but with more communication to warn the public. In Washington, there is a draft revision to the state smoke management plan currently under consideration by the USEPA, but the state must demonstrate that the proposed changes will not result in exceedances of the National Ambient Air Quality Standards (NAAQS).

What are the current perspectives of stakeholders and the general public about smoke and wildfire?

In general, the public does not like living in smoke, certainly not for three weeks or more. However, they do seem to accept smoke from wildfires, in contrast to smoke from prescribed burning.

Prescribed Fire and Smoke

What are the current uses of prescribed fire in different ecosystems? What are the goals for future use of prescribed fire and associated management actions (e.g., forest thinning)?

The Pacific Northwest Region uses prescribed fire in marine ecosystems to dispose of timber slash. In drier ecosystems, prescribed burning is used for slash disposal as a means of reducing hazardous fuels (i.e., to reduce wildfire risk) and enhancing forest health. Pile burning is the most common way in which prescribed burning is accomplished in the Region, although large landscape burns are also used where possible.

To what degree is smoke a barrier to use of prescribed fire (regulatory process, air quality concerns, conflicts with other burners, etc.)?

Weather conditions are the biggest barrier because they affect fuel conditions and periods of time during which burning can occur. Once fuels are in prescription, we need to share the airshed, because a good day for burning is usually a good day for all burners, not just the US Forest Service. Burning is limited by state smoke management rules, especially in areas where smoldering fuels follow along topographic drainages into communities overnight. Burning is also limited by human and financial resource availability, as well as public tolerance for smoke (e.g., following a bad summer for wildfire smoke).

What are the current perspectives of stakeholders and the general public about smoke and prescribed fire?

Stakeholders seem to understand the need for prescribed fire to reduce hazardous fuels and will tolerate it to some extent, but there is still a lot of negativity about smoke.

Smoke Research Needs and Scientific Efforts Applicable to the Region

Which new knowledge, research, modeling, and other tools are needed to predict and manage smoke from wildfire and prescribed fire in dominant ecosystems in the region?

We need better understanding of trade-offs around smoke to inform decisions about managing wildfires. For example, if a wildfire occurs in an area where smoke will be dispersed into nearby communities for weeks, is it better from a public and forest health perspective, to (1) use aggressive fire suppression to contain the fire in a few days, or (2) endure the smoke for two to four weeks?

A model is needed to more accurately predict smoke emissions and downwind impacts, especially overnight when smoke moves along topographic drainages into communities, in order to better understand whether we might exceed the NAAQS or an intrusion threshold.

Pacific Southwest Region

Donald Schweizer, Ricardo Cisneros, and Andrea Nick

Ecological and Social Context

What are the dominant forest/range/grassland ecosystems with which you work? What is the ecological role of fire in these systems? Which fire-related issues or widespread problems (forest health, invasive species, wildland–urban interface, etc.) are of greatest concern?

Dominant California ecosystems include desert, Mediterranean, forested mountain, and coastal forests. Mountainous forests, Mediterranean plant communities, and grasslands all experience wildfires to some extent. Species within each system are adapted to a specific fire regime. Smoke has been prevalent historically in California and benefits seed viability and growth of certain plants while hindering some fungi and pests.

Currently, some areas with short fire cycles have "missed" multiple fires, resulting in increased tree density and a higher accumulation of fuel which can contribute to intense fires. Historic suppression of nearly all wildfires has negatively affected forest resilience and health in the Sierra Nevada and southern Cascades where numerous fire cycles have been missed. In contrast, some areas with longer fire-return intervals have been experiencing more frequent fire. Chaparral has been replaced by invasive grasses in some areas that have burned frequently.

California has many rural communities and large metropolitan areas adjacent to federal lands. The WUI is a major concern and in many ecosystems invasive species, such as cheatgrass (*Bromus tectorum*) which is common in both desert and mountain forests, promote fire spread and growth. In addition, the forests of the Sierra Nevada have experienced high tree mortality, especially of ponderosa pine, from bark beetle infestations after a major drought in the early 2010's. These challenges, although seemingly contradictory—too little fire or too much fire—are locally significant and contribute to fire-related issues.

What have been the recent trends in wildfire occurrence? What are the projections for the future?

As of the end of the 2021 fire season, only two of the largest 20 wildfires on record in California occurred prior to the year 2000. Wildfire frequency, area burned, and fire

---

D. Schweizer
U.S. Forest Service, Pacific Southwest Region, Bishop, CA, USA

R. Cisneros
U.S. Forest Service, Pacific Southwest Region, Clovis, CA, USA

A. Nick
U.S. Forest Service, Pacific Southwest Region, Fawnskin, CA, USA

season length are expected to increase with climate warming. Where fuel loadings are elevated, fire intensity may increase as well. Over 95% of wildfires in California are human-caused.

During the past 20 years or so, which smoke events have been noteworthy (especially related to public health and safety)?

In the past 20 years, several large, high-intensity fires (e.g., Rim, Rough, McNally) created significant smoke impacts. Wildfires on state or private land or that were ignited by utility infrastructure (e.g., Camp, Paradise, Thomas, Tubbs fires) have resulted in unprecedented smoke episodes in urban areas (San Francisco, Los Angeles, etc.) unaccustomed to high concentrations of smoke.

What is the current regulatory environment? What are the major interactions with states and other governmental bodies?

Managers of the US Forest Service Pacific Southwest Region have been slowly shifting event smoke monitoring and management back to local air districts and air resources boards. In the early 2000s, fire managers and air regulators had serious conflicts. The USFS Air Resource Management Program responded by bringing different entities (California Air Resources Board, air districts, fire managers, non-governmental organizations, etc.) together and forming multiple groups to foster collaboration while expanding smoke monitoring. This includes continuous monitoring of background emissions and utilizing comparative analysis tools for a data-driven, science-based smoke management program. Currently, the Air Resource Management Program facilitates communication among local fire managers, tribes, and local, state, and elected officials. Air Resource Advisors (ARA) are the main contacts during a fire to provide a consistent interface with everyone. The ARA program has been successful in providing smoke monitoring and air quality information to fire personnel, and providing a direct interface between air regulatory specialists, fire response managers, and air resource managers. This allows USFS air resource managers to focus on fire tactics for long-term smoke reduction benefits instead of emergency smoke management.

Some of the smoke tools and products are easy to use and readily available. This allows many of the simpler tasks to be done by the ARAs and fire managers, allowing air resource managers to focus on more difficult and complex questions in smoke management. Strong collaboration is also required among stakeholders (e.g., US Forest Service, other agencies, California Air Resources Board, local air districts, tribes) to approve wilderness fires for ecological benefits that may reduce smoke exposure to neighboring communities and to provide consistent messages to the public.

What are the current perspectives of stakeholders and the general public about smoke and wildfire?

Public understanding of wildland fire in California is exceptional, and recent large fires have further helped to increase fire awareness and preparedness. However, the

Appendix A: Regional Perspectives on Smoke Issues and Management 297

oversimplified message of planned versus unplanned fire has in some cases obscured the complexity of fire as an ecological process.

Statewide, there are numerous prescribed fire councils working with ~37 agencies or groups under a common Memorandum of Understanding about smoke and wildfire. An active and vocal populace makes working at the local level critical, and we have some flexibility to work with communities where opinion diverges from full suppression to a desire to let wilderness fires go unless there are threats to life or property.

The utility-sparked fires have resulted in greater preparedness for future fires in California including increased deployment of fire management tools. For example, utilities have invested in meteorological networks to collect data in wildlands throughout the state, and fire-detection web cameras have been installed.

Prescribed Fire and Smoke

What are the current uses of prescribed fire in different ecosystems? What are the goals for future use of prescribed fire and associated management actions (e.g., forest thinning)?

USFS air resource managers support use of prescribed fire in the WUI to protect life and property and in wilderness or undeveloped areas to reduce fuel loads. The emphasis is typically on forest thinning and surface fuels reduction, as well as ecosystem function and ecological benefit. A long-term goal is for natural ignition fires to be left alone when possible. In addition, prescribed fire is used to control invasive plants and to enhance specific types of animal habitat.

To what degree is smoke a barrier to use of prescribed fire (regulatory process, air quality concerns, conflicts with other burners, etc.)?

Air quality is a challenging issue because background levels of human-caused pollution are high in many of the air districts in the southern half of California. Another issue is the small portion of the public that is vocally against all burning, making air regulators uneasy about allowing prescribed fire, and it has been difficult to convince regulators of the importance of burning for long-term public health objectives. In some cases, air regulators use models that overpredict smoke impacts, concluding that burning should not occur. In addition, fire season has expanded to cover most of the year in California, whereas staffing is still mostly seasonal, making it difficult to implement prescribed burning during some months. Finally, extended periods of high fire danger reduce times when areas are in prescription.

What are the current perspectives of stakeholders and the general public about smoke and prescribed fire?

The Pacific Southwest Region, stakeholders, and the public all have difficulty considering fire to be a binary event (planned versus unplanned) and prefer a more nuanced approach: There are big difference between a lightning-ignited wildfire at high elevation that could burn all summer, a small prescribed fire, a prescribed fire across a large landscape, and a wildfire like the Paradise Fire that involves many structures.

Smoke Research Needs and Scientific Efforts Applicable to the Region

Which new knowledge, research, modeling, and other tools are needed to predict and manage smoke from wildfire and prescribed fire in dominant ecosystems in the region?

The biggest knowledge gaps relate to quantifiable ways to assess emissions estimates, effectiveness of fire management tactics for reducing smoke exposure to communities, and smoke ecology where we need to know how changes to the timing and intensity of smoke may affect forest health. There is also a need for a national level of testing that can be implemented to determine if fire management tactics are reducing smoke.

Emissions estimates need to be validated with field measurements, as it often appears we are comparing "apples to oranges" with smoke estimates tending to be conservatively high, whereas agricultural and vehicle emissions have a financial incentive to be conservatively low. More high-quality rural and wilderness air quality monitoring is needed, as well as more research on the economic impacts of smoke. We are currently limited to data from visitor bureaus or similar entities for local impacts from smoke events. Anecdotally, it appears that in a big smoke event, tourists move to places where there is less or no smoke. This may decrease local economic benefits or allocate them to another location, though tourism in the year following fire may compensate for losses during the fire (see Chap. 7).

Some recent fires have been extreme in terms of fire behavior, duration, and interaction with the WUI, making it important to understand how high-intensity fires outside the natural range of variability may differ from historical fires, and how fires that involve infrastructure as a fuel differ from other wildfires.

Which current scientific efforts (federal, state, tribal, universities, NGO's, etc.) will help address these needs?

The Pacific Southwest Region works with universities throughout California and with Forest Service research scientists. Although we have benefited from ongoing research, differences in smoke and fire management within the Region may require research that accounts for this variability.

Rocky Mountain Region

Brian Keating

Ecological and Social Context

What are the dominant forest/range/grassland ecosystems with which you work? What is the ecological role of fire in these systems? Which fire-related issues or widespread problems (forest health, invasive species, wildland–urban interface, etc.) are of greatest concern?

The Rocky Mountain Region covers a broad range of forest and grassland ecosystems, including high-elevation subalpine fir (*Abies lasiocarpa*), lodgepole pine and mixed conifer, and lower-elevation ponderosa pine forest. There are also diverse and expansive shortgrass and tallgrass systems across all five states of the Region.

Fire plays an important role in maintaining the health and viability of these systems. Multiple land management issues exist due to 100+ years of fire exclusion, a departure from the historic fire regime. Most forest landscapes have higher tree densities and fuel loadings, and resultant potential for high-intensity wildfires. In addition, the Region has experienced widespread insect and disease outbreaks. Grasslands throughout the Region have experienced encroachment of western juniper (*Juniperus occidentalis*) and invasive species, partly due to lack of fire. The Region has also experienced significant increases in human population, creating fire management challenges on the wildland–urban interface.

What have been the recent trends in wildfire occurrence? What are the projections for the future?

Similar to most of the western USA, the Rocky Mountain Region has seen larger and higher-intensity wildfires in recent years, representing a departure from historic fire regimes. For example, 15 of Colorado's 20 largest wildfires on record occurred between 2011 and 2021. Climate warming is contributing to longer fire seasons and is expected to continue to affect the entire Region, with increasing frequency and extent of wildfires.

Over the past 10 years, the Region averaged 454 fires per year. During the past 16 years, the number of lightning-caused fires reported trended downward, although human-caused wildfires are trending upwards on national forests and other lands. Over the past 5 years, 91% of the fires reported were human-caused, accounting for 73% of the area burned.

Climate change is affecting the type and amount of fuels available for ignition and when those fuels are flammable. At lower elevations, where the length of the growing season is increasing, there is a potential for significant increases in

---

B. Keating
U.S. Forest Service, Rocky Mountain Region, Lakewood, CO, USA

wildfire frequency and extent which could contribute to increasing wildfire risks to neighboring communities.

During the past 20 years or so, which smoke events have been noteworthy (especially related to public health and safety)?

- The 416 Fire began on June 1, 2018 and continued for two months, burning 22,000 ha in Southwest Colorado. Durango, Colorado experienced poor air quality during this time, including several days with PM$_{2.5}$ greater than 60 µg m^{-3} and one day greater than 150 µg m^{-3}.
- In 2016, the Beaver Creek Fire (15,000 ha) produced significant amounts of smoke over a 3-month period, primarily affecting lightly populated portions of Wyoming. It was one of the first large fires in lodgepole pine forest that had been killed by mountain pine beetles (*Dendroctonus ponderosae*).
- In 2012–2013, numerous large fires in Colorado and Wyoming (e.g., High Park Fire, Waldo Canyon Fire, Fontenelle Fire, West Fork Complex) produced significant smoke episodes in populated areas, although durations were typically less than a week.
- In 2002, the Hayman Fire burned 56,000 ha in 20 days, causing significant smoke impacts in the Denver metropolitan area.
- In 2001, the Polhemus prescribed burn caused smoke impacts in the Denver metropolitan area. Discrepancies in communicated and actual burn details on the Polhemus fire triggered over a decade of distrust between land managers and the Colorado Air Pollution Control Division.

What is the current regulatory environment? What are the major interactions with states and other governmental bodies?

The US Forest Service works with state smoke and air quality Regulatory agencies across all five states in the Region. Not all states have the same air quality issues nor the same regulatory requirements. In recent years, as understanding of the important ecological role of fire has increased, relationships with states have transitioned from a regulatory stance to a cooperative effort with the US Forest Service to facilitate more prescribed fire while continuing to protect the public health and safety. The need to work together in maintaining ecosystem health and reducing hazardous fuels has been institutionalized, and all national forests in the Region have contributed to improving relationships through outreach and proactive communication with regulatory agencies and the public. These efforts have created room for smoke permits that allow for larger projects under broader environmental conditions. Interactions occur through the local prescribed fire council, annual stakeholder meetings, and periodic outreach by the Regional Smoke Management Coordinator for the Forest Service.

What are the current perspectives of stakeholders and the general public about smoke and wildfire?

The general public recognizes that little can be done about short-term smoke impacts from wildfires, but can be vocal and impatient when impacts extend beyond a few days. Incident management teams assigned to large wildfires routinely utilize ARAs who deploy air quality monitoring equipment in and near affected communities, and work with state regulatory and public health authorities to ensure affected populations are aware of the hazards associated with smoke exposure. ARAs also conduct modeling and provide data to the management team which are used to inform fire suppression-related actions. These efforts help mitigate smoke impacts to the public and wildland firefighters, and contribute to a greater understanding by stakeholders and the public about wildland fire management.

Prescribed Fire and Smoke

What are the current uses of prescribed fire in different ecosystems? What are the goals for future use of prescribed fire and associated management actions (e.g., forest thinning)?

The Rocky Mountain Region utilizes the full range of available prescribed fire actions including, but not limited to, pile burning (hand and machine piles), jackpot burning (for local concentrations of fuels), and broadcast burning (hand and aerial ignition). The prescribed fire methodology is dictated by the resource management objectives identified both in National Environmental Policy Act (NEPA) documents and project burn plans. Methods are further refined by factors such as fuel loading, topography, location, and size.

Prescribed fire is a critical management action for the Region to be successful in land stewardship activities, and an increased investment has been made to support prescribed fire activities. For example, the Region more than doubled its 10-year average of prescribed fire area for the 2019 reporting year. However, prescribed fire cannot on its own resolve all forest health and hazard issues. Mechanical thinning (both commercial and non-commercial) is critical to long-term success. Currently, the Region meets 50% of its hazardous fuels targets through commercial thinning (timber sale) activities (60% two years ago).

To what degree is smoke a barrier to use of prescribed fire (regulatory process, air quality concerns, conflicts with other burners, etc.)?

The air quality regulatory environment was frequently cited in the past as a significant barrier. However, improved relationships with regulators, proactive communication, more monitoring, and more science-based decision support models have reduced this barrier. Currently, the lack of a skilled and available workforce, as well as sociopolitical issues related to smoke and wildfire, are the largest obstacles.

What are the current perspectives of stakeholders and the general public about smoke and prescribed fire?

A majority of stakeholders in the Rocky Mountain Region understand the role that wildland fire, including prescribed fire, plays in maintaining forest health and support land management activities that use prescribed fire. Many agencies and other organizations are interested in partnerships that use prescribed fire across jurisdictional boundaries. Some private landowners apply fire on their lands in partnership with other agency efforts or as stand-alone treatments. Smoke is always a concern, especially adjacent to large population centers, although increased messaging, mitigation measures, and outreach have created an atmosphere of acceptance across most of the Region.

Smoke Research Needs and Scientific Efforts Applicable to the Region

Which new knowledge, research, modeling, and other tools are needed to predict and manage smoke from wildfire and prescribed fire in dominant ecosystems in the region?

There is a high priority for access to high-resolution meteorological forecast data to effectively predict smoke impacts in complex terrain. The most commonly available 4-km resolution is inadequate for smoke forecasting. A resolution of at least 1.33 km is required, but is often not available from the National Oceanic and Atmospheric Administration/National Weather Service (NOAA/NWS).

Which current scientific efforts (federal, state, tribal, universities, NGO's, etc.) will help address these needs?

NOAA/NWS is aware of our needs but has stated they do not currently have the resources to make fine-resolution forecasts available on a routine basis.

Southern Region

Michael D. Ward

Ecological and Social Context

What are the dominant forest/range/grassland ecosystems that you work with? What is the ecological role of fire in these systems? Which fire-related issues or widespread problems (forest health, invasive species, wildland–urban interface, etc.) are of greatest concern?

The Southern Region has a wide range of ecosystems:

- Longleaf pine/shortleaf pine (*Pinus palustris/P. echinata*)
  - Closed to semi-closed canopy with shrub–herbaceous understory
  - Open canopy with herbaceous understory
- Mixed pine flatwoods/plantation (basal area > 30 m^2 ac^{-1}) with shrub or litter understory
- Oak/hickory/pine Appalachian deciduous
- Mixed pine/hardwood Piedmont
- Hydric mixed hardwood (swamps and drainages)
- Xeric/mesic shrub/scrub.

All of these forest types are considered fire-dependent systems that require fire disturbance on a regular return interval to maintain at least one critical functional area. Return intervals vary between forest types. Variability in seasonality and intensity of fires creates conditions that support diverse plant and animal communities. Fire disturbance in these forest types typically maintains desired canopy closure, thus decreasing competition and encroachment by off-site species and facilitating seedbed preparation and nutrient cycling.

Issues of greatest concern for these forest types are:

- Fragmentation of forested areas due to human development, including agriculture and urban development.
- Lack of fire disturbance, which can alter the dominant forest type and result in critical habitat loss and increased risk of crown fires due to high fuel loading.
- Lack of fire or mechanical harvesting disturbance, resulting in dense stands that increase southern pine beetle (*Dendroctonus frontalis*) infestations with snags and dead and down fuels across thousands of hectares.

M. D. Ward
U.S. Forest Service, Southern Region, Atlanta, GA, USA

What have been the recent trends in wildfire occurrence? What are the projections for the future?

Approximately 45,000 wildfires and 400,000 ha burn every year in the southeastern USA. A recent study suggests that the risk of large-fire weeks will increase by 300% in this Region by mid-century (2041–2070).

During the past 20 years or so, which smoke events have been noteworthy (especially related to public health and safety)?

The following smoke events have had big impacts on visibility:

- Florida (January 2012)—Smoke-induced low visibility caused a multi-vehicle accident on Interstate 75, resulting in 18 injuries and 10 fatalities.
- Florida (January 2008)—Smoke-induced low visibility caused a 70-vehicle accident on Interstate 4, resulting in 38 injuries and 4 fatalities.
- Florida (2003)—Smoke-induced low visibility caused a 29-vehicle accident on Interstate 75, causing 3 fatalities.
- In 2000, smoke from wildfires drifting across Interstate 10 caused at least 10 fatalities.
- Many other non-fatal accidents have occurred throughout the Region in which wildland fire smoke contributed to low visibility.

In terms of public health, violations exceeding air quality index standards in Class 1 airsheds and non-attainment areas have occurred multiple times due to large wildfires (e.g., Okefenokee Swamp, Bradwell Bay, Green Swamp, Appalachian fires in 2016–2017).

What is the current regulatory environment? What are the major interactions with states and other governmental bodies?

Each state in the Southern Region has identified air quality monitoring and permitting requirements for prescribed fire. There are multiple prescribed fire councils, prescribed burn associations, fire learning networks, and other groups that work with regulatory agencies and stakeholders to develop standards for smoke management. Interactions with local, state, and federal agencies, and the public contribute to support of continued use of prescribed fire for ecosystem health and reduction of fuels and fire hazard.

What are the current perspectives of stakeholders and the general public about smoke and wildfire?

Some smoke is generally accepted in most communities and prescribed burning is prominent on private lands in the South. Some states burn more than 400,000 ha y^{-1} on private lands alone. The public generally understands prescribed fire and wildfire, especially in rural areas. The more populated urban areas have less understanding of the role of fire in forest landscapes.

Prescribed Fire and Smoke

What are the current uses of prescribed fire in different ecosystems? What are the goals for future use of prescribed fire and associated management actions (e.g., forest thinning)?

Prescribed fire is used to reduce hazardous fuels in order to prevent large crown fires and restore fire-dependent/resilient forests and ecosystems. Prescribed burn treatments are designed to reduce understory fuel loading of dead and downed woody debris and of live woody and herbaceous vegetation. Forest land management plans in the Southern Region identify nearly 8 million ha of fire-dependent forest lands in agency ownership that require fire-return intervals of 3 to 25 years, depending on location and forest type.

Prescribed fire is used to annually burn 2.6 million ha for forest management and 1.5 million ha for agriculture in the Southeast, which is considerably greater than in the rest of the USA. Burning on national forests accounts for 450,000 ha annually across all 12 states in the Southern Region, with 40,000 ha of timber-related fuelbed manipulation. Southern Region forest managers would like to burn 800,000 ha y^{-1} on national forests to meet land management plan goals.

To what degree is smoke a barrier to use of prescribed fire (regulatory process, air quality concerns, conflicts with other burners, etc.)?

More stringent $PM_{2.5}$ and ozone pollutant level regulations may reduce area burned within some airsheds. A rigorous planning process is required to gain authorization compliance within the constraints of NEPA procedures and cultural heritage impact investigations. Each federal agency has a stringent set of planning requirements that can make it challenging to use prescribed fire. Each state has specific guidelines for private landowners to follow for prescribed burning, some of which require training and continuing education. In addition, operational workforce capacity and related support funding are typically 50% less than what is required to safely implement the desired interagency prescribed fire workload.

What are the current perspectives of stakeholders and the general public about smoke and prescribed fire?

Most stakeholders and the general public understand that prescribed fire is an essential tool for managing healthy forests and wildlife habitat. There is also recognition that agencies and private landowners need more resources to implement prescribed fire in areas where treatments are needed.

Southwestern Region

Anita Rose, William Basye, Joshua Hall, Mary Lata, Charles Maxwell, Tessa Nicolet, and Ronald Sherron

Ecological and Social Context

What are the dominant forest/range/grassland ecosystems with which you work? What is the ecological role of fire in these systems? Which fire-related issues or widespread problems (forest health, invasive species, wildland–urban interface, etc.) are of greatest concern?

The Southwest Region has a wide range of ecosystems, including ponderosa pine forest, pinyon–juniper woodland, diverse shrub lands, diverse grasslands, and alpine: all are adapted to an arid climate. About 78% of the area covered by these systems is considered to have a frequent-fire regime. Wildfire risk is a major concern, especially in the wildland–urban interface. Droughts and the potential for climate change to increase the frequency and magnitude of droughts are also prominent concerns.

Most ecosystems associated with frequent-fire-return intervals are now outside their natural range of variation and adaptability with respect to fire and fuels (historical fire-return interval was ~20 years for most forests and many woodlands), and fires can now burn at much higher severity than they did historically. This causes high mortality of forest overstories, as well as flooding, soil erosion, and altered wildlife habitat.

What have been the recent trends in wildfire occurrence? What are the projections for the future?

Although the area burned by wildfire per year during the past 20 years has increased, we still have not reached historical burning levels. The only year during which more area burned compared to historical levels was 2011. Wildfire occurrence (especially large fires) has been low in recent years due in part to weak monsoonal weather patterns (less lightning), although some climate models suggest that monsoons will become stronger and last longer in the future. In general, fires are as likely to be caused by humans as by lightning. Managers have observed an increase in intra-season weather variability, with bursts of moisture and cooler temperatures. If more

---

A. Rose · W. Basye · C. Maxwell · T. Nicolet
U.S. Forest Service, Southwestern Region, Albuquerque, NM, USA

J. Hall
U.S. Forest Service, Southwestern Region, Santa Fe, NM, USA

M. Lata
U.S. Forest Service, Southwestern Region, Flagstaff, AZ, USA

R. Sherron
U.S. Forest Service, Southwestern Region, Phoenix, AZ, USA

Appendix A: Regional Perspectives on Smoke Issues and Management   307

variable spring weather, including cooler temperatures, becomes more prevalent, it could limit the extent of severe burning conditions.

During the past 20 years or so, which smoke events have been noteworthy (especially related to public health and safety)?

- Cerro Grande Fire (2000)
- Rodeo–Chediski Fire (2002)
- Las Conchas Fire (2011)
- Wallow Fire (2011)
- Whitewater–Baldy Complex Fire (2012).

What is the current regulatory environment? What are the major interactions with states and other governmental bodies?

Arizona and New Mexico both have enhanced smoke management programs. The main regulatory driver for smoke management in New Mexico is the Regional Haze Rule, whereas Arizona works under Title 18 (air quality) of the Arizona Administrative Code. There is significant coordination among the Forest Service, regulatory agencies, and health agencies in both states.

What are the current perspectives of stakeholders and the general public about smoke and wildfire?

Numerous groups work to help people in the Southwest Region understand the ecological role of fire (e.g., Southwest Fire Consortium, http://www.swfireconsortium.org; Southwest Fireclime, https://swfireclime.org). The public generally understands the benefits of wildland fire, including prescribed fire, in reducing fuels and promoting ecosystem resilience, and that smoke production is a necessary consequence. However, there is a small, well-organized minority of people opposed to smoke.

Prescribed Fire and Smoke

What are the current uses of prescribed fire in different ecosystems? What are the goals for future use of prescribed fire and associated management actions (e.g., forest thinning)?

All available tools are used to restore natural fire regimes and manage wildfire risk to protect and restore watersheds. The current annual goal for area burned in the Region is 382,000 ha, th 78% in forest and woodland ecosystems and a significant amount in semi-desert grasslands (36,000 ha).

To what degree is smoke a barrier to use of prescribed fire (regulatory process, air quality concerns, conflicts with other burners, etc.)?

The lack of observational smoke monitoring data is a significant limiting factor. A few monitors have been deployed on incidents, but a comprehensive monitoring network across the Region is lacking. The high-resolution weather models are quite accurate, but managers do not have a good quantitative handle on actual smoke and PM conditions.

A 2019 injunction currently restricts tree harvest in Arizona to protect the Mexican spotted owl (*Strix occidentalis lucida*). This is the biggest limiting factor for prescribed fire in the Southwest Region. Other regulatory issues are not a major concern. Climatically, the prescribed burning season is long and operational windows for conducting burns are wide. In some cases, national forests adjacent to each other compete for resources and burn windows in order to meet burning targets (typically area burned) without compromising air quality.

What are the current perspectives of stakeholders and the general public about smoke and prescribed fire?

The public generally understands the interactions among smoke, wildfire, and prescribed fire. A small but well-organized minority of people are opposed to smoke. Strong collaborations are helping people understand the ecological role of fire.

Smoke Research Needs and Current Efforts Applicable to the Region

Which new knowledge, research, modeling, and other tools are needed to predict and manage smoke from wildfire and prescribed fire in dominant ecosystems in the region?

The scales at which smoke is observed, predicted, and managed are not in alignment. We need high spatial and temporal resolution, because that is the scale at which decision making and application of smoke information occurs. We have high-resolution weather and smoke models that could help, but we lack observational PM data to calibrate and initialize the smoke models.

More ensemble forecasting, higher-resolution models that look farther into the future, enhanced satellite products, and robust fine-scale surface-wind observations would improve smoke predictions. Managing some locations for smoke would be easier with better nighttime surface-wind information. Currently, all smoke models use inputs from weather models which limits planning flexibility because models must be run within the timeframe set by the weather model. A model that focuses on seasonal weather statistics or that allows for the burner/planner to input their own weather data would be useful.

As most models are set to a pattern of active flaming followed by smoldering, they cannot account for areas in which flaming and smoldering are adjacent. However, in the Southwest, it is no longer safe to assume that fire will switch from flaming to smoldering, and it is becoming more difficult to forecast overnight smoke. Research and modeling of smoke movement overnight will become increasingly important as drought conditions may result in longer flaming periods, into the night and sometimes for 24 h.

Which current scientific efforts (federal, state, tribal, universities, NGO's, etc.) will help address these needs?

Recent research at the University of New Mexico is focused on smoke, climate change, and fuels treatments, and can inform planning documents. Recent advances in the use of Lidar for plume dynamics are promising. It would be helpful if some of the plume dynamics research would focus on smaller fires and nighttime smoke dispersion.

Appendix B
Smoke Monitoring Networks, Models, and Mapping Tools

Yvonne Y. Shih

This appendix summarizes monitoring networks, models, and mapping tools that are cited in Chaps. 1 through 8. Summary information was derived from web sites and other sources that are cited below. Scale refers to the geographic area for which each item is potentially relevant.

Air Information Report for Public Awareness and Community Tracking (AIRPACT)

Chapter(s): 4, 5

Sponsors(s): University of Washington.

Scale: Regional (Pacific Northwest U.S.).

Description: AIRPACT is a computerized system for predicting air quality (AQ) for the immediate future of one to three days. AIRPACT predicts air quality by calculating the chemistry and physics of air pollutants as determined by pollutant emissions within the context of the background, natural air chemistry and predicted meteorology.

Website/Reference: http://lar.wsu.edu/airpact.

Y. Y. Shih
U.S. Forest Service, Research and Development, Washington, DC, USA

Air Quality Index (AQI)

Chapter(s): 7

Sponsor(s): U.S. Environmental Protection Agency (USEPA).

Scale: National.

Description: AQI is the USEPA index for reporting air quality. It indicates how clean or polluted the air is, and which associated health effects might be a concern. The AQI focuses on health effects that may be experience within a few hours or days after breathing polluted air.

Website/Reference: https://www.airnow.gov/aqi/aqi-basics/.

Air Quality Sensor Performance Evaluation Center (AQ-SPEC)

Chapter(s): 7

Sponsor(s): South Coast Air Quality Management District (SCAQMD).

Scale: National.

Description: The AQ-SPEC program tests low-cost air monitoring sensors to establish performance standards by which sensors are evaluated. The program evaluates sensors in the field, under ambient conditions, and in the laboratory under controlled environmental conditions.

Website/Reference: http://www.aqmd.gov/aq-spec.

Air Resource Advisors (ARAs)

Chapter(s): 7, 8

Sponsor(s): U.S. Forest Service (USFS).

Scale: National.

Description: ARAs are technical specialists who work on incident management teams and coordinate with multiple agencies to address public health risks and concerns, risks to transportation safety, and fire personnel exposure (including in base camp). They have expertise in air quality science including air quality monitoring, smoke modeling, pollutant health thresholds, and communicating about smoke risks and mitigation. During wildfire incidents where smoke is a concern, their objective is to provide timely smoke forecast and impact information, and messages based on best available science.

Website/Reference: https://sites.google.com/firenet.gov/wfaqrp-external/air-resource-advisors.

Air Sensor Toolbox

Chapter(s): 7

Sponsor(s): USEPA.

Scale: National.

Description: The Air Sensor website provides the latest science on the performance, operation, and use of air sensor monitoring systems for technology developers, air quality managers, citizen scientists, and the public. The information can help the public learn more about air quality in their communities.

Website/Reference: https://www.epa.gov/air-sensor-toolbox.

ALERTWildfire

Chapter(s): 8

Sponsor(s): University of Nevada–Reno, University of California–San Diego, University of Oregon.

Scale: National.

Description: ALERTWildfire provides access to state-of-the-art Pan-Tilt-Zoom (PTZ) fire cameras and associated tools to help firefighters and first responders: (1) discover/locate/confirm fire ignition, (2) quickly scale fire resources up or down appropriately, (3) monitor fire behavior through containment, (4) help evacuations through enhanced situational awareness, and (5) ensure contained fires are monitored appropriately for as long as they are active.

Website/Reference: http://www.alertwildfire.org/about.html.

ARISense

Chapter(s): 6

Sponsors(s): Aerodyne Research.

Scale: Regional.

Description: ARISense aims to provide a robust, useful tool for researchers in atmospheric chemistry and environmental public health. Each ARISense system includes electrolytic sensors that measure gas phase pollutants: NO, NO_2, CO, and $O_3 + NO_2$, as well as a Non-Dispersive Infrared (NDIR) sensor to measure CO_2.

Website/Reference: https://www.aerodyne.com/product/arisense/.

Blended Global Biomass Burning Emissions Product (GBBEPx V3)

Chapter(s): 5

Sponsors(s): NOAA.

Scale: Global.

Description: The GBBEPx V3 system produces daily global biomass burning emissions ($PM_{2.5}$, BC, CO, CO_2, OC, and SO_2) blended with fire observations. It also produces hourly emissions from geostationary satellites at the level of individual fire pixels. Outputs also includes a fire detection record in an HMS format, quality flag in biomass burning emissions, spatial pattern of $PM_{2.5}$ emissions, and statistical $PM_{2.5}$ information at the continental scale.

Website/Reference: https://www.ospo.noaa.gov/Products/land/gbbepx/.

BlueSky Smoke Modeling Framework

Chapter(s): 2, 3, 4, 5, 8

Sponsors(s): USFS.

Scale: National.

Description: BlueSky is a modeling framework designed to predict cumulative impacts of smoke from forest, agricultural, and range fires. The BlueSky smoke modeling framework combines emissions, meteorology, and dispersion models to generate the best possible predictions of smoke impacts.

Website/Reference: https://www.fs.fed.us/bluesky/about/.

Appendix B: Smoke Monitoring Networks, Models, and Mapping Tools 315

CALPUFF

Chapter(s): 4

Sponsors(s): Atmospheric Studies Group at TRC Solutions.

Scale: Global.

Description: CALPUFF is a non-steady-state meteorological and air quality modeling system. The model has been listed by the USEPA as an alternative model for assessing long-range transport of pollutants and their impacts.

Website/Reference: http://www.src.com/.

Canopy Version of the Advanced Regional Prediction System (ARPS-Canopy)

Chapter(s): 4

Sponsor(s): USFS.

Scale: National.

Description: ARPS-Canopy is a meteorological modeling system designed to simulate atmospheric mesoscale and boundary-layer processes as well as atmospheric conditions within forest vegetation (canopy) layers. The meteorological output from ARPS-Canopy can be used to drive dispersion models for air-quality applications.

Website/Reference: https://doi.org/10.1002/jgrd.5049.

Clean Air Status and Trends Network (CASTNET)

Chapter(s): 6

Sponsors(s): USEPA, National Park Service (NPS), Bureau of Land Management (BLM), North American tribes.

Scale: National.

Description: CASTNET is a national monitoring network established to assess trends in pollutant concentrations, atmospheric deposition, and ecological effects caused by changes in air pollutant emissions.

Website/Reference: https://www.epa.gov/castnet.

Cloud-Aerosol Lidar and Infrared Pathfinder Satellite Observations (CALIPSO)

Chapter(s): 4

Sponsors(s): National Aeronautics and Space Administration (NASA), Center National D'Études Spatiales (CNES).

Scale: Global.

Description: CALIPSO was launched to study the roles of clouds and aerosols in climate and weather. The satellite comprises three instruments: Cloud-Aerosol Lidar with Orthogonal Polarization (CALIOP Lidar), Imaging Infrared Radiometer (IIR), and Wide Field Camera (WFC).

Website/Reference: https://eosweb.larc.nasa.gov/project/calipso/calipso_table.

Community Earth System Model (CESM)

Chapter(s): 4

Sponsors(s): National Center for Atmospheric Research (NCAR); National Science Foundation (NSF), Department of Energy (DOE).

Scale: Global.

Description: CESM is a fully coupled global climate model that provides computer simulations of the Earth's past, present, and future climate.

Website/Reference: http://www.cesm.ucar.edu/.

Community Health Vulnerability Index

Chapter(s): 7

Sponsor(s): USEPA.

Scale: National.

Description: The Community Health Vulnerability Index can be used to help identify communities at higher health risk from wildfire smoke. Health officials can use the tool, in combination with air quality models, to focus public health strategies on vulnerable populations living in areas where air quality is impaired, either by wildland fire smoke or other sources of pollution.

Website/Reference: https://www.epa.gov/air-research/community-health-vulnerability-index-provides-public-health-tool-protect-vulnerable.

Community Multiscale Air Quality Model (CMAQ)

Chapter(s): 4, 5, 6

Sponsors(s): USEPA.

Scale: Global (hemispheric).

Description: CMAQ is an active open-source development project that consists of a suite of programs for conducting air quality simulations. CMAQ combines current knowledge in atmospheric science and air quality modeling, multi-processor computing techniques, and an open-source framework to deliver fast, technically sound estimates of ozone, particulates, toxics, and acid deposition.

Website/Reference: https://www.epa.gov/cmaq.

Comprehensive Air Quality Model with Extensions (CAMx)

Chapter(s): 6

Sponsors(s): Ramboll Environ.

Scale: Global.

Description: CAMx is a photochemical grid model that comprises a "one-atmosphere" treatment of tropospheric air pollution over spatial scales ranging from neighborhoods to continents. Meteorological inputs are supplied to CAMx from separate weather prediction models (specifically WRF, MM5, and RAMS are supported). Emission inputs are supplied from external pre-processing systems (e.g., SMOKE and EPS3).

Website/Reference: http://www.camx.com/.

CONSUME 3.0

Chapter(s): 2, 5

Sponsors(s): USFS.

Scale: National.

Description: CONSUME 3.0 is designed to import data directly from the Fuel Characteristic Classification System (FCCS). Output is formatted to feed other models and provide usable outputs for burn plan preparation and smoke management requirements.

Website/Reference: https://www.fs.fed.us/pnw/fera/research/smoke/consume/.

Crisis Emergency Risk Communication (CERC)

Chapter(s): 7

Sponsors(s): Centers for Disease Control (CDC).

Scale: National.

Description: The CERC program provides trainings, tools, and resources to help health communicators, emergency responders, and leaders of organizations communicate effectively during emergencies.

Website/Reference: https://emergency.cdc.gov/cerc/.

DaySmoke

Chapter(s): 3, 4, 5

Sponsors(s): USFS.

Scale: Regional (southeastern U.S.).

Description: Daysmoke is a local smoke transport model and has been used to provide smoke plume rise information. It includes a large number of parameters describing the dynamic and stochastic processes of particle upward movement, burn emissions, fallout, and fluctuations.

Website/Reference: https://doi.org/10.5094/APR.2010.032 (Liu et al. 2010).

Environmental Benefits Mapping and Analysis Program–Community Edition (BenMAP-CE)

Chapter(s): 7

Sponsors(s): USEPA.

Scale: Global (USA and China).

Description: BenMAP-CE is an open-source computer program that calculates the number and economic value of air pollution-related deaths and illnesses. The software incorporates a database that includes many of the concentration–response relationships, population files, and health and economic data needed to quantify these impacts.

Website/Reference: https://www.epa.gov/benmap.

FARSITE

Chapter(s): 5

Sponsors(s): USFS.

Scale: National.

Description: FARSITE is a fire-growth simulation modeling system that uses spatial information on topography and fuels along with weather and wind files. It incorporates existing models for surface fire, crown fire, spotting, post-frontal combustion, and fire acceleration into a 2-D fire growth model. FARSITE computes wildfire growth and behavior for long time periods under heterogeneous conditions of terrain, fuels, and weather, and is used to simulate the spread of wildfires and fire use for resource benefit across the landscape.

Website/Reference: https://www.fs.usda.gov/rmrs/tools/farsite (Finney 2004).

FIRETEC

Chapter(s): 2, 3, 4, 5

Sponsors(s): USFS, DOE Los Alamos National Laboratory (LANL).

Scale: National.

Description: FIRETEC is a physics-based, 3-D computer code designed to simulate the constantly changing, interactive relationship between fire and its environment. It does so by representing the coupled interaction between fire, fuels, atmosphere, and topography across the landscape.

Website/Reference: https://www.frames.gov/firetec/home.

FireWork

Chapter(s): 5

Sponsors(s): Government of Canada.

Scale: Global (North America).

Description: FireWork is an air quality prediction system that indicates how smoke from wildfires is expected to move across North America over the next 48 h. The FireWork system makes it possible to include the effects of wildfire smoke in air quality forecasts by estimating the amount of pollution that will be added to the air. These smoke-forecast maps show how the air quality in a community may be affected by wildfire smoke.

Website/Reference: https://weather.gc.ca/firework/index_e.html.

Fire and Smoke Model Evaluation Experiment (FASMEE)

Chapter(s): 3, 4, 8

Sponsors(s): USFS, San Jose State University, Desert Research Institute (DRI), University of Idaho, Michigan Technological University, University of Washington.

Scale: National.

Description: FASMEE is a large-scale effort to identify how fuels, fire behavior, fire energy, and meteorology interact to determine the dynamics of smoke plumes, long-range transport of smoke, and local fire effects such as soil heating and vegetative response.

Website/Reference: https://doi.org/10.3390/atmos10020066 (Prichard et al. 2019).

Fire Effects Monitoring and Inventory System (FIREMON)

Chapter(s): 2

Sponsors(s): USFS, U.S. Geological Survey; NPS.

Scale: National.

Description: FIREMON is a plot-scale sampling system designed to characterize changes in ecosystem attributes over time. The system consists of standardized sampling methods, a sampling strategy manual, field forms, database access, and a data analysis program. The comprehensive sampling of fire effects is done so data can be assessed for significant impacts, shared across agencies, and used to update and improve fire management plans and prescriptions.

Website/Reference: https://www.frames.gov/firemon/home.

Fire Emission Production Simulator (FEPS)

Chapter(s): 5, 8

Sponsors(s): USFS.

Scale: National.

Description: FEPS is a user-friendly computer program designed for scientists and resource managers with some working knowledge of Microsoft Windows applications. The software manages data concerning consumption, emissions, and heat release characteristics of prescribed fires and wildfires.

Website/Reference: https://www.fs.fed.us/pnw/fera/feps/ (Ottmar et al. 2020).

Fire Energetics and Emissions Research v1.0 (FEER)

Chapter(s): 5

Sponsors(s): NASA.

Scale: Global.

Description: The FEER v1.0 product is a first version of a $1 \times 1°$ resolution map of coefficients of smoke emission. Emissions can be calculated for a given region and time period as the product of time-integrated fire radiative power (FRP) and smoke emission coefficients.

Website/Reference: https://feer.gsfc.nasa.gov/projects/emissions/.

Fire Influence on Regional to Global Environments and Air Quality (FIREX-AQ)

Chapter(s): 2, 3, 4, 6

Sponsors(s): National Oceanic and Atmospheric Administration (NOAA), NASA.

Scale: National.

Description: FIREX-AQ provides comprehensive observations to investigate the impact of wildfires and agricultural fires on air quality and climate. FIREX-AQ brings together scientists to explore the chemistry and fate of trace gases and aerosols in smoke using instrumented airplanes, satellites, unmanned aerial vehicles, and ground-based instrumentation.

Website/Reference: https://www.esrl.noaa.gov/csd/projects/firex-aq/.

Fire INventory from NCAR (FINN)

Chapter(s): 5

Sponsors(s): NCAR.

Scale: Global.

Description: The FINN model provides high resolution, global emission estimates from open burning. These emissions have been developed specifically to provide input needed for modeling atmospheric chemistry and air quality in a consistent framework at scales from local to global.

Website/Reference: https://www2.acom.ucar.edu/modeling/finn-fire-inventory-ncar.

Fire Learning Network (FLN)

Chapter(s): 8

Sponsor(s): The Nature Conservancy, USFS, Department of the Interior (DOI).

Scale: National.

Description: The FLN supports multi-agency, community-based projects to accelerate the restoration of landscapes that depend on fire to sustain native plant and animal communities. Collaborative planning, implementation, adaptive management, and sharing of lessons learned are at the core of the FLN.

Website/Reference: https://www.conservationgateway.org/ConservationPractices/FireLandscapes/FireLearningNetwork/Pages/fire-learning-network.aspx.

Fire Science Exchange Network (FSEN)

Chapter(s): 2, 8

Sponsors(s): USFS, BLM, USGS, NPS, U.S. Fish and Wildlife Service (USFWS), DOI.

Scale: National.

Description: The FSEN is a network of 15 regional science exchanges supported by the Joint Fire Science Program that focus on supporting active knowledge exchange. Key objectives include providing a range of stakeholders with current, regionally relevant, wildland fire science information and fostering dialogue in which scientists and managers frame questions and research needs that can be addressed during research efforts.

Website/Reference: https://www.firescience.gov/JFSP_exchanges.cfm.

First Order Fire Effects Model (FOFEM)

Chapter(s): 2, 5

Sponsors(s): USFS.

Scale: National.

Description: FOFEM is a computer program that predicts smoke production, fuel consumption, tree mortality, and soil heating caused by prescribed fire or wildfire.

Website/Reference: https://www.firelab.org/project/fofem-fire-effects-model.

The FLEXible PARTicle Dispersion Model (FLEXPART)

Chapter(s): 4

Sponsors(s): DOE Pacific Northwest National Library.

Scale: Global.

Description: FLEXPART is a Lagrangian transport and dispersion model suitable for the simulation of a large range of atmospheric transport processes. It can simulate dry and wet deposition, decay, and linear chemistry, and can be used in forward or backward mode with defined sources or in a domain-filling setting.

Website/Reference: https://www.flexpart.eu/.

Forest Vegetation Simulator (FVS)

Chapter(s): 2

Sponsors(s): USFS.

Scale: National.

Description: The FVS is a forest growth simulation model that can simulate forest vegetation change in response to natural succession, disturbances, and management. FVS includes all major tree species and can simulate most types of management or disturbance at any time during the simulation.

Website/Reference: https://www.fs.fed.us/fvs/.

Fuel and Fire Tools (FFT)

Chapter(s): 5

Sponsors(s): USFS.

Scale: National.

Description: Fuel and Fire Tools (FFT) is a software application that integrates several fire management tools, including the Fuel Characteristic Classification System (FCCS, version 3.0), CONSUME (version 4.2), Fire Emission Production Simulator (FEPS, version 2.0), Pile Calculator, and Digital Photo Series into a single user interface. The FFT suite of tools uses fuels data classified into fuelbeds to predict outputs related to prescribed fire and wildfire.

Website/Reference: https://www.fs.usda.gov/ccrc/tools/fuel-fire-tools-fft.

Fuel Characteristic Classification System (FCCS)

Chapter(s): 2, 4, 5

Sponsors(s): USFS, NPS, USFWS, BLM.

Scale: National.

Description: The FCCS provides fuelbeds, fuelbed characteristics, and associated predicted surface fire behavior, crown fire, and available fuel potentials to facilitate the mapping of fuelbed characteristics and fire hazard assessment.

Website/Reference: https://www.landfire.gov/fccs.php (https://www.fs.fed.us/pnw/fera/fft/fccsmodule.shtml).

Appendix B: Smoke Monitoring Networks, Models, and Mapping Tools 325

Geostationary Operational Environmental Satellite (GOES)

Chapter(s): 5

Sponsors(s): NASA, NOAA.

Scale: Global.

Description: GOES helps meteorologists observe and predict local weather events, including thunderstorms, tornadoes, fog, hurricanes, flash floods, and other severe weather. GOES observations also have proven helpful in monitoring dust storms, volcanic eruptions, and forest fires. GOES benefits that directly enhance the quality of human life and protection of Earth's environment include: supporting the search-and-rescue satellite aided system (SARSAT); contributing to the development of worldwide environmental warning services and enhancements of basic environmental services; improving the capability for forecasting and providing real-time warning of solar disturbances; and providing data that may be used to extend knowledge and understanding of the atmosphere and its processes.

Website/Reference: https://www.nasa.gov/content/goes.

Global Fire Assimilation System (GFAS)

Chapter(s): 5

Sponsors(s): Copernicus Atmosphere Monitoring Service (CAMS).

Scale: Global.

Description: The GFAS assimilates fire radiative power (FRP) observations from satellite-based sensors to produce daily emissions estimates from wildfire and biomass burning. It also provides information about injection heights derived from fire observations and meteorological information from the operational weather forecasts of the European Centre for Medium-Range Weather Forecasts (ECMWF).

Website/Reference: https://atmosphere.copernicus.eu/global-fire-emissions.

Global Fire Emissions Database (GFED)

Chapter(s): 5

Sponsors(s): NASA, Gordon and Betty Moore Foundation, The Netherlands Organization for Scientific Research (NWO).

Scale: Global.

Description: GFED combines satellite information on fire activity and vegetation productivity to estimate gridded monthly burned area and fire emissions, as well as scalars that can be used to calculate higher temporal resolution emissions.

Website/Reference: https://www.globalfiredata.org/.

Global Forecast System (GFS)

Chapter(s): 4

Sponsors(s): NOAA.

Scale: Global.

Description: GFS is a weather forecast model produced by the National Centers for Environmental Prediction (NCEP). Dozens of atmospheric and land-soil variables are available through this dataset, including temperatures, winds, precipitation, soil moisture, and atmospheric ozone concentration.

Website/Reference: https://www.ncdc.noaa.gov/data-access/model-data/model-datasets/global-forcast-system-gfs.

Hadley Centre Global Environment Model Version 2 (HadGEM2-ES)

Chapter(s): 4

Sponsors(s): United Kingdom Meteorological Office (Met Office).

Scale: Global.

Description: The HadGEM2 includes a coupled atmosphere–ocean configuration, a vertical extension in the atmosphere to include a well-resolved stratosphere, and an Earth-System configuration that includes dynamic vegetation, ocean biology, and atmospheric chemistry.

Website/Reference: https://www.metoffice.gov.uk/research/approach/modelling-systems/unified-model/climate-models/hadgem2 (Collins et al. 2008).

High-Resolution Rapid Refresh (HRRR) Model

Chapter(s): 4, 5

Sponsors(s): NOAA.

Scale: North American Continent.

Description: The HRRR is a real-time 3-km resolution, hourly updated, cloud resolving, convection-allowing atmospheric model.

Website/Reference: https://rapidrefresh.noaa.gov/hrrr/.

Hybrid Single-Particle Lagrangian Integrated Trajectory (HYSPLIT)

Chapter(s): 4, 8

Sponsors(s): NOAA, Australia Bureau of Meteorology.

Scale: Global.

Description: The HYSPLIT model is a complete system for computing simple air parcel trajectories, as well as complex transport, dispersion, chemical transformation, and deposition simulations. It has also been used in a variety of simulations describing the atmospheric transport, dispersion, and deposition of pollutants and hazardous materials.

Website/Reference: https://www.ready.noaa.gov/HYSPLIT.php.

The Interagency Fuel Treatment Decision Support System (IFTDSS)

Chapter(s): 2

Sponsors(s): DOI, USFS, NPS, USFWS, Bureau of Indian Affairs (BIA).

Scale: National.

Description: IFTDSS is a web-based application designed to make fuels treatment planning and analysis more efficient and effective by providing access to data and models through one simple user interface.

Website/Reference: https://iftdss.firenet.gov.

Light Detection and Ranging (Lidar)

Chapter(s): 2, 4

Sponsors(s): USFS, NOAA, NCAR, USGS, NASA, NPS.

Scale: Global.

Description: Lidar is a remote-sensing method that uses light in the form of a pulsed laser to variable distances to the Earth. These light pulses generate precise, 3-D information about the shape of the Earth and its surface characteristics.

Website/Reference: https://oceanservice.noaa.gov/facts/lidar.html.

Landscape Fire and Resource Management Planning Tools (LANDFIRE)

Chapter(s): 2, 5

Sponsors(s): USFS, DOI, USGS.

Scale: National.

Description: LANDFIRE provides geospatial products to support cross-boundary planning, operations, and management. This program produces consistent, comprehensive, geospatial data and databases that describe vegetation, wildland fuel, and fire regimes.

Website/Reference: https://www.landfire.gov/index.php.

Missoula Fire Lab Emission Inventory (MFLEI)

Chapter(s): 5

Sponsors(s): USFS.

Scale: National.

Description: The MFLEI is a retrospective, daily wildfire emission inventory for the contiguous United States with a spatial resolution of 250 m. MFLEI was produced using multiple datasets of fire activity and burned area, a newly developed wildland fuels map, and an updated emission factor database.

Website/Reference: https://doi.org/10.2737/RDS-2017–0039 (Urbanski et al. 2017).

Appendix B: Smoke Monitoring Networks, Models, and Mapping Tools   329

Multi-angle Imaging SpectroRadiometer (MISR)

Chapter(s): 4

Sponsors(s): NASA.

Scale: Global.

Description: MISR provides ongoing global coverage with high spatial detail. Its imagery is carefully calibrated to provide accurate measures of the brightness, contrast, and color of reflected sunlight by viewing the sunlit Earth simultaneously at nine widely spaced angles.

Website/Reference: https://misr.jpl.nasa.gov/.

National Emission Inventory (NEI)

Chapter(s): 5, 8

Sponsors(s): USEPA.

Scale: National.

Description: The NEI is a comprehensive and detailed estimate of air emissions of criteria pollutants, criteria precursors, and hazardous air pollutants. The NEI is released every three years based primarily on data provided by state, local, and tribal air agencies for sources in their jurisdictions, supplemented by data from the USEPA. The NEI is built using the Emissions Inventory System (EIS), first to collect data from state, local, and tribal air agencies, and then to blend those data with other data sources.

Website/Reference: https://www.epa.gov/air-emissions-inventories/national-emissions-inventory-nei.

The National Environmental Public Health Tracking Network

Chapter(s): 7

Sponsor(s): CDC.

Scale: National.

Description: The National Environmental Public Health Tracking Network brings together health data and environmental data from national, state, and city sources, including supporting information to make the data easier to understand. The Tracking Network has data and information on environments and hazards, health effects, and population health that individuals can access to identify areas and populations vulnerable to wildfire smoke hazards.

Website/Reference: https://ephtracking.cdc.gov/.

National Fire Danger Rating System (NFDRS)

Chapter(s): 2

Sponsors(s): USFS, National Wildfire Coordinating Group (NWCG).

Scale: National.

Description: The NFDRS allows fire managers to estimate today's or tomorrow's fire danger for a given area. It combines the effects of existing and expected levels of selected fire danger factors into one or more qualitative or numeric indices that reflect an area's fire protection needs.

Website/Reference: https://www.nwcg.gov/publications/pms437/fire-danger/nfdrs-system-inputs-outputs.

National Weather Service (NWS)

Chapter(s): 5

Sponsors(s): NOAA.

Scale: National.

Description: The NWS provides weather, water, and climate forecasts and warnings for the United States, its territories, adjacent waters, and ocean areas, for the protection of life and property and the enhancement of the national economy. The NWS helps to develop and implement operational air quality forecast guidance.

Website/Reference: https://airquality.weather.gov/.

Natural Fuels Photo Series

Chapter(s): 2

Sponsors(s): USFS, DOI, Department of Defense, Hawaii Department of Natural Resources, University of Brasilia, NPS.

Scale: National.

Description: The Natural Fuels Photo Series project is designed to help land managers appraise fuel and vegetation conditions in natural settings. Each group of photos in a series includes inventory information summarizing vegetation composition, structure and loading, woody material loading and density by size class, forest floor depth and loading, and other site characteristics.

Website/Reference: https://www.fs.fed.us/pnw/fera/research/fuels/photo_series/.

North American Mesoscale Forecast System (NAM)

Chapter(s): 4

Sponsors(s): NOAA.

Scale: North American Continent.

Description: The NAM is one of the major weather models run by the National Centers for Environmental Protection that produces weather forecasts. High-resolution forecasts are generated within the NAM using additional numerical weather models. These high-resolution forecast windows are generated over fixed regions and are occasionally run to follow significant weather events.

Website/Reference: https://www.ncdc.noaa.gov/data-access/model-data/model-datasets/north-american-mesoscale-forecast-system-nam.

NWCG Smoke Management Guide for Prescribed Fire

Chapter(s): 7

Sponsor(s): NWCG.

Scale: National.

Description: The NWCG Smoke Management Guide for Prescribed Fire contains information on smoke management techniques for prescribed fire, air quality regulations, smoke monitoring, modeling, communication, public perception of prescribed fire and smoke, climate change, practical meteorological approaches, and smoke tools. The primary focus of this document is to serve as the textbook in support of the NWCG RX-410, Smoke Management Techniques course.

Website/Reference: https://www.nwcg.gov/sites/default/files/publications/pms420-2.pdf.

PB-Piedmont (PB-P)

Chapter(s): 4

Sponsors(s): DRI.

Scale: National.

Description: PB-Piedmont is a smoke model for predicting nighttime smoke movement. The ability to make such predictions is useful for determining whether smoke from a prescribed burn will cause problems on local roadways. The modeling is based on North America Mesoscale (NAM) forecast data from NOAA/NCEP.

Website/Reference: https://piedmont.dri.edu/.

Photoload Sampling Technique

Chapter(s): 2

Sponsors(s): USFS.

Scale: National.

Description: The Photoload sampling technique is used to quickly and accurately estimate loadings for six common surface fuel components. This technique involves visually comparing fuel conditions in the field with Photoload sequences. Photoload sequences are a series of downward-looking and close-up oblique photographs depicting a sequence of graduated fuel loadings of synthetic fuelbeds for each of the six fuel components (the four size classes of downed dead woody, herbaceous fuels, and shrubs).

Website/Reference: https://doi.org/10.2737/RMRS-GTR-190 (Keane et al. 2007).

PurpleAir

Chapter(s): 6

Sponsors(s): Weather Underground, Coalition for Clean Air, Clean Air Carolina, EZSBC, SCAQMD.

Scale: National.

Description: PurpleAir sensors measure airborne particulate matter. PurpleAir uses the AQI breakpoints established by the USEPA to convert mass concentration into the AQI published on the PurpleAir map.

Website/Reference: https://www2.purpleair.com/.

QUIC-Fire

Chapter(s): 3, 5

Sponsors(s): DOE LANL, USFS.

Scale: National.

Description: QUIC-Fire enables prescribed fire planners to compare, evaluate, and design burn plans. QUIC-Fire couples the 3-D rapid wind solver QUIC-URB to the physics-based, cellular automata fire-spread model Fire-CA.

Website/Reference: https://doi.org/10.1016/j.envsoft.2019.104616 (Linn et al. 2020).

Quick Fire Emission Dataset v2.4 (QFED)

Chapter(s): 5

Sponsors(s): NASA.

Scale: Global.

Description: The QFED was developed to enable biomass-burning emissions of atmospheric constituents to be included in the NASA Goddard Earth Observing System (GEOS) modeling and data assimilation systems. QFED emissions are based on the fire radiative power (FRP) approach and draw on the cloud correction method developed in the Global Fire Assimilation System (GFAS), although QFED employs more sophisticated treatment of non-observed (e.g., obscured by clouds) land areas.

Website/Reference: https://gmao.gsfc.nasa.gov/research/science_snapshots/global_fire_emissions.php.

Smoke-Ready Toolbox for Wildfires

Chapter(s): 7

Sponsor(s): USEPA.

Scale: National.

Description: The USEPA Smoke-Ready Toolbox provides resources related to wildfire smoke and health in one place including USEPA resources, and links to other federal, state, and local webpages. Public health officials and others can use resources in the Toolbox to provide the public with information about the risks of smoke exposure and actions they can take to protect their health.

Website/Reference: https://www.epa.gov/smoke-ready-toolbox-wildfires.

Smoke Sense

Chapter(s): 7

Sponsor(s): USEPA.

Scale: National.

Description: Smoke Sense is a research project that enables citizen scientists to engage with a mobile phone application to explore current and forecast maps of air quality, learn how to protect their health from wildfire smoke, and record their smoke experiences, health symptoms, and behaviors taken to reduce their exposure to smoke.

Website/Reference: https://www.epa.gov/air-research/smoke-sense-study-citizen-science-project-using-mobile-app.

SPECIATE

Chapter(s): 6

Sponsors(s): USEPA.

Scale: National.

Description: SPECIATE is the USEPA repository of organic gas and particulate matter speciation profiles of air pollution sources.

Website/Reference: https://www.epa.gov/air-emissions-modeling/speciate.

VSMOKE

Chapter(s): 3, 4, 8

Sponsors(s): USFS.

Scale: Regional (southeastern US).

Description: The VSmoke model (Lavdas 1996) estimates downwind concentrations of particulate matter at 31 fixed distances, and how far and how clearly a person may see through the smoke plume at each distance. VSmoke also provides estimates of the dimensions of the plume above the ground at each of the 31 distances.

Appendix B: Smoke Monitoring Networks, Models, and Mapping Tools 335

Website/Reference: https://webcam.srs.fs.fed.us/tools/vsmoke/

Western Wildfire Experiment for Cloud Chemistry, Aerosol Absorption and Nitrogen (WE-CAN)

Chapter(s): 3, 4, 6

Sponsors(s): NSF, NCAR, NOAA, NASA, Colorado State University, University of Washington, University of Colorado, University of Montana, University of Wyoming.

Scale: Regional.

Description: The WE-CAN project characterizes the emissions and first day of evolution of western U.S. wildfire plumes, focusing on three sets of scientific questions related to fixed nitrogen, absorbing aerosols, and cloud activation and chemistry in wildfire plumes.

Website/Reference: https://www.eol.ucar.edu/field_projects/we-can.

Whole Atmosphere Community Climate Model (WACCM)

Chapter(s): 5

Sponsors(s): NCAR.

Scale: Global.

Description: WACCM is a comprehensive numerical model, spanning the range of altitude from the Earth's surface to the thermosphere. WACCM unifies certain aspects of upper-atmosphere modeling, middle-atmosphere modeling of atmospheric chemistry, and tropospheric modeling of climate and global dynamics, using the NCAR Community Earth System Model (CESM) as a common numerical framework.

Website/Reference: https://www2.acom.ucar.edu/gcm/waccm.

Wildfire Smoke: A Guide for Public Health Officials

Chapter(s): 7, 8

Sponsor(s): California Air Resources Board, California Department of Public Health, USEPA, CDC, USFS.

Scale: National.

Description: This guide is intended to provide state, tribal, and local public health officials with information they need to prepare for smoke events and communicate health risks and protection measures the public can take when wildfire smoke is present. Although developed for public health officials, the information in this document is useful to other groups including health professionals, air quality officials, and members of the public.

Website/Reference: https://www.airnow.gov/wildfire-smoke-guide-publications/.

Wildland-Urban Interface Fire Dynamics Simulator (WFDS)

Chapter(s): 2, 3, 4, 5

Sponsors(s): USFS, National Institute of Standards and Technology (NIST).

Scale: National.

Description: The WFDS extends the capabilities of the NIST structure fire code Fire Dynamics Simulator (FDS) to fires in vegetation and fire spread over outdoor domains.

Website/Reference: https://www.fs.fed.us/pnw/fera/wfds/simulation_models.shtml.

Wildland Fire Emissions Information System (WFEIS)

Chapter(s): 5

Sponsors(s): Michigan Tech Research Institute.

Scale: National.

Description: The WFEIS is a web-based tool that provides users a simple interface for computing wildland fire emissions across the continental United States and Alaska at large (up to regional) spatial scales. WFEIS integrates burned area maps along with corresponding fuel loading data layers and fuel consumption models to compute wildland and cropland fuel consumption and emissions for user-specified locations and date ranges.

Website/Reference: https://wfeis.mtri.org/.

WRF-SFIRE

Chapter(s): 2, 3, 4, 5

Sponsors(s): University of Colorado Denver, University of Utah.

Scale: National.

Description: WRF-SFIRE is a coupled atmosphere-wildfire model that combines the Weather Research and Forecasting Model (WRF) with a fire-spread model. A version from 2010 was released based on WRF 3.2 as WRF-Fire.

Website/Reference: https://www.openwfm.org/wiki/WRF-SFIRE_user_guide.

References

Collins WJ, Bellouin N, Doutriaux-Boucher M et al (2008) Evaluation of the HadGEM2 model. Hadley Centre Technical Note HCTN 74. Met Office, Exeter, United Kingdom

Finney MA (2004) FARSITE: Fire Area Simulator-model development and evaluation (Research Paper RMRS-RP-4). U.S. Forest Service, Rocky Mountain Research Station, Ogden

Keane RE, Dickinson LJ (2007) The photoload sampling technique: estimating surface fuel loadings from downward-looking photographs of synthetic fuelbeds (General Technical Report RMRS-GTR-190). U.S. Forest Service, Rocky Mountain Research Station, Fort Collins

Lavdas LG (1996) Program VSMOKE—users manual (General Technical Report SRS-6). U.S Forest Service, Southern Research Station, Asheville

Linn RR, Goodrick S, Brambilla S et al (2020) QUIC-fire: A fast-running simulation tool for prescribed fire planning. Environ Model Softw 125:104616

Liu L, Achtemeier GL, Goodrick SL, Jackson W (2010) Important parameters for smoke plume rise simulation with Daysmoke. Atmos Pollut Res 1:250–259

Ottmar RD, Prichard SJ, Drye B (2020) Fuel and fire tools (FFT). https://depts.washington.edu/fft. Accessed 16 July 2020

Prichard S, Larkin SN, Ottmar R et al (2019) The fire and smoke model evaluation experiment—a plan for integrated, large fire–atmosphere field campaigns. Atmosphere 10(2):66

Urbanski SP, Reeves MC, Corley RE et al (2017) Missoula Fire Lab Emission Inventory (MFLEI) for CONUS. Research data archive, updated 9 Jan 2018. U.S. Forest Service, Rocky Mountain Research Station, Fort Collins. https://doi.org/10.2737/RDS-2017-0039.

Appendix C
Abbreviations and Acronyms

ß	Burning efficiency or combustion factor (fraction of available fuel burned)
2D	Two dimensional
3D	Three dimensional
ALM	Air and land Managers
ARA	Air Resource Advisors
AQ	Air quality
AQI	Air quality index
BC	Black carbon
BIA	Bureau of Indian Affairs
BLM	Bureau of Land Management
C_2H_4	Ethene
CALFIRE	California Department of Forestry and Fire Protection
CARB	California Air Resources Board
CARPA	California Air Response Planning Alliance
CCN	Cloud condensation nuclei
CDC	Centers for Disease Control and Prevention
CFD	Computational fluid dynamics
CH_3COOH	Acetic acid
CH_3OH	Methanol
CH_4	Methane
CNES	National Center for Space Studies (France)
CO	Carbon monoxide
CO_2	Carbon dioxide
CONUS	Continental United States
CTM	Chemical transport model
CWD	Coarse woody debris, defined as logs or wood particles > 7.6 cm diameter
DNR	Department of Natural Resources
DOD	U.S. Department of Defense

DOE	U.S. Department of Energy
DOI	U.S. Department of the Interior
δX	Excess mixing ratio of species X
EBAM	Environmental beta attenuation monitor
EC	Elemental carbon
EF	Emission factor in grams of emissions per kilogram of fuel consumed
EFX	Emission factor for species X
FCCS	Fuel Characteristic Classification System
FEM	Federal equivalent monitor
FEPS	Fire Emission Production Simulator
FFI	FIREMON Feat Integrated
FFT	Fuel and Fire Tools
FiNN	Fire Inventory from NCAR
FLM	Fuel loading models
FMO	Fire Management Officer
FOFEM	First-Order Fire Effects Model
FRE	Fire radiative energy (Joules)
FRM	Federal Reference Method
FRP	Fire radiative power (Watts or Joules sec^{-1})
FWD	Fine woody debris, defined as logs or wood particles < 7.6 cm diameter
GBBEPx	Blended Global Biomass Burning Emissions Product
GFAS	Global Fire Assimilation System
GFED	Global Fire Emissions Database
HAPs	Hazardous air pollutants
HCHO	Formaldehyde
HCl	Hydrogen chloride
HNCO	Isocyanic acid
HONO	Nitrous acid
IASC	Interagency Air and Smoke Council
ICP	Incident command post
LBL	Land Between the Lakes
Lidar	Light Detection and Ranging laser scanning
MCE	Modified combustion efficiency
MFLEI	Missoula Fire Lab Emission Inventory
MODIS	Moderate Resolution Imaging Spectroradiometer
MOU	Memorandum of understanding
NAAQS	National Ambient Air Quality Standards
NASA	National Aeronautics and Space Administration
NBR	Normalized Burn Ratio
NCAR	National Center for Atmospheric Research
NDVI	Normalized Difference Vegetation Index
NEI	National Emission Inventory
NEPA	National Envronmental Policy Act
NH_3	Ammonia
NIOSH	National Institute for Occupational Safety and Health

Appendix C: Abbreviations and Acronyms

NIST	National Institute of Standards and Technology
NMOG	Non-methane organic gases
NO_X	Nitrogen oxide
NOAA	National Oceanic and Atmospheric Administration
NSF	National Science Foundation
O_3	Ozone
OA	Organic aerosol
OH	Hydroxyl radical
PAN	Peroxy acetyl nitrate
PEHSU	Pediatric Environmental Health Specialty Units
PFIRS	Prescribed Fire Information Reporting System
PM	Particulate matter
PM_1	Submicron particles with an aerodynamic diameter < 1 μm
$PM_{2.5}$	Fine particulate matter < 2.5 μm diameter
PNs	Peroxy nitrates
POA	Primary organic aerosol
RGB	Red, green, blue band assignments in true-color imagery
QFED	QUIC-Fire Emission Dataset
rBC	Refractory black carbon
RSC	Residual smoldering combustion
SAF	Society of American Foresters
SEP	Socioeconomic position
SfM	Structure-from-motion photogrammetry
SIP	State Implementation Plans
SO_2	Sulfur dioxide
SOA	Secondary organic aerosol
SOC	Secondary organic carbon
SRC	Smoke Ready Community
SRM	Society for Rangeland Management
SVOC	Semi-volatile organic aerosol
SVT	Stereoscopic vision technique
TLS	Terrestrial Lidar scanning (ground based)
UAS/UAV	Unmanned aerial system/unmanned aerial vehicle
UFP	Ultra-fine particles, aerosol with an aerodynamic diameter < 0.1 um
US EPA	United States Environmental Protection Agency
USDA	United States Department of Agriculture
USFWS	United States Fish and Wildlife Service
USGS	United States Geological Survey
UV	Ultraviolet
VIIRS	Visible Infrared Imaging Radiometer Suite
VOC	Volatile organic compounds
WFEIS	Wildland Fire Emissions Information System